T0320736

Statistical Theory

Designed for a one-semester advanced undergraduate or graduate statistical theory course, *Statistical Theory: A Concise Introduction, Second Edition* clearly explains the underlying ideas, mathematics, and principles of major statistical concepts, including parameter estimation, confidence intervals, hypothesis testing, asymptotic analysis, Bayesian inference, linear models, nonparametric statistics, and elements of decision theory. It introduces these topics on a clear intuitive level using illustrative examples in addition to the formal definitions, theorems, and proofs.

Based on the authors' lecture notes, the book is self-contained, which maintains a proper balance between the clarity and rigor of exposition. In a few cases, the authors present a "sketched" version of a proof, explaining its main ideas rather than giving detailed technical mathematical and probabilistic arguments.

Features:

- Second edition has been updated with a new chapter on *Nonparametric Estimation;* a significant update to the chapter on *Statistical Decision Theory;* and other updates throughout
- No requirement for heavy calculus, and simple questions throughout the text help students check their understanding of the material
- Each chapter also includes a set of exercises that range in level of difficulty
- Self-contained, and can be used by the students to understand the theory
- Chapters and sections marked by asterisks contain more advanced topics and may be omitted
- Special chapters on linear models and nonparametric statistics show how the main theoretical concepts can be applied to well-known and frequently used statistical tools

The primary audience for the book is students who want to understand the theoretical basis of mathematical statistics—either advanced undergraduate or graduate students. It will also be an excellent reference for researchers from statistics and other quantitative disciplines.

Felix Abramovich is a professor at the Department of Statistics and Operations Research at Tel Aviv University.

Ya'acov Ritov is a professor in the Department of Statistics at the University of Michigan, Ann Arbor. He is a professor emeritus of Statistics at the Hebrew University of Jerusalem.

CHAPMAN & HALL/CRC

Texts in Statistical Science Series

Joseph K. Blitzstein, *Harvard University, USA*
Julian J. Faraway, *University of Bath, UK*
Martin Tanner, *Northwestern University, USA*
Jim Zidek, *University of British Columbia, Canada*

For more information about this series, please visit: https://www.routledge.com/Chapman--HallCRC-Texts-in-Statistical-Science/book-series/CHTEXSTASCI

Statistical Theory
A Concise Introduction
Second Edition

Felix Abramovich
Tel Aviv University, Israel
Ya'acov Ritov
University of Michigan, USA

and
The Hebrew University of Jerusalem, Israel

CRC Press
Taylor & Francis Group
Boca Raton London New York

CRC Press is an imprint of the
Taylor & Francis Group, an **informa** business

A CHAPMAN & HALL BOOK

Second edition published 2023
by CRC Press
6000 Broken Sound Parkway NW, Suite 300, Boca Raton, FL 33487-2742

and by CRC Press
4 Park Square, Milton Park, Abingdon, Oxon, OX14 4RN

First edition published by CRC Press 2013

CRC Press is an imprint of Taylor & Francis Group, LLC

Library of Congress Cataloging-in-Publication Data

Names: Abramovich, Felix, author. | Ritov, Ya'acov, 1951- author.
Title: Statistical theory : a concise introduction / Felix Abramovich, Tel
   Aviv University, Israel, Ya'acov Ritov, University of Michigan, USA and
   The Hebrew University of Jerusalem, Israel.
Description: Second edition. | Boca Raton : Chapman & Hall, CRC Press,
   2023. | Series: Chapman and Hall/CRC texts in statistical science |
   Includes bibliographical references and index.
Identifiers: LCCN 2022024632 (print) | LCCN 2022024633 (ebook) | ISBN
   9781032007458 (hardback) | ISBN 9781032007472 (paperback) | ISBN
   9781003175407 (ebook)
Subjects: LCSH: Probabilities--Textbooks. | Statistics--Textbooks.
Classification: LCC QA276 .A23 2023 (print) | LCC QA276 (ebook) | DDC
   519.5--dc23/eng20221007
LC record available at https://lccn.loc.gov/2022024632
LC ebook record available at https://lccn.loc.gov/2022024633

ISBN: 978-1-032-00745-8 (hbk)
ISBN: 978-1-032-00747-2 (pbk)
ISBN: 978-1-003-17540-7 (ebk)

DOI: 10.1201/ 9781003175407

Typeset in Nimbus
by KnowledgeWorks Global Ltd.

Publisher's note: This book has been prepared from camera-ready copy provided by the authors.

Contents

List of Figures

List of Tables

Preface to the Second Edition

Similarly to the first edition, the second edition of the book is intended as a textbook for a course in statistical theory for advanced undergraduate and graduate students in statistics, mathematics, or other related fields. In addition to revisions and corrections, we extended its scope and added several new (mostly advanced) chapters and sections—see the list below. We believe that by selecting the corresponding sections, the second edition can be adapted to various course levels—from the first course in statistical theory to a more advanced one.

Our main motivation has remained to propose a "student-oriented" self-contained textbook on statistical theory that would introduce basic statistical concepts on a clear intuitive level with illustrative examples in addition to the (necessary!) formal definitions, theorems, and proofs. Following the lines of the first edition, we tried to keep a proper balance between the clarity and rigor of exposition. In a few cases, we preferred to present a "sketched" version of a proof explaining its main ideas or even to give it up at all rather than to follow detailed technical mathematical and probabilistic arguments. The interested reader can complete those proofs from other existing books on mathematical statistics (see the bibliography).

The book assumes a corresponding background in calculus and probability. An introductory course in statistics is desirable but not essential. We provide a short probabilistic appendix that presents (briefly) the main probabilistic material necessary for the book.

Throughout the book, we leave simple questions for readers. These questions do not require heavy calculus but are intended to check the understanding of the material. All but the two last chapters (Chapters 8 and 9) are provided with an (extended) series of exercises at different levels. Solutions to part of them can be found in the appendix. Chapter 8 on linear regression demonstrates how main theoretical concepts can be applied to this well-known and used statistical tool. Finally, Chapter 9 introduces basic ideas of nonparametric estimation. Chapters and sections marked by asterisks contain more advanced topics.

Here is the list of new chapters and sections added in the second edition:

- M-estimators (*Section 2.5) and their asymptotic properties (*Section 5.10).
- Multiple testing (*Section 4.6).
- Score (Rao) asymptotic tests and confidence regions (Section 5.11).
- Asymptotic properties of the posterior distribution (*Section 6.6).
- Elements of Statistical Decision Theory (*Chapter 7) has been significantly revised and extended.)
- Nonparametric Estimation (*Chapter 9).

Chapter 1

Introduction

1.1 Preamble

This is a book about statistical theory, but we should first understand what "statistics" means. As it happens, the answer to this question is probably not as obvious as it might seem. Most laypersons have some basic idea of what mathematics, chemistry or biology are about, but their perception of statistics is usually somewhat fuzzy and often misleading. Rather unfortunately, statistics would typically be associated by them only either with polls, surveys, sport statistics of records and scores, or with various tables and diagrams published regularly by central bureaus of statistics. This is not what is meant by statistics in the title of this book.

During the last decade, we have become witnesses to the Data Science revolution and the onset of the Big Data era. Being one of the foundation stones of Data Science, statistics nowadays plays a key role essentially in all areas of science, industry, and economics. It is not just a slogan but the reality. Modern statistics covers a wide range of disciplines, from the theoretical development of complex statistical and stochastic models to the most advanced applications of statistical methods in engineering, artificial intelligence, medicine, bioinformatics, finance, social sciences, and other fields. Typical statistical applications range from identification of risk factors for heart disease to design and analysis of experiments for improving a production process; from detecting differentially expressed genes in genome studies to estimating indices for tracking the economy; from signal and image processing to data mining. It is therefore not surprising that (at least) an introductory course in statistics has become an inherent part of essentially any undergraduate program.

So what is statistics? We can generally define statistics as a science dealing with methods for proper collection, analysis, and drawing conclusions from experimental data. Do we need statistics to tell us what the date will be tomorrow, in a week, or in a month? Evidently not: if today, say, is June 1st, we are *certain* that tomorrow will be the 2nd of June, in a week it will be the 8th, and July 1st will be in a month. On the other hand, if we know the price of a particular stock today, can we know for sure its price tomorrow, in a week, or in a month? Apparently not.... The principal difference in the latter case is that, due to a variety of (partially unknown) factors affecting stock prices, there will always be *uncertainty* in its future value. In fact, we can consider it as an outcome of a *random experiment*. If the random mechanism behind generating the stock price was known *exactly*, calculating probabilities of its possible values, expectations, etc. would be a purely probabilistic question with no need for data and statistical tools. However, such situations are rare in practice. To learn and understand an unknown underlying random mechanism, one can collect and then analyze experimental data, either by performing a series of experiments or, as in the stock market example, by utilizing past observations. To this end, statistical tools have become essential.

According to the above general definition of statistics, there are four main stages in any statistical study:

1. Finding reasonable probabilistic models for the investigated phenomena, experiments to be conducted, and the data to be collected.

2. Data collection.

3. Data analysis (initial descriptive statistics of the data, fitting an appropriate statistical model, and estimating its parameters).

4. Statistical inference.

If the data has been inappropriately collected, its further statistical analysis becomes problematic. For example, when conducting a poll for evaluating the chances of a candidate or party to win the elections, it is extremely important to have a representative sample of responders that should include people of different political views, ethnic groups, genders, ages, socio-economic status, etc. in the right proportions. Otherwise, the results of the poll might evidently be strongly biased. Testing the effect of a new drug by comparing the results between case and control groups could be misleading if the allocation of patients to the two groups was not random but effected by a health condition. The problems of data collection and the design of experiments are, however, beyond the scope of this book. Throughout the book, we will always assume that the observed data comes from a properly designed experiment.

The initial stage of the descriptive statistics of the collected data (graphical analysis, data summaries, etc.) is often useful for understanding the main features of the data and verifying the appropriate statistical modeling of the data generating process, which is a key part of the third stage. Incorrect modeling of the data makes its further analysis and inference meaningless. By statistical model, we generally mean the following: Suppose that the observed data is the sample $\mathbf{y} = (y_1, ..., y_n)$ of size n. We will model $\mathbf{y}$ as a realization of an n-dimensional random vector $\mathbf{Y} = (Y_1, ..., Y_n)$ with a joint distribution $f_{\mathbf{Y}}(\mathbf{y})$. As mentioned above, if the underlying distribution $f_{\mathbf{Y}}(\mathbf{y})$ was known, one would not need the data to predict future observations or to (precisely) calculate the expected value. However, the true distribution $f_{\mathbf{Y}}(\mathbf{y})$ of the data is rarely completely known. Nevertheless, it can often be reasonable to assume that it belongs to some *family of distributions $\mathscr{F}$*. It follows, for example, from the scientific consideration of the data collected, or past experience of similar data. Thus, count data of independent rare events, for example, typically follows a Poisson distribution although its mean will be unknown. The data distribution of continuous measurements influenced by many factors can usually be well approximated by a normal distribution although its mean and variance cannot be concluded from such general assumptions. Throughout most of the book, we will assume that $\mathscr{F}$ is a *parametric* family, that is, that we know the type of distribution $f_{\mathbf{Y}}(\mathbf{y})$ (for example, Poisson or normal) up to some unknown parameter(s) $\theta \in \Theta$, where Θ is a *parameter space*.

Typically, we will consider the case, where $y_1, ..., y_n$ are the results of independent identical experiments (e.g., a series of independent similar measurements over the same object). In this case, $Y_1, ..., Y_n$ can be treated as *independent, identically distributed* (i.i.d.) random variables with the common distribution $f_\theta(y)$ from a parametric family of distributions $\mathscr{F}_\theta$, $\theta \in \Theta$.

Example 1.1 A researcher is interested in the proportion of males and females among newborn babies. One of his questions, for example, is whether this proportion is indeed half–half? In order to collect sample data for his research, he asks a maternity unit of a local hospital for their records on the genders of newborn babies over the last week. During that period, 20 babies were born with the genders *F,M,F,F,M,F,M,M,F,F,F,F,M,M,F,M,F,F,M,F* (overall 12 girls and 8 boys). Define indicator variables $Y_i = 1$ if the i-th baby was girl, and $Y_i = 0$ for a boy, $i = 1, ..., 20$. The observed data can be written then as a binary sequence 10110100111100101101. What might be the reasonable statistical model for such data? If we assume that the gender of all deliveries are independent with the same chance for a girl, then we can model the data $y_1, ..., y_n$ as realizations of a series of Bernoulli trials, where $Y_i \sim \mathscr{B}(1, p)$, $i = 1, ..., 20$, and the probability $p \in (0, 1)$ of delivering a girl is the unknown parameter of the model.

Example 1.2 Mr. Skeptic went to a supermarket to buy cheese. He found a series of standard 200g cheese packages on a shelf but decided to check whether their weight was indeed 200g as claimed. He bought five randomly chosen packages and weighed them at home. The resulting weights were 200.3, 195.0, 192.4, 205.2, and 190.1g. Not only were all the weights different, but not even a single package weighed 200g! Is this justification for the skepticism of Mr. Skeptic? At first, Mr. Skeptic

was indeed somewhat puzzled, but after some thought, he agreed that, due to a variety of factors (e.g., the accuracy of the cutting machine and the weighing scales), measurement errors are always present. It would then be unreasonable to expect all the weights to be exactly the same, but to model the observed weights $y_1, ..., y_5$ as realizations of i.i.d. random variables $Y_1, ..., Y_5$, where

$$Y_i = \mu + \varepsilon_i, \quad i = 1, ..., 5, \tag{1.1}$$

μ is the true (unknown) weight and the random errors $\varepsilon_1, ..., \varepsilon_5$ are i.i.d. random variables with some common distribution. In many measurement models, it is reasonable to assume that the distribution of ε_i is normal with zero mean (all measurement devices are calibrated) and some (known or unknown) variance σ^2. Hence, $Y_i \sim \mathcal{N}(\mu, \sigma^2)$ and the main parameter of interest for Mr. Skeptic is the expected weight μ, in particular, whether $\mu = 200$g or not. He may wonder whether the supermarket is cheating (formally he would ask himself whether the data is consistent with the assumption that $\mu = 200$). Finally, knowing the value of σ^2 is secondary, but not without interest.

Generally, finding a "proper" model for the data cannot be completely "formalized" and may involve prior knowledge and/or past experience of the specific problem at hand (e.g., the measurement errors in Example 1.2 are known to be typically normally distributed), the "nature" of the data, common sense, descriptive analysis of the data, collaboration with a researcher that has requested for statistical consulting, and other factors. The appropriateness of the chosen model and its "goodness-of-fit" should also be validated. These (practically important!) issues are also not the subject of this book, where we will always consider the assumed statistical model for the data to be adequate.

Given a statistical model for the observed data, the first statistical goal is usually to estimate its parameters of interest. For example, in Example 1.1, a researcher would estimate the unknown proportion p of newborn girls in the entire population by the corresponding sample proportion $12/20 = 0.6$. Similarly, in Example 1.2, Mr. Skeptic would estimate the unknown true weight cheese packages μ by the sample average of measured weights $\bar{y} = \frac{1}{5} \sum_{i=1}^{5} y_i = 196.6$g.

The final stage of a statistical study is statistical inference, that is, drawing conclusions from the data that extend the results obtained from the sample to make a statement about the entire population. What can be said about the proportion p of newborn girls based on its estimate from a sample of size 20? Can it be concluded that girls are in the majority among newborn babies ($p > 0.5$), or can the difference between $p = 0.5$ and the sample proportion 0.6 still be explained by the random effect of sampling? Similarly, can Mr. Skeptic claim that the supermarket is cheating and that the true $\mu < 200$ or the difference between $\bar{y} = 196.6$ and $\mu = 200$ is within reasonable statistical error? These are typical examples of statistical inference questions.

In this book, we cover the theoretical ground for various statistical estimation and inference procedures, and investigate their properties and optimality. We introduce first several basic important concepts.

1.2 Likelihood

We start from the likelihood function that plays a fundamental role in statistical theory.

Suppose that we observe a sample $\mathbf{y} = (y_1, ..., y_n)$ of realizations of random variables $\mathbf{Y} = (Y_1, ..., Y_n)$ with a joint distribution $f_\theta(\mathbf{y})$, where the parameter(s) $\theta \in \Theta$. The set of all possible samples Ω is called the sample space.

Assume first that $\mathbf{Y}$ is discrete and consider the probability $P_\theta(Y_1 = y_1, ..., Y_n = y_n)$ to draw the specific, already observed sample $y_1, ..., y_n$. Note that we treat the values $y_1, ..., y_n$ as given and consider the above probability as a function of the unknown parameter θ. To emphasize this, define the likelihood function $L(\theta; \mathbf{y}) = P_\theta(\mathbf{y})$—the probability to observe the given data $\mathbf{y}$ for any possible value of $\theta \in \Theta$. The value $L(\theta; \mathbf{y})$, therefore, can be viewed as a measure of likeliness of θ to the observed data $\mathbf{y}$. If $L(\theta_1; \mathbf{y}) > L(\theta_2; \mathbf{y})$ for a given $\mathbf{y}$, we can say that the value θ_1 for θ is more suited to the data than θ_2. Such a statistical view of $P_\theta(\mathbf{y})$ is, in a way, opposite to the probabilistic one.

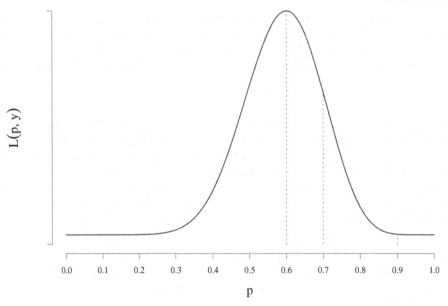

Figure 1.1: The likelihood function for baby data.

Within the probabilistic framework, we assume that the distribution of a random variable is known and are interested in calculating the probability to obtain its certain outcome. In statistics, we have already observed the realization of a random variable, but its distribution (or, at least, its parameters) is unknown. The goal is then to extract information about the parameters from the data.

As we have mentioned before, usually we will assume that $Y_1, \ldots, Y_n$ are i.i.d. with a probability function $f_\theta(y)$. In this case

$$L(\theta; \mathbf{y}) = P_\theta(\mathbf{y}) = \Pi_{i=1}^n P_\theta(Y_i = y_i) = \Pi_{i=1}^n f_\theta(y_i). \tag{1.2}$$

Example 1.3 (Bernoulli trials) Consider the newborn babies data from Example 1.1, where $Y_i \sim \mathscr{B}(1, p)$, $i = 1, \ldots, 20$. From (1.2) the likelihood $L(p, \mathbf{y})$ is

$$L(p; \mathbf{y}) = p^{y_1}(1-p)^{1-y_1} \cdot \ldots \cdot p^{y_{20}}(1-p)^{1-y_{20}} = p^{\Sigma_{i=1}^{20} y_i}(1-p)^{20 - \Sigma_{i=1}^{20} y_i} = p^{12}(1-p)^8 \tag{1.3}$$

(see Figure 1.1). We see that $p = 0.7$ is much more likely than $p = 0.9$, which means that the probability of having 12 females among 20 newborn babies if a true (unknown) p were 0.7 is much larger than if $p = 0.9$ (the likelihood ratio $L(0.7; \mathbf{y})/L(0.9; \mathbf{y}) \approx 322$). Intuitively, it is natural to seek the "most suitable" value of p for the data, that is, the one that maximizes $L(p; \mathbf{y})$ for a given $\mathbf{y}$. For our data, one can see (see also Example 2.1) that it is $12/20 = 0.6$. It turns out indeed to be a reasonable approach for estimating p, and we will discuss it in detail later in Section 2.2.

For a *continuous* random variable $\mathbf{Y}$ with a density function $f_\theta(\mathbf{y})$ the likelihood is defined in a similar way as $L(\theta; \mathbf{y}) = f_\theta(\mathbf{y})$. Although in this case the value $L(\theta; \mathbf{y})$ has no probabilistic meaning, it is still a measure of how likely a value θ is for the observed data sample $\mathbf{y}$. The likelihood ratio $L(\theta_1; \mathbf{y})/L(\theta_2; \mathbf{y})$ shows the strength of the evidence in favor of $\theta = \theta_1$ versus $\theta = \theta_2$. In fact, the likelihood $L(\theta; \mathbf{y})$ can be multiplied by any positive bounded function of $\mathbf{y}$ without losing its meaning.

For i.i.d. $Y_1, \ldots, Y_n \sim f_\theta(y)$ we again have

$$L(\theta; \mathbf{y}) = \Pi_{i=1}^n f_\theta(y_i).$$

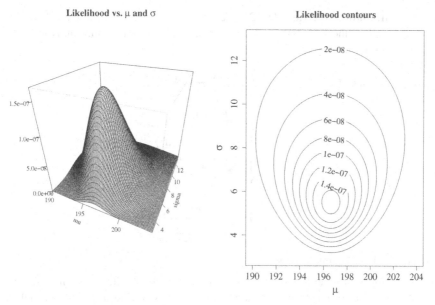

Figure 1.2: The likelihood function for Mr. Skeptic's data.

Example 1.4 (normal data) Consider the normal measurement data from Example 1.2, where both the expected weight μ and variance σ are the unknown parameters. We then have

$$L(\mu, \sigma; \mathbf{y}) = \Pi_{i=1}^{5} \frac{1}{\sqrt{2\pi}\sigma} e^{-\frac{(y_i - \mu)^2}{2\sigma^2}} = \left(\frac{1}{\sqrt{2\pi}\sigma} \right)^5 e^{-\frac{\Sigma_{i=1}^{5}(y_i - \mu)^2}{2\sigma^2}} \tag{1.4}$$

(see Figure 1.2).

1.3 Sufficiency

Consider again the newborn babies data from Example 1.3. Note that to calculate the likelihood $L(p; \mathbf{y})$ in (1.3) one essentially does not need to know the gender of *each* of the twenty newborn babies but the total number of girls among them is sufficient. In a sense, this number contains all available information about p and anything that can be learned about p from the data can be learned from the total number of girls.

To formulate the above half-heuristic arguments more rigorously, we start from a very general definition of a statistic:

Definition 1.1 (statistic) A statistic $T(\mathbf{Y})$ is any real or vector-valued function that can be computed using the data alone.

According to this definition, given a random sample $Y_1, ..., Y_n \sim f_\theta(y)$, $\Sigma_{i=1}^{n} Y_i$, $Y_{\max} = max(Y_1, ..., Y_n)$, $Y_1 - Y_3^{Y_2}$ are examples of statistics, while $Y_{\max}/\theta$ is not since it involves an unknown parameter θ.

We now want to define a *sufficient* statistic $T(\mathbf{Y})$ for θ that allows one to reduce the entire data $\mathbf{y}$ without any loss of information. In other words, given $T(\mathbf{Y})$ no other information on θ can be extracted from $\mathbf{y}$. A natural definition is then as follows:

Definition 1.2 (sufficient statistic) A statistic $T(\mathbf{Y})$ is sufficient for an unknown parameter θ if the conditional distribution of the all data $\mathbf{Y}$ given $T(\mathbf{Y})$ does not depend on θ.

Example 1.5 (continuation of Example 1.3) We show that the overall number of newborn girls $T(\mathbf{Y}) = \sum_{i=1}^{20} Y_i$ is indeed the sufficient statistic for p according to the definition. Note that $T(\mathbf{Y}) \sim \mathscr{B}(20, p)$ and, therefore, for any t,

$$
\begin{aligned}
P(\mathbf{Y}| \sum_{i=1}^{20} Y_i = t) &= \frac{P(\mathbf{Y} = \mathbf{y}, \sum_{i=1}^{20} Y_i = t)}{P(\sum_{i=1}^{20} Y_i = t)} \\
&= \begin{cases} \frac{P(\mathbf{Y} = \mathbf{y})}{P(\sum_{i=1}^{20} Y_i = t)} & \text{if } \sum_{i=1}^{20} y_i = t \\ 0 & \text{if } \sum_{i=1}^{20} y_i \neq t \end{cases} \\
&= \begin{cases} \frac{p^t(1-p)^{20-t}}{\binom{20}{t} p^t (1-p)^{20-t}} & \text{if } \sum_{i=1}^{20} y_i = t \\ 0 & \text{if } \sum_{i=1}^{20} y_i \neq t \end{cases} \\
&= \begin{cases} \frac{1}{\binom{20}{t}} & \text{if } \sum_{i=1}^{20} y_i = t \\ 0 & \text{if } \sum_{i=1}^{20} y_i \neq t \end{cases}
\end{aligned}
$$

that does not depend on the unknown p.

Definition 1.2 of a sufficient statistic allows one to check whether a *given* statistic $T(\mathbf{Y})$ is sufficient, but it does not provide one with a constructive way to find it. Recall our half-heuristic arguments at the beginning of this section on the newborn babies data from Example 1.3. We noted that the likelihood $L(p; \mathbf{y})$ depends on the data only through the total number of newborn girls that motivated us to think that it may be a sufficient statistic. Using Definition 1.2, we verified further that it is indeed so. However, do such arguments remain true in a general case? The following theorem gives the positive answer to this question and even extends this claim:

Theorem 1.1 (Fisher–Neyman factorization theorem) A statistic $T(\mathbf{Y})$ is sufficient for θ iff for all $\theta \in \Theta$, $L(\theta, \mathbf{y}) = g(T(\mathbf{y}), \theta) \cdot h(\mathbf{y})$, where the function $g(\cdot)$ depends on θ and the statistic $T(\mathbf{Y})$, while $h(\mathbf{y})$ does not depend on θ.

In particular, Theorem 1.1 implies that if the likelihood $L(\theta; \mathbf{y})$ depends on data only through $T(\mathbf{y})$ (like in Example 1.3), $T(\mathbf{Y})$ is a sufficient statistic for θ and $h(\mathbf{y}) \equiv 1$.

Proof. For simplicity of exposition we prove the theorem for discrete random variables.

Let $\mathbf{y}$ be a realization of $\mathbf{Y}$ and $T(\mathbf{y}) = t$. Suppose that $T(\mathbf{Y})$ is a sufficient statistic for θ. Then,

$$
L(\theta; \mathbf{y}) = P_\theta(\mathbf{Y} = \mathbf{y}) = P_\theta(\mathbf{Y} = \mathbf{y}, T(\mathbf{Y}) = t) = P_\theta(\mathbf{Y} = \mathbf{y}|T(\mathbf{Y}) = t) \cdot P_\theta(T(\mathbf{Y}) = t). \quad (1.5)
$$

Note that since $T(\mathbf{Y})$ is sufficient for θ, the first probability in the RHS of (1.5) does not depend on θ, while the second one depends on the data $\mathbf{Y}$ only through $T(\mathbf{Y})$. Thus, to complete the proof set $g(T(\mathbf{y}), \theta) = P_\theta(T(\mathbf{Y}) = T(\mathbf{y}))$ and $h(\mathbf{y}) = P_\theta(\mathbf{Y} = \mathbf{y}|T(\mathbf{Y}) = T(\mathbf{y}))$.

Conversely, if $L(\theta; \mathbf{y}) = g(T(\mathbf{y}), \theta) \cdot h(\mathbf{y})$, we show that $T(\mathbf{Y})$ is then sufficient for θ. For any t, consider the conditional probability

$$
\begin{aligned}
P_\theta(\mathbf{Y} = \mathbf{y}|T(\mathbf{Y}) = t) &= \frac{P_\theta(\mathbf{Y} = \mathbf{y}, T(\mathbf{Y}) = t)}{P_\theta(T(\mathbf{Y}) = t)} \\
&= \begin{cases} \frac{P_\theta(\mathbf{y} = \mathbf{y})}{\sum_{\mathbf{y}_j : T(\mathbf{y}_j) = t} P_\theta(Y = \mathbf{y}_j)} & \text{if } T(\mathbf{y}) = t \\ 0 & \text{if } T(\mathbf{y}) \neq t \end{cases} \\
&= \begin{cases} \frac{g(t, \theta) h(\mathbf{y})}{\sum_{\mathbf{y}_j : T(\mathbf{y}_j) = t} g(t, \theta) h(\mathbf{y}_j)} & \text{if } T(\mathbf{y}) = t \\ 0 & \text{if } T(\mathbf{y}) \neq t \end{cases} \\
&= \begin{cases} \frac{h(\mathbf{y})}{\sum_{\mathbf{y}_j : T(\mathbf{y}_j = t)} h(\mathbf{y}_j)} & \text{if } T(\mathbf{y}) = t \\ 0 & \text{if } T(\mathbf{y}) \neq t \end{cases}
\end{aligned}
$$

which does not depend on θ. Hence, by definition, $T(\mathbf{Y})$ is sufficient for θ. $\qquad \square$

Question 1.1 Show that any one-to-one function of a sufficient statistic is also a sufficient statistic.

Example 1.6 (Poisson data) Let $Y_1, \ldots, Y_n \sim Pois(\lambda)$. Then,

$$L(\lambda; \mathbf{y}) = \prod_{i=1}^{n} e^{-\lambda} \frac{\lambda^{y_i}}{y_i!} = e^{-n\lambda} \lambda^{\sum_{i=1}^{n} y_i} \left(\prod_{i=1}^{n} y_i! \right)^{-1}$$

and, according to Theorem 1.1, $T(\mathbf{Y}) = \sum_{i=1}^{n} Y_i$ is a sufficient statistic for λ, where $g(\sum_{i=1}^{n} y_i, \lambda) = e^{-n\lambda} \lambda^{\sum_{i=1}^{n} y_i}$ and $h(\mathbf{y}) = (\prod_{i=1}^{n} y_i!)^{-1}$.

Example 1.7 (continuation of Example 1.2 and Example 1.4) Consider again a random sample $Y_1, \ldots, Y_n \sim \mathcal{N}(\mu, \sigma^2)$ (in particular, $n = 5$ as in Example 1.4). Note that

$$\sum_{i=1}^{n} (y_i - \mu)^2 = \sum_{i=1}^{n} (y_i - \bar{y})^2 + 2(\bar{y} - \mu) \sum_{i=1}^{n} (y_i - \bar{y}) + n(\bar{y} - \mu)^2, \tag{1.6}$$

where the middle term in the RHS evidently vanishes since $\sum_{i=1}^{n} (y_i - \bar{y}) = \sum_{i=1}^{n} y_i - n\bar{y} = 0$. Hence, (1.4) yields

$$L(\mu, \sigma; \mathbf{y}) = \left(\frac{1}{\sqrt{2\pi}\sigma} \right)^n e^{-\frac{\sum_{i=1}^{n} (y_i - \bar{y})^2}{2\sigma^2}} e^{-\frac{n(\bar{y} - \mu)^2}{2\sigma^2}} \tag{1.7}$$

and from Theorem 1.1 it follows that $T(\mathbf{Y}) = (\bar{Y}, \sum_{i=1}^{n} (Y_i - \bar{Y})^2)$ is a sufficient statistic for $\theta = (\mu, \sigma)$.

What is its joint distribution? We now show that

a) $\bar{Y} \sim \mathcal{N}(\mu, \frac{\sigma^2}{n})$;

b) $\sum_{i=1}^{n} (Y_i - \bar{Y})^2 \sim \sigma^2 \chi_{n-1}^2$;

c) $\bar{Y}$ and $\sum_{i=1}^{n} (Y_i - \bar{Y})^2$ are independent.

Let A be any $n \times n$ orthogonal matrix (i.e., a matrix whose columns (rows) are orthonormal— have norm 1 and any two of them are orthogonal) with the first row of equal entries $a_{1j} = \frac{1}{\sqrt{n}}$, $j = 1, \ldots, n$. Define a random vector $\mathbf{U} = A(\mathbf{Y} - \mu \mathbf{1}_n)$, where $\mathbf{1}_n = (1, 1, \ldots, 1)^\top$. By the standard properties of random normal vectors (see Appendix A.4.16), we then have

$$\mathbf{U} \sim \mathcal{N}(\mathbf{0}, \sigma^2 I_n) \tag{1.8}$$

and, therefore, U_i are independent normal random variables $\mathcal{N}(0, \sigma^2)$.

In particular, (1.8) implies that $U_1 = \frac{1}{\sqrt{n}} \sum_{i=1}^{n} (Y_i - \mu) = \sqrt{n}(\bar{Y} - \mu)$ and, therefore, $\sqrt{n}(\bar{Y} - \mu) \sim \mathcal{N}(0, \sigma^2)$ or $\bar{Y} \sim \mathcal{N}(\mu, \frac{\sigma^2}{n})$.

Furthermore, since A is orthogonal, from (1.6) we have

$$\sum_{i=1}^{n} (Y_i - \bar{Y})^2 = \sum_{i=1}^{n} (Y_i - \mu)^2 - n(\bar{Y} - \mu)^2 = \mathbf{U}^\top \mathbf{U} - U_1^2 = \sum_{i=2}^{n} U_i^2$$

which is the sum of squares of $n - 1$ independent normal variables $U_i \sim \mathcal{N}(0, \sigma^2)$, $i = 2, \ldots, n$ and hence, $\sum_{i=1}^{n} (Y_i - \bar{Y})^2 \sim \sigma^2 \chi_{n-1}^2$.

Finally, since $\bar{Y}$ depends only on U_1, while $\sum_{i=1}^{n} (Y_i - \bar{Y})^2 = \sum_{i=2}^{n} U_i^2$, independency of U_i's yields independency of $\bar{Y}$ and $\sum_{i=1}^{n} (Y_i - \bar{Y})^2$.

1.4 *Minimal sufficiency

A sufficient statistic is not unique (even not up to its one-to-one transformations — see Section 1.3). As an obvious example, if $T(\mathbf{Y})$ is sufficient, then for any other statistic $S(\mathbf{Y})$, $(T(\mathbf{Y}), S(\mathbf{Y}))$ is also sufficient (why?). Moreover, the entire sample $\mathbf{Y} = (Y_1, ..., Y_n)$ is always a (trivial) sufficient statistic! We may seek a *minimal* sufficient statistic implying the maximal reduction of the data. If $T_1(\mathbf{Y})$ and $T_2(\mathbf{Y})$ are two sufficient statistics and $T_1(\mathbf{Y})$ is a function of $T_2(\mathbf{Y})$, then some redundancy in $T_2(\mathbf{Y})$ is lost in the transformation to $T_1(\mathbf{Y})$. We would naturally prefer $T_1(\mathbf{Y})$ since it more parsimoniously captures information about the unknown parameter θ within the data. Such considerations lead us to the following definition of a minimal sufficient statistic:

Definition 1.3 (minimal sufficient statistic) A sufficient statistic $T(\mathbf{Y})$ is called *a minimal sufficient* if it is a function of any other sufficient statistic.

Except for several very special examples, a minimal sufficient statistic always exists and can be explicitly constructed in the following way. Note that any statistic $T(\mathbf{Y})$ (a sufficient statistic in particular) divides the sample space Ω into (non-overlapping) equivalence classes, where each class contains a subset of samples $\mathbf{y} \in \Omega$ with the same value $T(\mathbf{y})$. For example, consider a (sufficient) statistic $T(\mathbf{Y}) = \sum_{i=1}^n Y_i$ for a series of Bernoulli trials (Example 1.5). It induces the partition of the sample space of 2^n binary sequences of length n into $n + 1$ equivalence classes, where all $\binom{n}{k}$ sequences with the same total number k of ones will fall within the same equivalence class ($k = 0, ..., n$). Let $T_1(\mathbf{Y}), T_2(\mathbf{Y})$ be two sufficient statistics, where $T_1(\mathbf{Y})$ is a function of $T_2(\mathbf{Y})$. Then, evidently, if two elements of a sample space belong to the same equivalence class induced by $T_2(\mathbf{Y})$, they will necessarily fall within the same equivalence class induced by $T_1(\mathbf{Y})$ as well (explain why). In this sense, the partition of Ω by $T_1(\mathbf{Y})$ is coarser. The minimal sufficient statistic therefore should yield the coarsest possible partition of Ω among all sufficient statistics. This observation will be the key point in constructing the minimal sufficient statistic below.

Assume the existence of a minimal sufficient statistic and consider partitioning the sample space Ω, where $\mathbf{y}_1, \mathbf{y}_2 \in \Omega$ are assigned to the same equivalence class iff the likelihood ratio $L(\theta; \mathbf{y}_1)/L(\theta; \mathbf{y}_2)$ does not depend on θ. Define a statistic $T(\mathbf{Y})$ in such a way that $T(\mathbf{y}_1) = T(\mathbf{y}_2)$ if $\mathbf{y}_1$ and $\mathbf{y}_2$ belongs to the same equivalence class and $T(\mathbf{y}_1) \neq T(\mathbf{y}_2)$ otherwise (obviously, $T(\mathbf{Y})$ is defined in this way for up to one-to-one transformations).

Theorem 1.2 The statistic $T(\mathbf{Y})$ defined above is the minimal sufficient statistic for θ.

Proof. First we show that $T(\mathbf{Y})$ is sufficient. For simplicity, consider the discrete case. Similar to the proof of the Fisher–Neyman factorization theorem, one has

$$P_\theta(\mathbf{Y} = \mathbf{y}|T(\mathbf{Y}) = t) = \frac{P_\theta(\mathbf{Y} = \mathbf{y}, T(\mathbf{Y}) = t)}{P_\theta(T(\mathbf{Y}) = t)}$$

$$= \begin{cases} \frac{P_\theta(\mathbf{y} = \mathbf{y})}{\sum_{\mathbf{y}_j : T(\mathbf{y}_j) = t} P_\theta(Y = \mathbf{y}_j)} & \text{if } T(\mathbf{y}) = t \\ 0 & \text{if } T(\mathbf{y}) \neq t \end{cases} \qquad (1.9)$$

$$= \begin{cases} \left(\sum_{\mathbf{y}_j : T(\mathbf{y}_j) = t} \frac{L(\theta; \mathbf{y}_j)}{L(\theta; \mathbf{y})} \right)^{-1} & \text{if } T(\mathbf{y}) = t \\ 0 & \text{if } T(\mathbf{y}) \neq t \end{cases} .$$

Since $T(\mathbf{y}) = T(\mathbf{y}_j) = t$ in (1.9), all $\mathbf{y}_j$ and $\mathbf{y}$ belong to the same equivalence class induced by $T(\mathbf{Y})$ and, therefore, the likelihood ratios $L(\theta; \mathbf{y}_j)/L(\theta; \mathbf{y})$ do not depend on θ (see the definition of $T(\mathbf{Y})$). Hence, the conditional distribution of $\mathbf{Y}|T(\mathbf{Y}) = t$ does not depend on θ as well and $T(\mathbf{Y})$ is sufficient for θ.

To prove the minimality of $T(\mathbf{Y})$ consider any other sufficient statistic $S(\mathbf{Y})$ and the corresponding partitioning of Ω. Let $\mathbf{y}_1, \mathbf{y}_2 \in \Omega$ belong to the same equivalence class of that partition. According to the factorization theorem,

$$\frac{L(\theta; \mathbf{y}_1)}{L(\theta; \mathbf{y}_2)} = \frac{g(S(\mathbf{y}_1), \theta)h(\mathbf{y}_1)}{g(S(\mathbf{y}_2), \theta)h(\mathbf{y}_2)} = \frac{h(\mathbf{y}_1)}{h(\mathbf{y}_2)}$$

which does not depend on θ. By definition, $\mathbf{y}_1, \mathbf{y}_2$ then fall within the same equivalence class induced by $T(\mathbf{Y})$ as well, and, therefore, $T(\mathbf{Y})$ is a function of $S(\mathbf{Y})$. $\square$

Example 1.8 (continuation of Example 1.5) We now construct the minimal sufficient statistic $T(\mathbf{Y})$ and verify that indeed $T(\mathbf{Y}) = \sum_{i=1}^{n} Y_i$. For any two Bernoulli sequences, $\mathbf{Y}_1$ and $\mathbf{Y}_2$, consider the likelihood ratio

$$\frac{L(p;\mathbf{y}_1)}{L(p;\mathbf{y}_2)} = p^{\sum_{i=1}^{n} y_{1i} - \sum_{i=1}^{n} y_{2i}} (1-p)^{\sum_{i=1}^{n} y_{2i} - \sum_{i=1}^{n} y_{1i}}$$

that does not depend on p iff $\sum_{i=1}^{n} y_{1i} = \sum_{i=1}^{n} y_{2i}$. Thus, Theorem 1.2 implies that $T(\mathbf{Y}) = \sum_{i=1}^{n} Y_i$ is the minimal sufficient statistic.

Example 1.9 (continuation of Example 1.7) For any two random samples $\mathbf{Y}_1$ and $\mathbf{Y}_2$ from the normal distribution $\mathcal{N}(\mu, \sigma^2)$, the corresponding likelihood ratio is

$$\frac{L(\mu,\sigma;\mathbf{y}_1)}{L(\mu,\sigma;\mathbf{y}_2)} = \exp\left\{ -\frac{n(\bar{\mathbf{y}}_1 - \mu)^2 - n(\bar{\mathbf{y}}_2 - \mu)^2 + \sum_{i=1}^{n}(y_{1i} - \bar{\mathbf{y}}_1)^2 - \sum_{i=1}^{n}(y_{2i} - \bar{\mathbf{y}}_2)^2}{2\sigma^2} \right\}$$

(see (1.7)) that does not depend on (μ, σ) iff $\bar{\mathbf{y}}_1 = \bar{\mathbf{y}}_2$ and $\sum_{i=1}^{n}(y_{1i} - \bar{\mathbf{y}}_1)^2 = \sum_{i=1}^{n}(y_{2i} - \bar{\mathbf{y}}_2)^2$. Thus, $T(\mathbf{Y}) = (\bar{\mathbf{Y}}, \sum_{i=1}^{n}(Y_i - \bar{\mathbf{Y}})^2)$ is the minimal sufficient statistic for (μ, σ).

It is clear that the minimal sufficient statistic is unique up to one-to-one transformations.

Question 1.2 For the previous Example 1.9, show that $(\sum_{i=1}^{n} Y_i, \sum_{i=1}^{n} Y_i^2)$ is also a minimal sufficient statistic for (μ, σ).

1.5 *Completeness

Another important property of a statistic is *completeness*:

Definition 1.4 (complete statistic) Let $Y_1, .., Y_n \sim f_\theta(y)$, where $\theta \in \Theta$. A statistic $T(\mathbf{Y})$ is complete if no statistic $g(T)$ exists (except $g(T) \equiv 0$) such that $E_\theta g(T) = 0$ for all $\theta \in \Theta$. In other words, if $E_\theta g(T) = 0$ for all $\theta \in \Theta$, then necessarily $g(T) \equiv 0$.

Example 1.10 (Bernoulli trials) Let $Y_i \sim \mathscr{B}(1,p)$, $i = 1, ..., n$ and show that the (minimal) sufficient statistic $T(\mathbf{Y}) = \sum_{i=1}^{n} Y_i$ is complete. Note that T is a binomial random variable $\mathscr{B}(n,p)$. An arbitrary function $g(T)$ can obtain at most $n+1$ different values $g(0), ..., g(n)$. Thus,

$$E_p g(T) = \sum_{k=0}^{n} g(k) \binom{n}{k} p^k (1-p)^{n-k} = (1-p)^n \sum_{k=0}^{n} g(k) \binom{n}{k} \left(\frac{p}{1-p}\right)^k,$$

which is zero for all values of p only if there is an n-th degree polynomial function of $\frac{p}{1-p}$, which is identically zero. But the only polynomial function which is 0 over an interval is the trivial polynomial with all coefficients equal 0. Hence, $g(k) = 0$ for all $k = 0, ..., n$ and, therefore, $g(T) \equiv 0$.

To verify completeness for a general distribution can be a nontrivial mathematical problem. Fortunately, as we shall show in the following Section 1.6, it is much simpler for the exponential family of distributions that includes many of the "common" distributions.

Completeness is often used to prove the uniqueness of various estimators (see, e.g., Section 2.7.5). Completeness of a *sufficient* statistic yields its minimal sufficiency, as the following theorem proves.

Theorem 1.3 Assume the existence of a minimal sufficient statistic (see Section 1.4). If a sufficient statistic $T(\mathbf{Y})$ is complete, then it is also minimal sufficient.

Proof. Let S be a minimal sufficient statistic. Then, by definition, S should be a function of T, that is, $S = g_1(T)$ for some function g_1.

On the other hand, define $g_2(S) = E_\theta(T|S)$. Since S is a sufficient statistic, the conditional distribution of $T|S$ and the conditional expectation $E(T|S)$, in particular, do not depend on the unknown parameter $\theta \in \Theta$, and thus $g_2(S)$ is a statistic. We have $ET = E(E(T|S)) = E(g_2(S))$ and, therefore,

$E_\theta(T - g_2(S)) = 0$ for any $\theta \in \Theta$. But $g_2(S) = g_2(g_1(T))$, hence $E_\theta(T - g_2(g_1(T))) = 0$, that due to the completeness of T yields $T = g_2(g_1(T)) = g_2(S)$. Thus, T is a one-to-one transformation of a minimal sufficient statistic and, hence, also minimal. $\square$

However, a minimal sufficient statistic may not necessarily be complete as demonstrated in the following counterexample:

Example 1.11 Let $Y_1, ..., Y_n \sim \mathcal{N}(\mu, a^2\mu^2)$ and $a > 0$ is known. Such a model describes, for example, a series of measurements on an object using a measuring device (say, a series of weights), whose accuracy (standard deviation) is relative to the true (unknown) measure of the object rather than being constant.

It is easy to show that $T(\mathbf{Y}) = (\sum_{i=1}^n Y_i, \sum_{i=1}^n Y_i^2)$ is a minimal sufficient statistic for μ (see Exercise 1.5). $\sum_{i=1}^n Y_i \sim \mathcal{N}(n\mu, na^2\mu^2)$ and, hence, $E_\mu(\sum_{i=1}^n Y_i)^2 = n^2\mu^2 + na^2\mu^2$. On the other hand, $E_\mu(\sum_{i=1}^n Y_i^2) = n(\mu^2 + a^2\mu^2)$. An immediate calculus shows then that

$$E_\mu\left\{\frac{n+a^2}{1+a^2}\sum_{i=1}^n Y_i^2 - \left(\sum_{i=1}^n Y_i\right)^2\right\} = 0$$

for all μ and, therefore, there is a function $g(T)$ with expectation identically zero and T is not complete.

1.6 Exponential family of distributions

In this section, we introduce an important family of distributions that include many "standard" distributions (e.g., exponential, binomial, Poisson, and normal) and discuss its main common properties that will be used throughout the book.

We start from the one parameter case.

Definition 1.5 (one-parameter exponential family) A (generally, multivariate) family of distributions $\{f_\theta(\mathbf{y}) : \theta \in \Theta\}$ is said to be an (one parameter) exponential family if

(i) Θ is an open interval

(ii) the support of the distribution, $\text{supp}(f_\theta)$, does not depend on θ

(iii) $f_\theta(\mathbf{y}) = \exp\{c(\theta)T(\mathbf{y}) + d(\theta) + S(\mathbf{y})\}$, $\mathbf{y} \in \text{supp}(f_\theta)$ where $c(\cdot), T(\cdot), d(\cdot)$ and $S(\cdot)$ are known functions; $c(\theta)$ is usually called the *natural* parameter of the distribution

Note that the functions $c(\cdot), T(\cdot), d(\cdot), S(\cdot)$ are not uniquely defined.

Consider several examples of distributions from the exponential family.

Example 1.12 (exponential distribution) Recall that the density $f_\theta(y)$ of the exponential distribution $\exp(\theta)$ is $f_\theta(y) = \theta e^{-\theta y}, y \geq 0$. Thus, $\text{supp}(f_\theta) = [0, \infty)$ does not depend on θ and the density f_θ evidently has the required form with

$$c(\theta) = \theta, \quad T(y) = -y, \quad d(\theta) = \ln\theta, \quad S(y) \equiv 0.$$

Example 1.13 (binomial distribution) Suppose $Y \sim \mathcal{B}(n, p)$, where $0 < p < 1$. Then, $\text{supp}(f_p) = \{0, 1, ..., n\}$ for all p and

$$f_p(y) = \binom{n}{y}p^y(1-p)^{n-y} = \exp\left\{y\ln\left(\frac{p}{1-p}\right) + n\ln(1-p) + \ln\binom{n}{y}\right\}.$$

Therefore, f_p belongs to the exponential family with the natural parameter $c(p) = \ln\frac{p}{1-p}$, $T(y) = y$, $d(p) = n\ln(1-p)$ and $S(y) = \ln\binom{n}{y}$.

Definition 1.5 of the one-parameter exponential family can be naturally extended to the multi-parameter case:

Definition 1.6 (*k*-**parameter exponential family**) Let $\mathbf{Y} \sim f_\theta(\mathbf{y})$, where $\theta = (\theta_1, ..., \theta_p)$, where θ belongs to an open set. We say that f_θ belongs to the *k*-parameter exponential family if

(i) $\text{supp}(f_\theta)$ does not depend on θ

(ii) $f_\theta(\mathbf{y}) = \exp\{\sum_{j=1}^k c_j(\theta) T_j(\mathbf{y}) + d(\theta) + S(\mathbf{y})\}$, $\mathbf{y} \in \text{supp}(f_\theta)$ for some known functions $c_j(\cdot), T_j(\cdot)$, $j = 1, ..., k$; $d(\cdot)$ and $S(\cdot)$.

The functions $c(\theta) = (c_1(\theta), ..., c_k(\theta))$ are the natural parameters of the distribution.

Note that the dimensionality p of the original parameter θ is not necessarily the same as the dimensionality k of the natural parameter $c(\theta)$.

Example 1.14 (**normal distribution with unknown mean and variance**) Consider the normal distribution $\mathcal{N}(\mu, \sigma^2)$, where both μ and σ are unknown, $\theta = (\mu, \sigma)$, and $dim(\theta) = 2$. Its support is the entire real line for any μ and σ, while

$$f_\theta(y) = \frac{1}{\sqrt{2\pi}\sigma} \exp\left\{-\frac{(y-\mu)^2}{2\sigma^2}\right\} = \exp\left\{y\frac{\mu}{\sigma^2} - y^2\frac{1}{2\sigma^2} - \frac{1}{2}\left(\frac{\mu^2}{\sigma^2} + \ln(2\pi\sigma^2)\right)\right\}$$

that corresponds to a two-parameter exponential family with

$$c_1(\theta) = \frac{\mu}{\sigma^2}, \ c_2(\theta) = -\frac{1}{2\sigma^2},$$

$$T_1(y) = y, \ T_2(y) = y^2,$$

$$d(\theta) = -\frac{1}{2}\left(\frac{\mu^2}{\sigma^2} + \ln(2\pi\sigma^2)\right), \ S(y) \equiv 0.$$

Example 1.15 Consider the normal distribution $\mathcal{N}(\mu, a^2\mu^2)$ with a known $a > 0$ from Example 1.11, where the standard deviation is proportional to the mean. This is a *one*-parameter distribution whose support is the entire real line. Following the calculus in the previous Example 1.14, we have

$$f_\mu(y) = \frac{1}{\sqrt{2\pi}a\mu} \exp\left\{-\frac{(y-\mu)^2}{2a^2\mu^2}\right\}$$

$$= \exp\left\{y\frac{1}{a^2\mu} - y^2\frac{1}{2a^2\mu^2} - \frac{1}{2}\left(\frac{1}{a^2} + \ln(2\pi a^2\mu^2)\right)\right\}.$$

Hence, the considered distribution belongs to a *two*-parameter exponential family, where

$$c_1(\mu) = \frac{1}{a^2\mu}, \ c_2(\mu) = -\frac{1}{2a^2\mu^2},$$

$$T_1(y) = y, \ T_2(y) = y^2,$$

$$d(\theta) = -\frac{1}{2}\left(\frac{1}{a^2} + \ln(2\pi a^2\mu^2)\right), \ S(y) \equiv 0.$$

Other examples of distributions from the exponential family are given in Exercise 1.2 and Exercise 1.3. An immediately obvious counterexample of a distribution that does not belong to this family is the uniform distribution $\mathcal{U}(0, \theta)$, since its support $(0, \theta)$, does depend on θ.

We now study some basic common properties of distributions within the exponential family.

Theorem 1.4 Consider a random sample $Y_1, ..., Y_n$, where $Y_i \sim f_\theta(y)$ and f_θ belongs to a *k*-parameter exponential family of distributions with natural parameters $c(\theta) = (c_1(\theta), ..., c_k(\theta))$, that is, $f_\theta(y_i) = \exp\left\{\sum_{j=1}^k c_j(\theta) T_j(y_i) + d(\theta) + S(y_i)\right\}$, $y_i \in \text{supp}(f_\theta)$. Then,

(i) the joint distribution of $\mathbf{Y} = (Y_1, ..., Y_n)$ also belongs to the *k*-parameter exponential family with the same natural parameters $c(\theta)$

(ii) $T(\mathbf{Y}) = (\sum_{i=1}^n T_1(Y_i), ..., \sum_{i=1}^n T_k(Y_i))$ is the sufficient statistic for $c(\theta)$ (and, therefore, for θ)

Proof. The joint distribution $f_\theta(\mathbf{y})$ of $\mathbf{Y}$ is

$$
\begin{aligned}
f_\theta(\mathbf{y}) &= \prod_{i=1}^n f_\theta(y_i) = \prod_{i=1}^n \exp\left\{\sum_{j=1}^k c_j(\theta)T_j(y_i) + d(\theta) + S(y_i)\right\} \\
&= \exp\left\{\sum_{j=1}^k c_j(\theta)\sum_{i=1}^n T_j(y_i) + nd(\theta) + \sum_{i=1}^n S(y_i)\right\},
\end{aligned}
$$

which is of the form of the exponential family with the natural parameters $c(\theta) = (c_1(\theta), ..., c_k(\theta))$. The sufficiency of $T(\mathbf{Y}) = (\sum_{i=1}^n T_1(Y_i), ..., \sum_{i=1}^n T_k(Y_i))$ follows immediately by the Fisher–Neyman factorization theorem. □

For example, applying Theorem 1.4 for Bernoulli data implies that $T(\mathbf{Y}) = \sum_{i=1}^n Y_i$ is a sufficient statistic for p (see Example 1.13). We have already obtained this result directly from the definition of a sufficient statistic in Example 1.5 previously. Similarly, for the normal sample with the unknown mean μ and variance σ^2 in Example 1.14, a sufficient statistic is $(\sum_{i=1}^n Y_i, \sum_{i=1}^n Y_i^2)$ (compare with the sufficient statistic $(\bar{Y}, \sum_{i=1}^n (Y_i - \bar{Y})^2)$ derived from the factorization theorem in Example 1.7).

Theorem 1.5 * Let $Y_1, ..., Y_n$ be a random sample from a distribution $f_\theta(y)$, $\theta = (\theta_1, ..., \theta_p) \in \Theta$ belonging to the k-parameter exponential family with the natural parameters $(c_1(\theta), ..., c_k(\theta))$, that is, $f_\theta(y) = \exp\left\{\sum_{j=1}^k c_j(\theta)T_j(y) + d(\theta) + S(y)\right\}$, $y \in \text{supp}(f_\theta)$. Assume that the set $\{(c_1(\theta), ..., c_k(\theta)),\ \theta \in \Theta\}$ contains an open k-dimensional rectangle. Then, the sufficient statistic $T(\mathbf{Y}) = (\sum_{i=1}^n T_1(Y_i), ..., \sum_{i=1}^n T_k(Y_i))$ is complete (and, therefore, minimal sufficient if the latter exists).

The proof of Theorem 1.5 can be found, for example, in Lehmann and Romano (2005, Theorem 4.3.1).

Example 1.16 * Consider again a normal sample from $\mathcal{N}(\mu, \sigma^2)$, where both parameters are unknown. We showed in Example 1.14 that this distribution belongs to a two-parameter exponential family with the natural parameters $c_1(\mu, \sigma) = \frac{\mu}{\sigma^2}$ and $c_2(\mu, \sigma) = -\frac{1}{2\sigma^2}$. The set $\{(\frac{\mu}{\sigma^2}, -\frac{1}{2\sigma^2}),\ \mu \in \mathbb{R}, \sigma \in \mathbb{R}^+\}$ is the lower half plane and certainly contains a two-dimensional rectangle. Thus, the corresponding sufficient statistic $T(\mathbf{Y}) = (\sum_{i=1}^n Y_i, \sum_{i=1}^n Y_i^2)$ is complete and, hence, minimal sufficient as we have already seen in Example 1.9.

On the other hand, for the normal distribution $\mathcal{N}(\mu, a^2\mu^2)$ with a known $a > 0$ from Example 1.15, the natural parameters are $c_1(\mu) = \frac{1}{a^2\mu}$ and $c_2(\mu) = -\frac{1}{2a^2\mu^2}$. One verifies immediately that the set $\{(\frac{1}{a^2\mu}, -\frac{1}{2a^2\mu^2}),\ \mu \in \mathbb{R}\}$ consists only of a single parabola and does not contain a two-dimensional rectangle. In fact, we showed in Example 1.11 that the corresponding sufficient statistic $T(\mathbf{Y}) = (\sum_{i=1}^n Y_i, \sum_{i=1}^n Y_i^2)$ is not complete in this case. Nevertheless, it is still minimal sufficient (see Exercise 1.5).

1.7 Exercises

Exercise 1.1 Let $Y_1, ..., Y_n$ be a random sample from a Poisson distribution $Pois(\lambda)$. In Example 1.6, we found that $T(\mathbf{Y}) = \sum_{i=1}^n Y_i$ is a sufficient statistic for λ using the Fisher–Neyman factorization theorem.

1. Obtain this result directly using the definition of a sufficient statistic. What is the distribution of $T(\mathbf{Y})$?

2. *Show that $T(\mathbf{Y})$ is a minimal sufficient statistic for λ.

3. *Show that $T(\mathbf{Y})$ is complete using

   (a) the definition of a complete statistic.

   (b) the properties of the exponential family of distributions.

Exercise 1.2 Find sufficient statistics for the parameters of the following distributions based on a random sample of size n:

1. geometric distribution $Geom(p)$

2. uniform distribution $\mathscr{U}(\theta, \theta+1)$

3. uniform distribution $\mathscr{U}(-\theta, \theta)$

4. uniform distribution $\mathscr{U}(\theta_1, \theta_2)$

5. shifted exponential distribution $f_{\theta,\tau}(y) = \theta e^{-\theta(y-\tau)}$, $y \geq \tau$, $\theta > 0$, where

   (a) τ is known

   (b) τ is unknown

6. double exponential distribution $f_\theta(y) = \frac{\theta}{2}e^{-\theta|y|}$, $-\infty < y < \infty$, $\theta > 0$

7. Gamma distribution $Gamma(\alpha, \beta)$

8. Beta distribution $Beta(p, q)$

9. log-normal distribution $f_{\mu,\sigma}(y) = \frac{1}{\sqrt{2\pi}\sigma y}e^{-\frac{(\ln y - \mu)^2}{2\sigma^2}}$

10. Weibull distribution $f_{\alpha,\beta}(y) = \beta\alpha y^{\alpha-1}\exp\{-\beta y^\alpha\}$, $y > 0$, $\alpha > 0$ $\beta > 0$, where

    (a) α is known

    (b) α is unknown

11. Pareto distribution $f_{\alpha,\beta}(y) = \frac{\alpha}{\beta}\left(\frac{\beta}{y}\right)^{\alpha+1}$, $y \geq \beta$, $\alpha, \beta > 0$, where

    (a) β is known

    (b) β is unknown

12. inverse Gaussian distribution $f_{\theta_1,\theta_2}(y) = \sqrt{\frac{\theta_1}{2\pi y^3}}\exp\left\{-\frac{\theta_1(y-\theta_2)^2}{2\theta_2^2 y}\right\}$, $y > 0$, $\theta_1 > 0$ $\theta_2 > 0$, where

    (a) θ_1 is known

    (b) θ_2 is known

    (c) both parameters are unknown

Exercise 1.3 Which of the distributions in Exercise 1.2 belong to the exponential family? Find their natural parameters. In addition, consider also the following distributions:

1. Poisson distribution $Pois(\lambda)$

2. Cauchy distribution $f_\theta(y) = \frac{1}{\pi(1+(y-\theta)^2)}$, $-\infty < y < \infty$

3. $f_\theta(y) = \frac{\theta}{(1+y)^{\theta+1}}$, $y \geq 0$, $\theta > 0$

4. $f_\theta(y) = \theta y^{\theta-1}$, $0 < y < 1$, $\theta > 0$

5. $f_\theta(y) = \frac{2(y+\theta)}{1+2\theta}$, $0 < y < 1$, $\theta > 0$

Exercise 1.4 Consider a population with three kinds of individuals labeled, say, 1, 2, and 3 occurring in the following proportions: $p_1 = p^2$, $p_2 = 2p(1-p)$, $p_3 = (1-p)^2$ $(0 < p < 1)$. In a random sample of size n chosen from this population, n_1 individuals have been labeled 1, n_2—2 and n_3—3 respectively $(n_1 + n_2 + n_3 = n)$.

1. Write down the likelihood function.

2. Does the distribution of the data belong to the exponential family? What are the natural parameter and the sufficient statistic?

Exercise 1.5 *Let $Y_1, ..., Y_n \sim \mathcal{N}(\mu, a^2\mu^2)$, where $a > 0$ is known (see Example 1.11). Show that $T(\mathbf{Y}) = (\sum_{i=1}^n Y_i, \sum_{i=1}^n Y_i^2)$ is a minimal sufficient statistic for μ.

Chapter 2

Point Estimation

2.1 Introduction

As we have discussed in Section 1.1, estimation of the unknown parameters of distributions from the data is one of the key issues in statistics.

Suppose that $Y_1, ..., Y_n$ is a random sample from a distribution $f_\theta(y)$, where $f_\theta(y)$ is a probability or a density function for a discrete or continuous random variable, respectively, assumed to belong to a parametric family of distributions $\mathscr{F}_\theta$, $\theta \in \Theta$. In other words, the data distribution f_θ is supposed to be known up to the unknown parameter(s) $\theta \in \Theta$. For example, the birth data in Example 1.1 is a random sample from a Bernoulli $B(1, p)$ distribution with the unknown parameter p. The goal is to estimate the unknown θ from the data. We start from a general definition of an estimator:

Definition 2.1 (**estimator**) A (point) *estimator* $\hat{\theta} = \hat{\theta}(\mathbf{Y})$ of an unknown parameter θ is any statistic used for estimating θ. The value of $\hat{\theta}(\mathbf{y})$ evaluated for a given sample is called an *estimate*.

Thus, $\bar{Y}$, $Y_{max} = \max(Y_1, ..., Y_n)$, and $Y_3 \log(|Y_1|) - Y_2^{Y_5}$ are examples of estimators of θ. This is a general, somewhat trivial definition that still does not say anything about the goodness of estimation. One would evidently be interested in "good" estimators. In this chapter, we firstly present several methods of estimation and then define and discuss their goodness.

2.2 Maximum likelihood estimation

This is probably the most used method of estimation of parameters in parametric models. Its underlying idea is simple and intuitively clear. Recall that given a random sample $Y_1, ..., Y_n \sim f_\theta(y)$, $\theta \in \Theta$, the likelihood function $L(\theta; \mathbf{y})$ defined in Section 1.2 is the joint probability (for a discrete random variable) or density (for a continuous random variable) of the observed data as a function of an unknown parameter(s) θ: $L(\theta; \mathbf{y}) = \prod_{i=1}^{n} f_\theta(y_i)$.

As we have discussed, $L(\theta; \mathbf{y})$ is the measure of likeliness of a parameter's values θ for the observed data $\mathbf{y}$. It is only natural then to seek "the most likely" value of θ, and this is the idea behind the *maximum likelihood* estimation.

Definition 2.2 (**maximum likelihood estimator**) The maximum likelihood estimator (MLE) $\hat{\theta}$ of θ is $\hat{\theta} = \arg\max_{\theta \in \Theta} L(\theta; \mathbf{y})$—the value of θ, which maximizes the likelihood.

Finding MLEs is, therefore, essentially an exercise in calculus. In practice, it is usually easier to find a maximum of a *log*-likelihood function $l(\theta; \mathbf{y}) = \ln L(\theta; \mathbf{y}) = \sum_{i=1}^{n} \ln f_\theta(y_i)$.

Question 2.1 Explain why a maximizer of the log-likelihood function is an MLE of θ.

Question 2.2 Show that an MLE is necessarily a function of a sufficient statistic for θ.

A series of examples below demonstrate various possible scenarios and describe the main properties of MLEs.

Example 2.1 (**binomial distribution**) Consider again the birth data example, where $Y_i \sim \mathscr{B}(1, p)$, $i = 1, ..., n$ and p is the probability of a newborn girl. As we have seen in Example 1.3, the likelihood function $L(p; \mathbf{y})$ is

$$L(p; \mathbf{y}) = p^{\sum_{i=1}^{n} y_i} (1 - p)^{n - \sum_{i=1}^{n} y_i}$$

and the log-likelihood

$$l(p;\mathbf{y}) = \sum_{i=1}^{n} y_i \ln p + (n - \sum_{i=1}^{n} y_i) \ln(1-p).$$

Differentiating $l(p;\mathbf{y})$ w.r.t. p yields

$$l_p'(p;\mathbf{y}) = \frac{\sum_{i=1}^{n} y_i}{p} - \frac{n - \sum_{i=1}^{n} y_i}{1-p} = \frac{\sum_{i=1}^{n} y_i - np}{p(1-p)}.$$

Solving $l_p'(p;\mathbf{y}) = 0$ one obtains the MLE of p:

$$\hat{p} = \frac{\sum_{i=1}^{n} Y_i}{n}, \tag{2.1}$$

where one can easily check (e.g., by verifying that the second derivative $l_p''(p;\mathbf{y}) < 0$) that the $\hat{p}$ in (2.1) is indeed a maximum. For the considered example, where there were 12 girls among a total of 20 newborn babies, the corresponding estimate is $\hat{p} = 0.6$ (see also Figure 1.1).

Note that the MLE $\hat{p}$ is the proportion of successes (girls) in a series of n trials, which is a natural estimator for the probability of success in a population.

Example 2.2 (uniform distribution $\mathcal{U}(0,\theta)$) Suppose $Y_1, ..., Y_n \sim \mathcal{U}(0,\theta)$, where θ is unknown and we want to find its MLE. The uniform density is $f_\theta(y) = \frac{1}{\theta}$, $0 \le y \le \theta$ and the likelihood is therefore

$$L(\theta;\mathbf{y}) = \frac{1}{\theta^n}, \; \theta \ge y_{\max}, \tag{2.2}$$

where $y_{\max} = \max(y_1, ..., y_n)$, and 0 otherwise. Differentiating $L(\theta;\mathbf{y})$ in (2.2) to find its maximum does not work now since the likelihood is not differentiable at $\theta = y_{\max}$. However, it is strictly decreasing for $\theta \ge y_{\max}$ (see Figure 2.1) and the MLE is therefore $\hat{\theta} = Y_{\max}$.

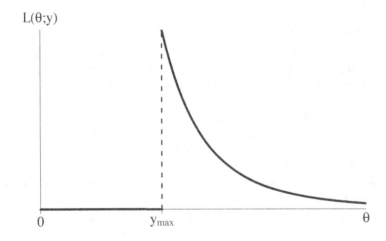

Figure 2.1: Likelihood function for $\mathcal{U}(0,\theta)$.

Example 2.3 (uniform distribution $\mathcal{U}(\theta,\theta+1)$) Now we are given observations $y_1, ..., y_n$ from a uniform distribution on a unit length interval but the location of the interval is unknown: $f_\theta(y) = 1$, $\theta \le y \le \theta + 1$. Let $y_{\min} = \min(y_1, ..., y_n)$ and $y_{\max} = \max(y_1, ..., y_n)$. Simple straight-forward calculus implies then that $L(\theta;\mathbf{y}) = 1$, if $y_{\max} - 1 \le \theta \le y_{\min}$ (note that $y_{\max} - 1 \le y_{\min}$ since otherwise the sample range is more than one, which contradicts the assumptions on the data distribution), and is zero otherwise (see Figure 2.2). The MLE of θ is therefore the entire interval $(Y_{max} - 1, Y_{min})$. This example demonstrates that an MLE is not necessarily unique and in some cases may even be an infinite set of equally likely values.

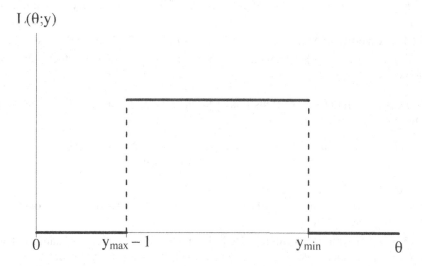

Figure 2.2: Likelihood function for $\mathscr{U}(\theta, \theta + 1)$.

Example 2.4 (**exponential distribution** $\exp(\theta)$) Every day John takes a bus from his home to the University campus. Over a period of 30 days he recorded how much time he waited for a bus. In fact, his average waiting time was 8 minutes. It would be reasonable to model the recorded waiting times $t_1, ..., t_{30}$ as sampled from an exponential distribution $\exp(\theta)$ with unknown θ with the density $f_\theta(t) = \theta e^{-\theta t}$, $t \geq 0$. John wants to find an MLE of θ.

The likelihood is

$$L(\theta; \mathbf{y}) = \prod_{i=1}^{n} \theta e^{-\theta t_i} = \theta^n e^{-\theta \sum_{i=1}^{n} t_i}$$

and the log-likelihood is

$$l(\theta; \mathbf{y}) = n \ln \theta - \theta \sum_{i=1}^{n} t_i. \qquad (2.3)$$

Differentiating $l(\theta; \mathbf{y})$ w.r.t. θ and solving $l'_\theta(\theta; \mathbf{y}) = 0$ yields the MLE $\hat{\theta} = n / \sum_{i=1}^{n} t_i = \frac{1}{\bar{t}} = 0.125$ ($l''_\theta(\theta; \mathbf{y}) = -\frac{n}{\theta^2} < 0$ and, therefore, it is indeed maximum).

In fact, it is likely that John would be more interested in estimating the expected waiting time for a bus $\mu = E(t) = \frac{1}{\theta}$ rather than θ itself, which is the mean rate of bus arrivals per minute. A first attempt to find the corresponding MLE would be to re-write the log-likelihood $l(\theta; \mathbf{y})$ in (2.3) in terms of μ:

$$l(\mu; \mathbf{y}) = -n \ln \mu - \frac{\sum_{i=1}^{n} t_i}{\mu}$$

and to find its maximum $\hat{\mu} = \frac{\sum_{i=1}^{n} t_i}{n} = \bar{t} = 8$ min. Note that $\mu = \frac{1}{\theta}$ and exactly the same relation holds for their MLEs: $\hat{\mu} = \frac{1}{\hat{\theta}}$. In fact, once the MLE $\hat{\theta}$ was obtained, it was not necessary to do the calculus again to find $\hat{\mu}$!

Does the invariance property of an MLE estimator demonstrated in Example 2.4 hold only in some specific cases or is it also true in the general case? Let θ be an unknown parameter and define a new parameter of interest $\xi = g(\theta)$, which is a function of θ (e.g., $\xi = \mu = \theta^{-1}$ in the previous Example 2.4). Essentially, it is simply a reparametrization of the original problem. One needs to be somewhat more careful when re-writing the likelihood in terms of a new parameter ξ when $g(\cdot)$ is not a one-to-one function, e.g., $g(\theta) = \theta^2$. In this case, define the function $\widetilde{L}(\xi; \mathbf{y}) = \sup_{\theta:g(\theta)=\xi} L(\theta; \mathbf{y})$ (aka *profile likelihood*) and the MLE $\hat{\xi} = \arg\max_\xi \widetilde{L}(\xi; \mathbf{y})$. Then the following theorem holds:

Theorem 2.1 (invariance of MLE) Let $\hat{\theta}$ be an MLE of θ, $\xi = g(\theta)$ and define $\hat{\xi} = \arg\max_\xi \widetilde{L}(\xi; \mathbf{y})$, where $\widetilde{L}(\xi; \mathbf{y}) = \sup_{\theta:g(\theta)=\xi} L(\theta; \mathbf{y})$. Then, $\hat{\xi} = g(\hat{\theta})$. In other words, an MLE of a function of a parameter is the same function of a parameter's MLE.

Proof. Theorem 2.1 is intuitively obvious-it is just a matter of reparametrization of the likelihood. Formally, for any value of ξ, we have

$$\widetilde{L}(\xi; \mathbf{y}) = \sup_{\theta:g(\theta)=\xi} L(\theta; \mathbf{y}) \leq L(\hat{\theta}; \mathbf{y}) = \sup_{\theta:g(\theta)=g(\hat{\theta})} L(\theta; \mathbf{y}) = \widetilde{L}(g(\hat{\theta}); \mathbf{y})$$

and, therefore, $g(\hat{\theta})$ is the maximizer of $\widetilde{L}(\cdot; \mathbf{y})$, i.e. $\hat{\xi} = g(\hat{\theta})$. $\qquad\square$

Theorem 2.1 establishes an extremely helpful property of MLEs that allows one to estimate easily various parameters of interest.

Example 2.5 (continuation of Example 2.4) Today John is in a hurry for an exam. He knows that if he waits for a bus for more than 5 minutes, he will be late. Based on the collected data, he wants to estimate the chance to be late. He would take a taxi instead if it is sufficiently high.

Let p be the probability of waiting for more than 5 minutes, that is, $p = P(t > 5) = e^{-5\theta}$. Following Theorem 2.1, the MLE for p can be easily calculated as $\hat{p} = e^{-5\hat{\theta}} = e^{-5/8} = 0.535$. The estimated chance of being late is definitely quite high, and John would be advised not to wait for a bus but to take a taxi.

Example 2.6 There are two cashiers in a shop: the experienced employee Alice, and Kate, who was hired very recently. It is not surprising that on average, Alice serves clients twice as fast as Kate. Clients choose a cashier at random. To estimate the expected service times per client of both cashiers, an owner of the shop registered the service times of their clients during one working day. Alice served n clients for $t_1, ..., t_n$ minutes, while Kate served m clients for $\tau_1, ..., \tau_m$ minutes respectively.

We should first find a proper model for the data. It is reasonable to assume that the service time of each of the two cashiers is exponentially distributed. Hence, $t_i, ..., t_n$ and $\tau_1, ..., \tau_m$ are independent random samples from $\exp(2\theta)$ and $\exp(\theta)$, respectively, where $\mu_A = 1/(2\theta)$ and $\mu_K = 2\mu_A = 1/\theta$ are the expected service times of Alice and Kate.

We start from finding an MLE for θ. This is an example where not all the data is identically distributed, but both samples nevertheless contain relevant information about the unknown common parameter θ.

Following the definition of the likelihood, one has

$$L(\theta; \mathbf{y}) = \prod_{i=1}^{n} \left(2\theta e^{-2\theta t_i}\right) \prod_{j=1}^{m} \left(\theta e^{-\theta \tau_j}\right) = 2^n \theta^{n+m} e^{-\theta\left(2\sum_{i=1}^{n} t_i + \sum_{j=1}^{m} \tau_j\right)}.$$

The log-likelihood is then

$$l(\theta; \mathbf{y}) = n \ln 2 + (n+m) \ln \theta - \theta \left(2\sum_{i=1}^{n} t_i + \sum_{j=1}^{m} \tau_j\right)$$

and

$$l'_\theta(\theta; \mathbf{y}) = \frac{n+m}{\theta} - \left(2\sum_{i=1}^{n} t_i + \sum_{j=1}^{m} \tau_j\right).$$

Solving $l'_\theta(\theta; \mathbf{y}) = 0$ and verifying that $l''_\theta(\theta; \mathbf{y}) < 0$ implies that the MLE $\hat{\theta} = (n + m)/(2\sum_{i=1}^n t_i + \sum_{j=1}^m \tau_j)$, while $\hat{\mu}_A = 1/(2\hat{\theta})$ and $\hat{\mu}_K = 2\hat{\mu}_A = 1/\hat{\theta}$.

So far, we have considered examples with a single unknown parameter. However, maximum likelihood estimation can be directly applied in a multiparameter case as well.

Example 2.7 (normal distribution) Consider a sample $Y_1, ..., Y_n$ from a normal distribution $\mathcal{N}(\mu, \sigma^2)$, where both μ and σ^2 are unknown. The log-likelihood (see (1.4))

$$l(\mu, \sigma; \mathbf{y}) = -n \ln \sqrt{2\pi} - n \ln \sigma - \frac{\sum_{i=1}^n (y_i - \mu)^2}{2\sigma^2}.$$

To find MLEs for μ and σ^2 (or, σ—see Theorem 2.1), we differentiate $l(\mu, \sigma; \mathbf{y})$ w.r.t. both parameters

$$\frac{\partial l}{\partial \mu} = \frac{\sum_{i=1}^n (y_i - \mu)}{\sigma^2} = 0$$

$$\frac{\partial l}{\partial \sigma} = -\frac{n}{\sigma} + \frac{\sum_{i=1}^n (y_i - \mu)^2}{\sigma^3} = 0$$

and solve the above system of two simultaneous equations to get

$$\hat{\mu} = \bar{Y}, \quad \hat{\sigma}^2 = \frac{\sum_{i=1}^n (Y_i - \bar{Y})^2}{n}. \tag{2.4}$$

We leave the reader to verify that (2.4) is the maximal point.

In particular, for Example 1.2, the MLE estimates of Mr. Skeptic are $\hat{\mu} = 196.6g$ and $\hat{\sigma}^2 = 30.02g^2$ ($\hat{\sigma} = 5.48g$).

Example 2.8 (Cauchy distribution) A Cauchy distribution $Cauchy(\theta)$ with the parameter θ has a density

$$f_\theta(y) = \frac{1}{\pi(1 + (y - \theta)^2)}, \quad -\infty < y < \infty$$

(see Appendix A.4.10). It arises, in particular, in a series of physical problems and can be obtained as a distribution of the ratio of two independent normal variables. The Cauchy distribution is an example of a distribution with undefined mean, variance, and higher moments. The density is symmetric around θ, which is the mode and the median of the distribution.

The log-likelihood of a sample $Y_1, ..., Y_n$ from $Cauchy(\theta)$ is then

$$l(\theta; \mathbf{y}) = -n \ln \pi - \sum_{i=1}^n \ln\left((1 + (y_i - \theta)^2\right).$$

Differentiating and setting $l'_\theta(\theta; \mathbf{y})$ to zero leads to the following equation:

$$l'_\theta(\theta; \mathbf{y}) = 2\sum_{i=1}^n \frac{y_i - \theta}{1 + (y_i - \theta)^2} = 0 \tag{2.5}$$

However, unlike the previous examples, the solution of (2.5) does not exist in a closed form, and the MLE $\hat{\theta}$ can be obtained only approximately by various numerical procedures.

Although the conception of MLE was motivated by an intuitively clear underlying idea, the justification for its use is much deeper. It is a really "good" method of estimation, and we will present the theoretical ground for this claim later in Section 5.8.

2.3   Method of moments

Another popular method of estimation is the method of moments. Its main idea is based on expressing the *population* moments of the distribution of data in terms of its unknown parameter(s) and equating them to their corresponding *sample* moments. The parameter(s) are then estimated by the solutions of the resulting equations.

Suppose a random variable Y has a distribution $f_\theta(y)$, which is a function of p (unknown) parameters $\theta = (\theta_1, ..., \theta_p)$. The j-th moment of Y is $\mu_j(\theta) = E(Y^j) = \int y^j f_\theta(y) dy$ if the integral exists. In particular, $\mu_1 = EY$, $\mu_2 = EY^2 = (EY)^2 + Var(Y)$. Consider the p first moments. Given a random sample $Y_1, ..., Y_n \sim f_\theta(y)$ and equating each $\mu_j(\theta)$ to the corresponding *sample* moment $m_j = \frac{1}{n} \sum_{i=1}^n Y_i^j$, one obtains the system of p simultaneous equations $\mu_j(\theta) = m_j$, $j = 1, ..., p$. Solving this system implies the method of moments estimators (MME) $\hat{\theta} = (\hat{\theta}_1, ..., \hat{\theta}_p)$.

Example 2.9 (estimating of the mean and the variance by the method of moments)   Assume that the two first moments of Y are finite and we want to estimate $\mu = EY$ and $\sigma^2 = Var(Y)$. Applying the method of moments yields the following system of two simultaneous equations:

$$\mu = \bar{Y}$$
$$\mu^2 + \sigma^2 = \frac{1}{n} \sum_{i=1}^n Y_i^2$$

that implies,

$$\hat{\mu}_{MME} = \bar{Y}$$
$$\hat{\sigma}^2_{MME} = \frac{1}{n} \sum_{i=1}^n Y_i^2 - \bar{Y}^2 = \frac{1}{n} \sum_{i=1}^n (Y_i - \bar{Y})^2.$$

Example 2.10 (uniform distribution $\mathcal{U}(0, \theta)$)   Recall that the MLE of θ is $\hat{\theta}_{MLE} = Y_{\max}$ (see Example 2.2). Now we estimate θ by the method of moments: $\mu_1 = \theta/2$, $m_1 = \bar{Y}$ and solving $\mu_1 = m_1$ one has

$$\hat{\theta}_{MME} = 2\bar{Y}.$$

The two estimators are different. At this stage we do not know how to compare between them, but we will return to this example later (see Exercise 2.12).

Question 2.3 (invariance of MME)   Show that MMEs are invariant w.r.t. one-to-one transformations, that is, if $g(\theta)$ is a one-to-one function of θ, then $\widehat{g(\theta)}_{MME} = g(\hat{\theta}_{MME})$.

We should point out some possible problems that arise with method of moments estimators. Consider again Example 2.10 and assume that we observed a sample $0.2, 0.6, 2, 0.4$ of size $n = 4$ from $U(0, \theta)$. The MME $\hat{\theta}_{MME} = 2\bar{y} = 1.6$. Such an estimate evidently does not make much sense since one of the observations $y_3 = 2 > \hat{\theta}_{MME}$.

Although it is conventional to use the first moments for estimating the unknown parameters, one can nevertheless work with higher moments and the resulting MMEs will generally be different. We can estimate θ, for example, by μ_2 rather than by μ_1. We have $\mu_2 = (EY)^2 + Var(Y) = \theta^2/4 + \theta^2/12 = \theta^2/3$. Solving $\mu_2 = m_2 = \frac{1}{n} \sum_{i=1}^n Y_i^2$ yields another MME $\tilde{\theta}_{MME}$:

$$\tilde{\theta}_{MME} = \sqrt{\frac{3}{n} \sum_{i=1}^n Y_i^2}.$$

For the considered sample $\tilde{\theta}_{MME} \approx 1.85$, which is still nonsensical.

Due to the above reasons, the MME is less used than their MLE counterparts that also enjoy various asymptotic optimality properties (see Section 5.8). However, unlike the latter, they are usually simpler to compute and can be used, for example, as reasonable initial values in numerical iterative procedures for MLEs. On the other hand, the method of moments does not require knowledge of the entire distribution of the data (up to the unknown parameters) but only its moments and thus may be less sensitive to a possible misspecification of a model.

2.4 Method of least squares

The method of least squares plays a key role in regression and analysis of variance. We briefly present here its main ideas.

In a typical regression setup, we are given n observations $(\mathbf{x}_i, y_i), i = 1, ..., n$ over m explanatory variables (predictors) $\mathbf{x} = (x_1, ..., x_m)$ and the response variable Y. We assume that

$$y_i = g_\theta(\mathbf{x}_i) + \varepsilon_i, \quad i = 1, ..., n,$$

where the response function $g_\theta(\cdot) : \mathbb{R}^m \to \mathbb{R}$ has a known parametric form and depends on $p \le n$ unknown parameters $\theta = (\theta_1, ..., \theta_p)$, and the noise ε_i's are i.i.d. random variables with zero means and a finite variance.

We look for a $\hat{\theta}$ that yields the best fit $g_{\hat{\theta}}(\mathbf{x})$ to the observed $\mathbf{y}$ w.r.t. the Euclidean distance, that is, which minimizes

$$||\mathbf{y} - \mathbf{g}_\theta(\mathbf{x})||_2^2 = \sum_{i=1}^n (y_i - g_\theta(\mathbf{x}_i))^2.$$

Provided that $g_\theta(\cdot)$ is differentiable w.r.t. θ, the least squares estimator (LSE) $\hat{\theta}$ satisfies the system of *normal* equations

$$\frac{\partial}{\partial \theta_j} \sum_{i=1}^n (y_i - g_\theta(\mathbf{x}_i))^2 = -2 \sum_{i=1}^n (y_i - g_\theta(\mathbf{x}_i)) \frac{\partial g_\theta(\mathbf{x}_i)}{\partial \theta_j} = 0, \quad j = 1, ..., p. \tag{2.6}$$

Question 2.4 Show that if, in addition, ε_i's are normally distributed, the LSE coincide with the MLE.

For a *linear* regression, where $g_\theta(\cdot)$ is linear in θ, that is, $g_\theta(\mathbf{x}) = \sum_{j=1}^p \theta_j g_j(\mathbf{x})$ and functions $g_j(\cdot) : \mathbb{R}^m \to \mathbb{R}, j = 1, ..., p$ are known, the system of normal equations (2.6) is linear, and its solution is available in the closed form. For a *nonlinear* regression, however, it can generally be only found numerically.

We discuss the least squares estimation for linear regression in more detail in Chapter 8.

2.5 *M-estimators

Recall that the motivation for MLE and LSE was to find estimators that maximize/minimize a certain goodness-of-fit criterion. Thus, the MLE maximizes the likelihood $L(\theta, \mathbf{y}) = \Pi_{i=1}^n f_\theta(y_i)$ or, equivalently, the log-likelihood $l(\theta, \mathbf{y}) = \sum_{i=1}^n \ln f_\theta(y_i)$. Similarly, the LSE minimizes the sum of squares $\sum_{i=1}^n (y_i - g_\theta(\mathbf{x}_i))^2$. More generally, one can consider any function $\rho(\theta, y)$ as a measure of the goodness-of-fit and look for an estimator that maximizes/minimizes $\sum_{i=1}^n \rho(\theta, y_i)$ w.r.t. θ. Such estimators are called *M-estimators* ("M" for "maximum" or "minimum")[1]:

Definition 2.3 (*M-estimator*) For a given function $\rho(\theta, y)$, the *M*-estimator $\hat{\theta}$ of θ is $\hat{\theta} = \arg\min_{\theta \in \Theta} \sum_{i=1}^n \rho(\theta, Y_i)$.

Hence, MLE and LSE are *M*-estimators corresponding respectively to $\rho(\theta, y) = -\ln f_\theta(y)$ and $\rho(\theta, y) = (y - g_\theta(\mathbf{x}))^2$.

Assume that ρ is differentiable w.r.t. θ and let $\psi(\theta, y) = \rho'_\theta(\theta, y)$. The *M*-estimator $\hat{\theta}$ then is the solution of the equation $\sum_{i=1}^n \psi(\theta, Y_i) = 0$.

---

[1] Maximizing ρ is obviously equivalent to minimizing $-\rho$ so in what follows we can talk about minimization without loss of generality.

It turns out that various well-known estimators can be viewed as M-estimators for a particular $\rho(\theta, y)$ or, equivalenlty, $\psi(\theta, y)$.

Example 2.11 (sample mean) Consider $\rho(\theta, y) = (y - \theta)^2$ and, hence, $\psi(\theta, y) = 2(\theta - y)$. The resulting M-estimator $\hat{\theta}$ then satisfies $\sum_{i=1}^{n}(\hat{\theta} - Y_i) = 0$ yielding $\hat{\theta} = \bar{Y}$. In other words, a sample mean $\bar{Y}$ is an M-estimator of θ corresponding to $\rho(\theta, y) = (y - \theta)^2$.

Example 2.12 (sample median) Consider now $\rho(\theta, y) = |y - \theta|$ so that $\psi(\theta, y) = sign(y - \theta)$. Solving

$$\sum_{i=1}^{n} sign(Y_i - \theta) = \#\{i : Y_i > \theta\} - \#\{i : Y_i < \theta\} = 0$$

implies that $\hat{\theta}$ is a sample median $Y_{(0.5)}$.

In fact, any sample α-quantile $Y_{(\alpha)}$ is an M-estimator corresponding to $\rho(\theta, y) = \alpha(y - \theta)_+ + (1 - \alpha)(\theta - y)_+$, where $x_+ = x$ for positive x and zero otherwise (see Exercise 2.22). Example 2.12 is a particular case for $\alpha = 0.5$.

Question 2.5 Show that $Y_{\min}$ and $Y_{\max}$ are also M-estimators.

Similar to MLEs, multiparameter extensions of M-estimators are straightforward. The family of M-estimators includes, therefore, a variety of different types of estimators. Nevertheless, as we will see in Section 5.10, they share common important asymptotic properties.

2.6   Goodness-of-estimation: mean squared error

So far, we have considered several methods of estimation. A natural question is how to compare between various estimators. First, we should define a measure of goodness-of-estimation.

Suppose that we want to estimate an unknown parameter θ of the distribution $f_\theta(y)$, $\theta \in \Theta$ from a sample of observations $Y_1, ..., Y_n \sim f_\theta(y)$. Recall that any estimator $\hat{\theta} = \hat{\theta}(Y_1, ..., Y_n)$ is a function of a random sample and, therefore, is a random variable by itself with a certain distribution, expectation, variance, etc.

Example 2.13 Let $Y_1, ..., Y_n \sim \mathcal{N}(\mu, \sigma^2)$. Consider the MLE $\hat{\mu} = \bar{Y}$ of μ (which is also the MME). Then, $\hat{\mu} \sim \mathcal{N}(\mu, \sigma^2/n)$.

A first, somewhat naive attempt to measure the goodness-of-estimation of $\hat{\theta}$ would be to consider the error $|\hat{\theta} - \theta|$. However, θ is unknown and, as we have mentioned, an estimator $\hat{\theta}$ is a random variable and hence, the value $|\hat{\theta} - \theta|$ will vary from sample to sample. It may be "small" for some of the samples, while "large" for others and, therefore, cannot be used as a proper criterion for goodness-of-estimation of an estimator $\hat{\theta}$. A more reasonable measure would then be an *average* distance over all possible samples, that is, the *mean absolute error* $E|\hat{\theta} - \theta|$, where the expectation is taken w.r.t. joint distribution of $\mathbf{Y} = (Y_1, ..., Y_n)$. It indeed can be used as a measure of goodness-of-estimation but usually, mostly due to convenience (for example, $E|\hat{\theta} - \theta|$ is not differentiable), a more conventional measure is the *mean squared error* (MSE) given by

$$MSE(\hat{\theta}, \theta) = E(\hat{\theta} - \theta)^2.$$

The mean squared error $MSE(\hat{\theta}, \theta)$ can be decomposed into two components:

$$
\begin{aligned}
MSE(\hat{\theta}, \theta) &= E(\hat{\theta} - \theta)^2 = E\left((\hat{\theta} - E\hat{\theta}) + (E\hat{\theta} - \theta)\right)^2 \\
&= E(\hat{\theta} - E\hat{\theta})^2 + (E\hat{\theta} - \theta)^2 \\
&= Var(\hat{\theta}) + b^2(\hat{\theta}, \theta),
\end{aligned}
\tag{2.7}
$$

where the cross-product term vanishes (why?) and $b(\hat{\theta}, \theta) = E\hat{\theta} - \theta$ is called the *bias*. The decomposition (2.7) shows that the overall $MSE(\hat{\theta}, \theta)$ includes a *stochastic* error (the variance term) and a *systematic* or *deterministic* error (the bias term).

To better understand the meanings of both components in (2.7) consider again Example 1.2 from Chapter 1, where Mr. Skeptic estimates the unknown mean weight μ of a standard cheese package by measuring the weight Y of a randomly chosen package. As we have discussed there, repeating the experiment several times with a series of different packages, the resulting weights (estimates) will be different due to the measurement errors of the weighing scale (ignoring for simplicity other possible factors). In fact, we can consider these weights as realizations of a random variable Y, whose variance depends on the device's accuracy. The inherent randomness of an estimator is reflected than in the variance term of the MSE decomposition (2.7). However, it is not the only possible error source. Assume an ideal situation in which Mr. Skeptic uses an absolutely noiseless weighing scale in the sense that all the weights are identical with zero variance (in a more realistic scenario it would mean that its precision is so high that the variance of measurements can be neglected). Does it necessarily mean that his estimator is free of error in this case and he now observes the true μ? It does not, since the weighing scale might not be properly calibrated! For example, its baseline might not be set to a zero level and, therefore, the (same) observed weight will systematically be biased. This systematic error enters (2.7) via the bias term.

Example 2.14 (estimating the expectation by the sample mean) Given a random sample $Y_1,...,Y_n$ we want to estimate the unknown $\mu = EY$ by the sample mean $\bar{Y}$, which is also the MME of μ (see Example 2.9). $E\bar{Y} = \mu$ and, therefore, $\bar{Y}$ is an *unbiased* estimator of μ (the bias $b(\bar{Y},\mu) = 0$), while $Var(\bar{Y}) = \sigma^2/n$. Thus, regardless of the specific distribution of the data (we require only the finiteness of its first two moments), $MSE(\bar{Y},\mu) = \sigma^2/n$. For comparison, consider another trivial estimator $\tilde{\mu} = Y_1$. It would be a nonsensical estimator since it is based on a single observation and ignores all other available data. However, it is still unbiased since $EY_1 = \mu$. But its variance and, therefore, MSE is $Var(\tilde{\mu}) = \sigma^2$, which is n times larger than that of $\hat{\mu} = \bar{Y}$. This essentially explains why averaging is done when μ is estimated.

Having defined the goodness-of-estimation measure by MSE, one can compare different estimators and choose the one with the smallest MSE. However, since the $MSE(\hat{\theta},\theta)$ typically depends on the unknown θ, it is a common situation where no estimator is *uniformly* superior for all $\theta \in \Theta$.

Example 2.15 Suppose we observe a series of Bernoulli trials $Y_1,...,Y_n \sim \mathscr{B}(1,p)$ and want to estimate the unknown p. Consider the following three estimators:

$$\hat{p}_1 = \frac{\sum_{i=1}^n Y_i}{n}, \quad \hat{p}_2 = \frac{\sum_{i=1}^n Y_i + 1}{n+2} \quad \text{and} \quad \hat{p}_3 = Y_1.$$

The natural estimator $\hat{p}_1$ is the MLE and the MME. Since $\hat{p}_1 = \bar{Y}$, it is unbiased, $Var(\hat{p}_1) = \frac{p(1-p)}{n}$ and

$$MSE(\hat{p}_1,p) = \frac{p(1-p)}{n}.$$

The justification for $\hat{p}_2$ can be clearer if we interpret it as the proportion of successes in a series of $n+2$ Bernoulli trials with two artificially added trials: one success and one failure. These imaginary observations represent a prior belief that the unknown p is around 0.5. Thus (at this stage on the intuitive level only), $\hat{p}_2$ may be somewhat better than $\hat{p}_1$ if the true (unknown!) p is indeed close to 0.5. $E(\hat{p}_2) = (np + 1)/(n+2)$ and $Var(\hat{p}_2) = np(1 - p)/(n+2)^2$. Unlike $\hat{p}_1$, $\hat{p}_2$ is a biased estimator but with a smaller variance, where

$$MSE(\hat{p}_2,p) = (E\hat{p}_2 - p)^2 + Var(\hat{p}_2) = \frac{(1-2p)^2 + np(1-p)}{(n+2)^2}.$$

Finally, the trivial estimator $\hat{p}_3$ is unbiased with a variance $Var(\hat{p}_3) = p(1-p)$ and

$$MSE(\hat{p}_3,p) = p(1-p).$$

Figure 2.3: MSE for three estimators of p in $\mathscr{B}(10,p)$: $MSE(\hat{p}_1,p)$ (solid line), $MSE(\hat{p}_2,p)$ (dashed line), and $MSE(\hat{p}_3,p)$ (dotted line).

We plot all three MSEs as functions of p on Figure 2.3. One immediately realizes that $\hat{p}_1$ uniformly (for all p) outperforms $\hat{p}_3$. In fact, it is not so surprising and we have already discussed in Example 2.14 that the sample mean is always better for estimating the expectation than a single observation. It is more intriguing to compare $\hat{p}_1$ with $\hat{p}_2$. However, we see that neither of them is *uniformly* better! The latter outperforms when p is around 0.5, while for other values of p the former is preferable. This is unfortunately a quite common situation in comparing estimators. Moreover, we cannot even say that $\hat{p}_2$ uniformly outperforms the trivial $\hat{p}_3$ since very close to the boundaries of the unit interval, the $MSE(\hat{p}_3,p)$ is still smaller.

2.7   Unbiased estimation

2.7.1   Definition and main properties

As we have argued in the previous Section 2.6, ideally, a good estimator with a small MSE should have both low variance and low bias. However, it might be hard to "kill these two birds with one stone." For example, consider a trivial constant estimator $\hat{\theta} = \theta_0$ of an unknown parameter θ, where θ_0 is any fixed value. Evidently, $Var(\hat{\theta}) = 0$ but its bias $E\hat{\theta} - \theta = \theta_0 - \theta$ is large when the true θ is far from θ_0. One of the common approaches is to first control the bias component of the overall MSE and to consider *unbiased* estimators.

Definition 2.4 (unbiased estimator) $\hat{\theta}$ is an unbiased estimator of the unknown parameter $\theta \in \Theta$ if $E\hat{\theta} = \theta$ for all possible values of $\theta \in \Theta$.

There is no general rule or algorithm for deriving an unbiased estimator. In fact, unbiasedness is a property of an estimator rather than a method of estimation. One usually checks an MLE or any other estimator for bias. Sometimes one can then modify the original estimator to "correct" its bias.

We have already seen that the sample mean is always an unbiased estimator of the expectation (see Example 2.14). Now we find an unbiased estimator for the variance.

Example 2.16 (unbiased estimator of the variance) Consider a sample variance $\hat{\sigma}^2 = \frac{1}{n}\sum_{i=1}^n (Y_i - \bar{Y})^2$ of the random sample $Y_1, ..., Y_n$ and check it for unbiasedness:

$$E\hat{\sigma}^2 = \frac{E\sum_{i=1}^n (Y_i - \bar{Y})^2}{n} = \frac{E(\sum_{i=1}^n Y_i^2 - n\bar{Y}^2)}{n},$$

where $EY_i^2 = (EY)^2 + Var(Y) = \mu^2 + \sigma^2$ and similarly, $E\bar{Y}^2 = (E\bar{Y})^2 + Var(\bar{Y}) = \mu^2 + \frac{\sigma^2}{n}$. Thus, applying simple calculus, we have

$$E\hat{\sigma}^2 = \frac{n-1}{n}\sigma^2 \neq \sigma^2. \tag{2.8}$$

Therefore, the sample variance $\hat{\sigma}^2$ is a biased estimator for the variance σ^2. However, one immediately realizes from (2.8) that multiplying $\hat{\sigma}^2$ by the constant $\frac{n}{n-1}$ yields an unbiased estimator

$$s^2 = \frac{n}{n-1}\hat{\sigma}^2 = \frac{1}{n-1}\sum_{i=1}^n (Y_i - \bar{Y})^2. \tag{2.9}$$

What does unbiasedness of an estimator $\hat{\theta}$ for θ actually mean? Suppose we were observing not a single sample but *all possible* samples of size n from a sample space and were calculating the estimates $\hat{\theta}_j$ for each one of them. The unbiasedness would mean then that the *average* value of $\hat{\theta}$ over the entire sample space is θ, but it does not guarantee yet that $\hat{\theta}_j \approx \theta$ for *each* particular sample. The dispersion of $\hat{\theta}_j$'s around their average value θ might be large and, since in reality we have only a single sample, its particular value of $\hat{\theta}$ might be quite away from θ. To ensure with high confidence that $\hat{\theta}_j \approx \theta$ for *any* sample we need in addition for the variance $Var(\hat{\theta})$ to be small. Thus, whilst unbiasedness is an appealing property of an estimator, it is not sufficient without a "small" variance (see also the discussion at the beginning of the section).

As we have already seen (e.g., Example 2.14), an unbiased estimator is not necessarily unique. Moreover, if there are two unbiased estimators $\hat{\theta}_1$ and $\hat{\theta}_2$ of θ, any linear combination $\alpha\hat{\theta}_1 + (1-\alpha)\hat{\theta}_2$ is also an unbiased estimator for θ (see Exercise 2.8).

On the other hand, an unbiased estimator may not exist as we demonstrate in the following counterexample:

Example 2.17 Let $Y_1, ..., Y_n \sim \mathscr{B}(1, p)$. We show that there is no unbiased estimator for the odds $\theta = \frac{p}{1-p}$.

Suppose that there exists an unbiased estimator $\hat{\theta} = \hat{\theta}(\mathbf{Y})$ for θ, that is, $E\hat{\theta} = \frac{p}{1-p}$ for all $0 \leq p \leq 1$. On the other hand, there are 2^n various samples $\mathbf{y}_j$ of zeros and ones of size n and, therefore,

$$E\hat{\theta} = \sum_{j=0}^{2^n} \hat{\theta}(\mathbf{y}_j) p^{\sum_{i=1}^n y_{ji}} (1-p)^{n-\sum_{i=1}^n y_{ji}},$$

which is a polynomial function of p of (finite) degree n. However, the function $\frac{p}{1-p}$ cannot be expanded in a *finite* Taylor series.

Unlike maximum likelihood estimation, unbiasedness is not invariant under nonlinear transformations of the original parameter: if $\hat{\theta}$ is an unbiased estimator of θ, $g(\hat{\theta})$ is generally a biased estimator for $g(\theta)$ (see Exercise 2.9 for a simple counterexample). In particular, s from (2.9) is a biased estimator for the standard deviation σ.

2.7.2   Uniformly minimum variance unbiased estimators: The Cramer–Rao lower bound

As we have discussed, one should attempt to find an unbiased estimator with a "small" variance; ideally, an unbiased estimator with a uniformly minimal variance:

Definition 2.5 (**uniformly minimum variance unbiased estimator**) $\hat{\theta}$ is called a uniformly minimum variance unbiased estimator (UMVUE) of θ, if

1. $\hat{\theta}$ is an unbiased estimator of θ, that is, $E\hat{\theta} = \theta$ for all $\theta \in \Theta$

2. for any other unbiased estimator $\tilde{\theta}$ of θ, $Var(\hat{\theta}) \leq Var(\tilde{\theta})$ uniformly for all $\theta \in \Theta$

The UMVUE (if it exists) is necessarily unique (see Exercise 2.10).

Question 2.6 Is an UMVUE (if one exists) necessarily the best estimator of θ in terms of the MSE?

As we have mentioned, there is no general algorithm to obtain unbiased estimators in general and UMVUE in particular. Given an unbiased estimator and its variance, can we say that it is an UMVUE or at least close to one and cannot be much improved? In fact, there exists a lower bound for a variance of an unbiased estimator, which can be used as a benchmark for evaluating its goodness.

Assume that $\ln f_\theta(\mathbf{y})$ is differentiable w.r.t. θ and define the *Fisher information number* $I(\theta) = E((\ln f_\theta(\mathbf{y}))'_\theta)^2$. The derivative $(\ln f_\theta(\mathbf{y}))'_\theta$ is sometimes called *a score function*.

Theorem 2.2 (**Cramer–Rao lower bound**) Let $Y_1, ..., Y_n$ have a joint distribution $f_\theta(\mathbf{y})$, where $f_\theta(\mathbf{y})$ satisfies the following two regularity conditions:

1. the support supp $f_\theta(\mathbf{y})$ does not depend on θ;

2. for any statistic $T = T(Y_1, ..., Y_n)$ with a finite $E_\theta|T|$, integration w.r.t. $\mathbf{y}$ and differentiation w.r.t. θ in calculating $(ET)'_\theta$ can be interchanged, that is,

$$(ET)'_\theta = \frac{d}{d\theta} \int T(\mathbf{y})f_\theta(\mathbf{y})d\mathbf{y} = \int T(\mathbf{y})f'_\theta(\mathbf{y})d\mathbf{y}.$$

Let T be an unbiased estimator for θ with finite variance. Then,

$$Var(T) \geq \frac{1}{I(\theta)}.$$

More generally, if T is an unbiased estimator for $g(\theta)$, where $g(\cdot)$ is differentiable, then,

$$Var(T) \geq \frac{(g'(\theta))^2}{I(\theta)}.$$

Proof. Since T is an unbiased estimator for $g(\theta)$, we have

$$g(\theta) = ET = \int T(\mathbf{y})f_\theta(\mathbf{y})d\mathbf{y},$$

and under the regularity conditions of the theorem,

$$g'(\theta) = \int T(\mathbf{y})f'_\theta(\mathbf{y})d\mathbf{y} = \int T(\mathbf{y})(\ln f_\theta(\mathbf{y}))'_\theta f_\theta(\mathbf{y})d\mathbf{y} \tag{2.10}$$
$$= E\left\{T(\mathbf{Y})(\ln f_\theta(\mathbf{Y}))'_\theta\right\}.$$

On the other hand,

$$E(\ln f_\theta(\mathbf{Y}))'_\theta = \int (\ln f_\theta(\mathbf{y}))'_\theta f_\theta(\mathbf{y})d\mathbf{y} = \int f'_\theta(\mathbf{y})d\mathbf{y} = \frac{d}{d\theta}\int f_\theta(\mathbf{y})d\mathbf{y} = 0 \tag{2.11}$$

since $\int f_\theta(\mathbf{y})d\mathbf{y} = 1$, and (2.10) implies

$$\left|E\left\{T(\mathbf{Y})(\ln f_\theta(\mathbf{y}))'_\theta\right\}\right| = \left|Cov\left\{T(\mathbf{Y}), (\ln f_\theta(\mathbf{Y}))'_\theta\right\}\right| \leq \sqrt{Var(T) \cdot Var((\ln f_\theta(\mathbf{Y}))'_\theta)}, \tag{2.12}$$

where due to (2.11),
$$Var((\ln f_\theta(\mathbf{Y}))'_\theta) = E(\ln f_\theta(\mathbf{Y}))'_\theta)^2 = I(\theta). \tag{2.13}$$

Combining (2.10), (2.12), and (2.13) completes the proof. □

The regularity conditions required in Theorem 2.2 are satisfied for most "standard" distributions. In particular, they hold for distributions from the exponential family, where $f_\theta(\mathbf{y}) = \exp\{c(\theta)T(\mathbf{y}) + d(\theta) + S(\mathbf{y})\}$, $y \in \text{supp}(f_\theta)$, and $\text{supp} f_\theta$ do not depend on θ, if $c'_\theta(\theta)$ does not vanish. On the other hand, an immediate counterexample is the uniform distribution $\mathscr{U}(0, \theta)$, whose support depends on the unknown parameter.

Throughout the book we are especially interested in the case where $Y_1, ..., Y_n$ is a random sample from a distribution $f_\theta(y)$. The joint distribution then is $f_\theta(\mathbf{y}) = \prod_{i=1}^{n} f_\theta(y_i)$ or $\ln f_\theta(\mathbf{y}) = \sum_{i=1}^{n} \ln f_\theta(y_i)$, and the following corollary holds:

Corollary 2.1 Let $Y_1, ..., Y_n \sim f_\theta(y)$ that satisfies the regularity conditions of Theorem 2.2. Then,

$$I(\theta) = nI^*(\theta),$$

where $I^*(\theta) = E((\ln f_\theta(y))'_\theta)^2$ is the Fisher information number of $f_\theta(y)$, and, therefore, for any unbiased estimator T of $g(\theta)$,

$$Var(T) \geq \frac{(g'_\theta(\theta))^2}{nI^*(\theta)}.$$

Proof. From (2.13) one has

$$I(\theta) = Var((\ln f_\theta(\mathbf{Y}))'_\theta) = \sum_{i=1}^{n} Var((\ln f_\theta(Y_i))'_\theta) = n\, Var((\ln f_\theta(Y_1))'_\theta) = nI^*(\theta).$$

$\square$

Example 2.18 (Cramer–Rao lower bound for the normal mean) Consider a random sample $Y_1, ..., Y_n$ from a normal distribution $\mathscr{N}(\mu, \sigma^2)$ and find the Cramer–Rao lower bound for an unbiased estimator of μ. We have

$$\ln f(y) = -\ln\sqrt{2\pi} - \ln\sigma - \frac{(y-\mu)^2}{2\sigma^2} \tag{2.14}$$

and

$$(\ln f(y))'_\mu = \frac{y-\mu}{\sigma^2}. \tag{2.15}$$

Thus,

$$I^*(\mu) = E(\ln f(Y))'^2_\mu = \frac{E(Y-\mu)^2}{\sigma^4} = \frac{1}{\sigma^2}$$

and the Fisher information number is $I(\mu) = nI^*(\mu) = n/\sigma^2$. The Cramer–Rao lower bound for unbiased estimators of μ is, therefore, σ^2/n. As we have seen in Example 2.14, the sample mean $\bar{Y}$ is unbiased for μ and $Var(\bar{Y}) = \sigma^2/n$. Thus, for the normal data, $\bar{Y}$ is an UMVUE for μ.

There is another, usually more convenient formula for calculating the Fisher information number $I(\theta)$ other than its direct definition:

Proposition 2.1 If $\ln f_\theta(\mathbf{y})$ is twice differentiable w.r.t θ and the regularity conditions of Theorem 2.2 hold for the second derivative $f''_\theta(\mathbf{y})$ as well, then

$$I(\theta) = -E\,(\ln f_\theta(\mathbf{Y}))''_\theta. \tag{2.16}$$

Proof. We have

$$\begin{aligned}
(\ln f_\theta(\mathbf{y}))''_\theta &= ((\ln f_\theta(\mathbf{y})'_\theta)'_\theta = \left(\frac{f'_\theta(\mathbf{y})}{f_\theta(\mathbf{y})}\right)'_\theta \\
&= \frac{f''_\theta(\mathbf{y})f_\theta(\mathbf{y}) - (f'_\theta(\mathbf{y}))^2}{f_\theta(\mathbf{y})^2} = \frac{f''_\theta(\mathbf{y})}{f_\theta(\mathbf{y})} - (\ln f_\theta(\mathbf{y})'_\theta)^2,
\end{aligned} \tag{2.17}$$

The proof follows by taking the expectations of both sides in (2.17) and noting that

$$E \frac{f_\theta''(\mathbf{Y})}{f_\theta(\mathbf{Y})} = \int \frac{f_\theta''(\mathbf{y})}{f_\theta(\mathbf{y})} f_\theta(\mathbf{y}) dy = \frac{d^2}{d\theta^2} \int f_\theta(\mathbf{y}) dy = 0.$$

$\square$

Example 2.19 (continuation of Example 2.18) We have already found by the definition of the Fisher information number that for a normal sample $I(\mu) = \frac{n}{\sigma^2}$. Now apply Proposition 2.1 and calculate it by (2.16). Differentiating (2.15) again w.r.t. μ implies $(\ln f(y))_\mu'' = -\frac{1}{\sigma^2}$ and $I(\mu) = -nE(\ln f(y))_\mu'' = \frac{n}{\sigma^2}$.

It is important to emphasize that Theorem 2.2 holds in one direction only: if the variance of an unbiased estimator does not achieve the Cramer–Rao lower bound, one still cannot claim that it is not an UMVUE. Nevertheless, it can be used as a benchmark for measuring the goodness of an unbiased estimator.

Example 2.20 (Cramer–Rao lower bound for the normal variance) We continue Example 2.18 for the normal sample and now find the Cramer–Rao lower bound for estimating the variance σ^2. Differentiating (2.14) w.r.t. σ, we have

$$(\ln f(y))_\sigma' = -\frac{1}{\sigma} + \frac{(y-\mu)^2}{\sigma^3}$$

and

$$(\ln f(y))_\sigma'' = \frac{1}{\sigma^2} - 3\frac{(y-\mu)^2}{\sigma^4}.$$

Thus, $I(\sigma) = -nE(\ln f(y))_\sigma'' = \frac{2n}{\sigma^2}$ and the Cramer–Rao lower bound for estimating $g(\sigma) = \sigma^2$ is $(2\sigma)^2/I(\sigma) = \frac{2\sigma^4}{n}$.

Consider the unbiased estimator $s^2 = \frac{1}{n-1}\sum_{i=1}^n (Y_i - \bar{Y})^2$ from (2.9). From Exercise 2.11, $Var(s^2) = \frac{2\sigma^4}{n-1}$. It does not achieve the lower bound although it closely approaches it. Meanwhile we can only say that it is "nearly the best" unbiased estimator of σ^2. We will show in Section 2.7.4 that if μ is unknown, it is "exactly the best" unbiased estimator, that is, an UMVUE. Note that for the *known* μ, the UMVUE for σ^2 is $\hat{\sigma}^2 = \frac{1}{n}\sum_{i=1}^n (Y_i - \mu)^2$, which does achieve the Cramer–Rao lower bound—see Exercise 2.14.

In fact, the Cramer–Rao lower bound is achieved only for distributions from the exponential family:

Theorem 2.3 Assume that $Y_1, \ldots, Y_n$ has a joint distribution $f_\theta(\mathbf{y})$ that belongs to the (one-parameter) exponential family with $f_\theta(\mathbf{y}) = \exp\{c(\theta)T(\mathbf{y}) + d(\theta) + S(\mathbf{y})\}$, $\mathbf{y} \in \text{supp}(f_\theta)$ and $\text{supp}(f_\theta)$ does not depend on θ, where $c'(\theta)$ is continuous and does not vanish. Then, $T(\mathbf{Y})$ is an UMVUE of $E_\theta(T(\mathbf{Y}))$ that achieves the Cramer–Rao lower bound.

Conversely, if $f_\theta(\mathbf{y})$ satisfies somewhat stronger conditions than the regularity conditions of Theorem 2.2, and there exists an unbiased estimator $T(\mathbf{Y})$ of $g(\theta)$ that achieves the Cramer–Rao lower bound for every θ, then the distribution of $\mathbf{Y}$ belongs to the one-parameter exponential family with the sufficient statistic $T(\mathbf{Y})$.

Proof. Assume that $f_\theta(\mathbf{y})$ belongs to the one-parameter exponential family. Then, $\ln f_\theta(\mathbf{y}) = c(\theta)T(\mathbf{y}) + d(\theta) + S(\mathbf{y})$, $\mathbf{y} \in \text{supp}(f_\theta)$ and, therefore, $(\ln f_\theta(\mathbf{y}))_\theta' = c'(\theta)T(\mathbf{y}) + d'(\theta)$. From (2.11) we have

$$E(\ln f_\theta(\mathbf{Y}))_\theta' = c'(\theta)ET(\mathbf{Y}) + d'(\theta) = 0$$

implying $ET(\mathbf{Y}) = -\frac{d'(\theta)}{c'(\theta)}$.

Furthermore, from (2.13),

$$I(\theta) = Var\left((\ln f_\theta(\mathbf{Y}))'_\theta\right) = Var\left(c'(\theta)T(\mathbf{Y}) + d'(\theta)\right) = c'(\theta)^2 Var(T(\mathbf{Y}))$$

and, therefore,

$$Var(T(\mathbf{Y})) = \frac{I(\theta)}{c'(\theta)^2} \tag{2.18}$$

On the other hand,

$$I(\theta) = -E(\ln f_\theta(\mathbf{Y}))''_\theta = -E\left(c''(\theta)T(\mathbf{Y}) + d''(\theta)\right) = -c''(\theta)ET(\mathbf{Y}) - d''(\theta)$$
$$= \frac{c''(\theta)d'(\theta) - c'(\theta)d''(\theta)}{c'(\theta)}$$

Let $g(\theta) = -\frac{d'(\theta)}{c'(\theta)}$ so that $T(\mathbf{Y})$ is an unbiased estimator of $g(\theta)$ and show that $Var(T(\mathbf{Y}))$ in (2.18) achieves the Cramer–Rao lower bound. The latter is

$$\frac{(g'(\theta))^2}{I(\theta)} = \left(-\frac{c'(\theta)d''(\theta) - c''(\theta)d'(\theta)}{c'(\theta)^2}\right)^2 \frac{1}{I(\theta)} = \left(\frac{I(\theta)}{c'(\theta)}\right)^2 \frac{1}{I(\theta)} = \frac{I(\theta)}{c'(\theta)^2}$$

which is exactly (2.18).

Conversely, from the proof of the Cramer–Rao lower bound it follows that the inequality appears only in (2.12), where we used

$$\left|Cov\left\{T(\mathbf{Y}), (\ln f_\theta(\mathbf{Y}))'_\theta\right\}\right| \le \sqrt{Var(T) \cdot Var((\ln f_\theta(\mathbf{Y}))'_\theta)} \tag{2.19}$$

Note that the equality in (2.19) holds if $T(\mathbf{Y})$ is a linear function of $(\ln f_\theta(\mathbf{Y}))'_\theta$, that is, if there exist $a(\theta)$ and $b(\theta)$ such that $(\ln f_\theta(\mathbf{Y}))'_\theta = a(\theta)T(\mathbf{Y}) + b(\theta)$. In this case $\ln f_\theta(\mathbf{Y}) = c(\theta)T(\mathbf{Y}) + d(\theta) + S(\mathbf{Y})$, where $c'(\theta) = a(\theta)$, $d'(\theta) = b(\theta)$ and $S(\mathbf{Y})$ is a normalizing term to ensure that $f_\theta(\mathbf{y})$ is a density. Thus, $f_\theta(\mathbf{y})$ belongs to the one-parameter exponential family with the sufficient statistic $T(\mathbf{Y})$. $\qquad\square$

2.7.3   *The Cramer–Rao lower bound for multiparameter case

In this section, we extend the Cramer–Rao lower bound (see Theorem 2.2) for multivariate θ. We first define the Fisher information matrix—a multivariate extension of the Fisher information number:

Definition 2.6 (Fisher information matrix) Let $Y_1, ..., Y_n$ have a joint distribution $f_\theta(\mathbf{y})$, where $\theta = (\theta_1, ..., \theta_p)$ is a p-dimensional parameter. The Fisher information *matrix*

$$I(\theta) = E\left(\nabla \ln f_\theta(\mathbf{y}) \cdot \nabla \ln f_\theta(\mathbf{y})^\top\right),$$

where the gradient vector $\nabla \ln f_\theta(\mathbf{y}) = \left(\frac{\partial \ln f_\theta(\mathbf{y})}{\partial \theta_1}, ..., \frac{\partial \ln f_\theta(\mathbf{y})}{\partial \theta_p}\right)^\top$, i.e.

$$I_{jk}(\theta) = E\left(\frac{\partial \ln f_\theta(\mathbf{y})}{\partial \theta_j} \cdot \frac{\partial \ln f_\theta(\mathbf{y})}{\partial \theta_k}\right), \quad j, k = 1, ..., p.$$

Theorem 2.4 (multivariate Cramer–Rao lower bound) Let the joint distribution $f_\theta(\mathbf{y})$ of $\mathbf{Y} = (Y_1, ..., Y_n)$ with the p-dimensional parameter θ satisfy the following regularity conditions:

1. supp $f_\theta(\mathbf{y})$ does not depend on θ;

2. for any statistic $T = T(\mathbf{Y})$ with a finite $E|T|$, integration w.r.t. $\mathbf{y}$ and differentiation w.r.t. θ in calculating the gradient of ET can be interchanged, that is,

$$\frac{\partial}{\partial \theta_j} \int T(\mathbf{y})f_\theta(\mathbf{y})d\mathbf{y} = \int T(\mathbf{y}) \left(\frac{\partial}{\partial \theta_j} f_\theta(\mathbf{y}) \right) d\mathbf{y} \quad \text{for all } j = 1,...,p.$$

Let T be an unbiased estimator of a scalar-valued function $g(\theta) : \mathbb{R}^p \to \mathbb{R}$ with finite variance. Then,

$$Var(T) \geq \nabla g(\theta)^\top I^{-1}(\theta) \nabla g(\theta), \tag{2.20}$$

where the gradient vector $\nabla g(\theta) = (\frac{\partial g(\theta)}{\partial \theta_1}, ..., \frac{\partial g(\theta)}{\partial \theta_p})^\top$.

Proof. Repeating the arguments from the proof of Theorem 2.2 for the univariate case, it is easy to show that under the regularity conditions, $E\frac{\partial \ln f_\theta(\mathbf{y})}{\partial \theta_j} = 0$ for all $j = 1,...,p$ and, therefore,

$$I_{jk}(\theta) = Cov\left(\frac{\partial \ln f_\theta(\mathbf{y})}{\partial \theta_j}, \frac{\partial \ln f_\theta(\mathbf{y})}{\partial \theta_k} \right), \quad j,k = 1,...,p. \tag{2.21}$$

In particular, $I(\theta)$ is a positive semi-definite matrix. Since $T(\mathbf{Y})$ is an unbiased estimator of $g(\theta)$, for any $j = 1,...,p$, we have

$$\begin{aligned}
\frac{\partial g(\theta)}{\partial \theta_j} &= \frac{\partial}{\partial \theta_j} \int T(\mathbf{y})f_\theta(\mathbf{y}d\mathbf{y} \\
&= \int T(\mathbf{y}) \left(\frac{\partial f_\theta(\mathbf{y})}{\partial \theta_j} \right) d\mathbf{y} = \int T(\mathbf{y}) \left(\frac{\partial \ln f_\theta(\mathbf{y})}{\partial \theta_j} \right) f_\theta(\mathbf{y})d\mathbf{y} \\
&= Cov\left(T(\mathbf{Y}), \frac{\partial \ln f_\theta(\mathbf{Y})}{\partial \theta_j} \right).
\end{aligned} \tag{2.22}$$

Consider now the covariance matrix V of the $(p+1)$-dimensional random vector $(T(\mathbf{Y}), \nabla \ln f_\theta(\mathbf{Y}))^\top$. Then, (2.21) and (2.22) imply

$$V = \left(\begin{array}{cc} Var(T(\mathbf{Y})) & \nabla g(\theta)^\top \\ \nabla g(\theta) & I(\theta) \end{array} \right).$$

Since V is the covariance matrix, it is positive semi-definite and $\det(V) \geq 0$. On the other hand, using standard linear algebra,

$$\det(V) = \det(I(\theta)) \left(Var(T(\mathbf{Y})) - \nabla g(\theta)^\top I^{-1}(\theta) \nabla g(\theta) \right)$$

which yields (2.20).   □

Similarly to the univariate case, one can prove the following proposition (the proof is left as Exercise 2.23 for readers):

Proposition 2.2 Under the regularity conditions of Theorem 2.4,

1. If $Y_1,...,Y_n$ is a random sample from $f_\theta(y)$, then $I(\theta) = nI^*(\theta)$, where $I^*(\theta)$ is the Fisher information matrix of $f_\theta(y)$.

2. If the differentiation and integration can be exchanged up to second derivative, then $I(\theta)$ can alternatively be calculated as

$$I(\theta) = -E\left(\frac{\partial^2 \ln f_\theta(\mathbf{Y})}{\partial \theta \partial \theta^\top} \right).$$

Example 2.21 (Fisher information matrix for the normal distribution) Let $Y_1, ..., Y_n \sim \mathcal{N}(\mu, \sigma^2)$, where both μ and σ are unknown, and calculate the Fisher information matrix $I(\mu, \sigma)$. In fact, we have already obtained its diagonal elements $I_{\mu\mu} = \frac{n}{\sigma^2}$ and $I_{\sigma\sigma} = \frac{2n}{\sigma^2}$ in Example 2.18 and Example 2.20 respectively. To derive $I_{\mu\sigma}$, recall that

$$\ln f(y) = -\ln\sqrt{2\pi} - \ln\sigma - \frac{(y-\mu)^2}{2\sigma^2}.$$

Hence,

$$\frac{\partial^2 \ln f(y)}{\partial\mu\partial\sigma} = -\frac{2(y-\mu)}{\sigma^3},$$

$I_{\mu\sigma}^* = -E\left(\frac{\partial^2 \ln f(y)}{\partial\mu\partial\sigma}\right) = 0$ and the Fisher information matrix is

$$I(\mu, \sigma) = \begin{pmatrix} \frac{n}{\sigma^2} & 0 \\ 0 & \frac{2n}{\sigma^2} \end{pmatrix}.$$

The same result could be obtained directly from Definition 2.6 of the Fisher information matrix after slightly more complicated calculus.

2.7.4   Rao–Blackwell theorem

The Cramer–Rao lower bound allows one only to evaluate the goodness of a proposed unbiased estimator but does not provide any constructive way to derive it. In fact, as we have argued, there is no such general rule at all. However, if one manages to obtain any initial (even crude) unbiased estimator, it may be possible to improve it. The Rao–Blackwell theorem below shows that if there is an unbiased estimator that is not a function of a sufficient statistic W, one can construct another unbiased estimator based on W with an MSE not larger than the original one. Thus, in terms of MSE, only unbiased estimators based on a sufficient statistic are of interest. This demonstrates again a strong sense of the notion of sufficiency.

Theorem 2.5 (Rao–Blackwell theorem) Let T be an unbiased estimator of θ and W be a sufficient statistic for θ. Define $T_1 = E(T|W)$. Then, for all $\theta \in \Theta$

1. T_1 is an unbiased estimator of θ, that is, $ET_1 = \theta$.
2. $Var(T_1) \leq Var(T)$.

Proof. The proof is remarkably simple. We show first that T_1 is indeed an estimator, that is, it does not depend on the unknown θ. Here we exploit the sufficiency of W that guarantees that the conditional distribution $T|W$ (and, therefore, $E(T|W)$ in particular) does not depend on θ. To verify the unbiasedness of T_1 note that for any $\theta \in \Theta$, we have

$$\theta = ET = E(E(T|W)) = ET_1.$$

Similarly,

$$Var(T) = Var(E(T|W)) + E(Var(T|W)) \geq Var(T_1)$$

since the second term is evidently nonnegative.   □

Question 2.7 Show that:

1. $T_1 = E(T|W) = T$ and "Rao–Blackwellization" does not improve the initial unbiased estimator iff T itself is a function of a sufficient statistic W;
2. there is no sense in applying "Rao–Blackwellization" more than once, since $T_2 = E(T_1|W) = T_1$.

Example 2.22 (**Bernoulli trials**) We start from a simple example. Let $Y_1, ..., Y_n$ be a series of Bernoulli trials $\mathscr{B}(1, p)$ for which we want to find an unbiased estimator of p.

To apply the Rao–Blackwell theorem we need to find a sufficient statistic W and an initial unbiased estimator T. As we know, $W = Y_1 + ... + Y_n$ (see Example 1.5). It is usually more difficult to construct T. Note that in the considered example, the unknown parameter p is the probability of a certain event. We can then employ a useful "trick" by defining T as an indicator of that event: one if it occurs and zero otherwise. In this case, obviously, $ET = p$. For example, in our case p is the probability of success in a single (say, the first) trial and T is thus simply Y_1. Then, $T|W \sim \mathscr{B}(1, P(Y_1 = 1|Y_1 + ... + Y_n = W))$ and

$$
\begin{aligned}
T_1 &= E(T|W) = P(Y_1 = 1|Y_1 + ... + Y_n = W) = \frac{P(Y_1 = 1, Y_1 + ... + Y_n = W)}{P(Y_1 + ... + Y_n = W)} \\
&= \frac{P(Y_1 = 1)P(Y_2 + ... + Y_n = W - 1)}{P(Y_1 + Y_2 + ... + Y_n = W)},
\end{aligned} \tag{2.23}
$$

where $Y_2 + ... + Y_n \sim \mathscr{B}(n-1, p)$ and $Y_1 + Y_2 + ... + Y_n \sim \mathscr{B}(n, p)$. Substituting the corresponding probabilities into (2.23), we have

$$
T_1 = \frac{p\binom{n-1}{W-1}p^{W-1}(1-p)^{n-W}}{\binom{n}{W}p^W(1-p)^{n-W}} = \frac{\binom{n-1}{W-1}}{\binom{n}{W}} = \frac{W}{n} = \bar{Y},
$$

which is known to be UMVUE (see Exercise 2.13).

Example 2.23 This is a less intuitive example. Suppose we have records $Y_1, ..., Y_n$ on the annual numbers of earthquakes in a certain seismic region over the last n years. It is reasonable to assume that the Y_i is i.i.d. $Pois(\lambda)$. We want to estimate the probability that the following year will not be "quiet" (seismically active), and there will be at least one earthquake in the region.

Let $p = P(Y \geq 1) = 1 - e^{-\lambda}$. The sufficient statistic $W = Y_1 + ... + Y_n$ is the total number of earthquakes over n years. Since the unknown parameter is again the probability of a certain event (seismically active year), the corresponding indicator variable $T = 1$ if $Y_1 \geq 1$ and zero otherwise, is an unbiased estimator for p. Then, $T|W \sim B(1, P(Y_1 \geq 1|Y_1 + ... + Y_n = W))$ and

$$
\begin{aligned}
T_1 = E(T|W) &= 1 - P(Y_1 = 0|Y_1 + ... + Y_n = W) \\
&= 1 - \frac{P(Y_1 = 0, Y_1 + ... + Y_n = W)}{P(Y_1 + ... + Y_n = W)} \\
&= 1 - \frac{P(Y_1 = 0)P(Y_2 + ... + Y_n = W)}{P(Y_1 + Y_2 + ... + Y_n = W)} \\
&= 1 - \frac{e^{-\lambda}e^{-(n-1)\lambda}((n-1)\lambda)^W/W!}{e^{-n\lambda}(n\lambda)^W/W!} \\
&= 1 - \left(\frac{n-1}{n}\right)^W = 1 - \left(1 - \frac{1}{n}\right)^{\sum_{i=1}^n Y_i}.
\end{aligned}
$$

Unlike the previous example, it would be quite hard to obtain the above-unbiased estimator by "intuitive" arguments.

Compare T_1 with the MLE $\hat{p}$ of p. The MLE of λ is $\hat{\lambda} = \bar{Y}$ (see Exercise 2.16), and therefore $\hat{p} = 1 - e^{-\bar{Y}}$. The two estimators are different but note that for large n

$$
T_1 = 1 - \left(1 - \frac{1}{n}\right)^{\sum_{i=1}^n Y_i} = 1 - \left\{\left(1 - \frac{1}{n}\right)^n\right\}^{\bar{Y}} \sim 1 - e^{-\bar{Y}} = \hat{p}.
$$

Does "Rao–Blackwellization" necessarily yield an UMVUE? Generally not. To guarantee UMVUE an additional requirement of *completeness* on a sufficient statistic W is needed. For interested and advanced readers, we discuss it in more detail in Section 2.7.5. However, under mild conditions, it can be shown that if the distribution of the data belongs to the exponential family (see Definition 1.6), where $\dim(\boldsymbol{\theta}) = \dim(c(\boldsymbol{\theta}))$, and an unbiased estimator is a function of the corresponding sufficient statistic $(\sum_{i=1}^{n} T_1(y_i), ..., \sum_{i=1}^{n} T_p(y_i))$, it is an UMVUE (see Section 2.7.5 for more detail). Thus, in particular, the unbiased estimator $(1 - \frac{1}{n})^{\sum_{i=1}^{n} Y_i}$ for $p = P(Y \geq 1)$ for the Poisson data in Example 2.23 is an UMVUE. A normal distribution with the unknown mean and variance $(\dim(\boldsymbol{\theta}) = 2)$ belongs to the two-dimensional exponential family (see Example 1.14), and it can be shown that $s^2 = \frac{1}{n-1} \sum_{i=1}^{n} (Y_i - \bar{Y})^2$, which is a function of a sufficient statistic, is an UMVUE for the variance although it does not achieve the Cramer–Rao lower bound (see Example 2.24).

Finally, note that despite its elegance, the applications of the Rao–Blackwell theorem in more complicated cases are quite limited. The two main obstacles are in finding an initial unbiased estimator T and calculating the conditional expectation $E(T|W)$.

2.7.5 *Lehmann–Scheffé theorem

As we have mentioned in the previous Section 2.7.4, the Rao–Blackwell theorem still does not guarantee that the resulting unbiased estimator is an UMVUE. The following theorem establishes that it is true under the additional *completeness* condition (see Section 1.5) of a sufficient statistic:

Theorem 2.6 (Lehmann–Scheffé theorem) Let T be an unbiased estimator of θ and W a *complete* sufficient statistic for θ. Then, $T_1 = E(T|W)$ is the unique UMVUE of θ.

Proof. Consider any unbiased estimator S of θ. If S is not a function of a sufficient statistic W, it can be necessarily improved by Rao–Blackwellization (see Question 2.7) and, therefore, the UMVUE should be a function of W. On the other hand, the completeness of W guarantees that T_1 is the *unique* unbiased estimator of θ, which is a function of W. Indeed, suppose there is another unbiased estimator $T_2 = T_2(W)$. Then,

$$E(T_1(W) - T_2(W)) = E(T_1(W)) - E(T_2(W)) = 0$$

and, due to the completeness of W, $T_1(W) = T_2(W)$. $\square$

From the Lehmann–Scheffé theorem it follows that if T is an unbiased estimator of θ and a function of a complete sufficient statistic, it is the unique UMVUE of θ. For distributions from the exponential family, completeness of a sufficient statistic can be established by Theorem 1.5 from Section 1.6.

Example 2.24 (unbiased estimate for the normal variance—continuation of Example 2.20) In Example 2.20, we found an unbiased estimator

$$s^2 = \frac{1}{n-1} \sum_{i=1}^{n} (Y_i - \bar{Y})^2$$

for the variance of a normal sample with unknown mean μ and variance σ^2. We saw that s^2 does not achieve the Cramer–Rao lower bound (although it is very close). Now we can show that it is an UMVUE. Recall that the sufficient statistic for $\theta = (\mu, \sigma)$ is $W = (\bar{Y}, \sum_{i=1}^{n} (Y_i - \bar{Y})^2)$ (see Section 1.6), and W is also complete (see Example 1.16). Since s^2 is a function of W, it is an UMVUE.

2.8 Exercises

Exercise 2.1 Let $Y_1, ..., Y_n$ be a random sample from the following distributions with the unknown parameter(s). Estimate them by maximum likelihood and by the method of moments.

1. double exponential distribution $f_\theta(y) = \frac{\theta}{2} e^{-\theta|y|}$, $-\infty < y < \infty$, $\theta > 0$

2. shifted exponential distribution $f_{\theta,\tau}(y) = \theta e^{-\theta(y-\tau)}$, $y \geq \tau$, $\theta > 0$, where

   (a) τ is known

   (b) both θ and τ are unknown

   (*Hint*: to calculate the moments of a shifted exponential distribution one can easily show that $Y = T + \tau$, where $T \sim \exp(\theta)$)

3. Pareto distribution $f_{\alpha,\beta}(y) = \frac{\alpha}{\beta}\left(\frac{\beta}{y}\right)^{\alpha+1}$, $y \geq \beta$, $\alpha, \beta > 0$

4. $f_\theta(y) = \theta y^{\theta-1}$, $0 \leq y \leq 1$, $\theta > 0$

Exercise 2.2 Let $Y_1, ..., Y_n \sim Gamma(\alpha, \beta)$.

1. Can you obtain the MLE of α and β in the closed form?

2. Estimate α and β by the method of moments.

Exercise 2.3 An electronic device contains k identical, independently working sensitive elements whose lifetimes are exponentially distributed $\exp(\theta)$. Even if one element does not operate, the entire device is rendered out of order and must be sent to a laboratory for repair. n devices with lifetimes $t_1, ..., t_n$ hours have been sent to a laboratory. Based on this data, find the MLEs of θ, the expected lifetimes of a single element and an entire device, and the probability that a device will operate at least a hours.
(*Hint*: show first that $t_i \sim \exp(k\theta)$.)

Exercise 2.4 A start-up company recruits new employees among graduates. Hiring requirements include passing an IQ test with a score larger than 120. Within a population of graduates, IQ is normally distributed with an unknown mean μ and a standard deviation of 5. According to the company policy, candidates do not obtain the scores of their tests, but only notification of whether they have passed it or not. Only 10 candidates among 100 successfully passed the test.

1. Find the MLE of μ.

2. The company was disappointed by the low number of new employees it managed to recruit and decided in the future that they would reduce the required IQ score to 115. Find the MLE of the chance for a randomly chosen graduate to be hired by the company after passing the test.

Exercise 2.5 A counter records a daily number of cars passing across a junction that is known to have a Poisson distribution with an unknown λ. The counter transmits its daily record to the control center every midnight and resets to restart to count data for the next day. However, with an (unknown) probability p the counter has a random fault and does not restart counting but records a zero count instead that day.

1. Given the counts $Y_1, ..., Y_n$ transmitted by the counter during a period of n days, find the sufficient statistic for (λ, p).

2. Can you obtain the MLE for λ and p in the closed form?

3. Estimate λ and p by the methods of moments.

Exercise 2.6 Let $Y_1, ..., Y_n \sim \exp(\theta)$, where it is known that $\theta \leq \theta_0$. Find the MLE of θ.

Exercise 2.7 Two random samples of n men and m women participated in a political poll to evaluate the support for Candidate A in the population. Out of those participated in the poll, X men and Y women supported his candidacy. In addition, it is also known that in the entire population, he is more popular among women rather than men. What are the MLE for the percentage of support for Candidate A among men and among women?

Exercise 2.8 Let T_1 and T_2 be two unbiased estimators of θ.

1. Show that for any α, the linear combination $\alpha T_1 + (1 - \alpha) T_2$ is also unbiased.

2. Assume, in addition, that T_1 and T_2 are independent with variances V_1 and V_2, respectively. Find the unbiased estimator with the minimal variance among all unbiased estimators from the first paragraph. Is it necessarily an UMVUE?

Exercise 2.9 Let T be an unbiased estimator of θ. Show that T^2 is a biased estimator of θ^2. What is its bias?

Exercise 2.10 Show that if T_1 and T_2 are both UMVUE with finite variances, then $T_1 = T_2$ (uniqueness of UMVUE).
(*Hint*: consider, for example, $T = (T_1 + T_2)/2$.)

Exercise 2.11 Let $Y_1, ..., Y_n$ be a normal sample where both μ and σ are unknown. Consider a family of estimators for σ^2 of the form

$$\hat{\sigma}^2_{a_n} = \frac{\sum_{i=1}^n (Y_i - \bar{Y})^2}{a_n},$$

where a_n may depend on n but not on the data.

1. Do the MLE $\hat{\sigma}^2 = \frac{1}{n} \sum_{i=1}^n (Y_i - \bar{Y})^2$ (see Example 2.7) and the UMVUE $s^2 = \frac{1}{n-1} \sum_{i=1}^n (Y_i - \bar{Y})^2$ (see Example 2.20) belong to this family?

2. What is the MSE of $\hat{\sigma}^2_{a_n}$?

3. Compare the MLE $\hat{\sigma}^2$ and the UMVUE s^2 by MSE.

4. Find an estimator with the minimal MSE within the family.

(*Hint*: recall that $\sum_{i=1}^n (Y_i - \bar{Y})^2 \sim \sigma^2 \chi^2_{n-1}$.)

Exercise 2.12

1. Show that the MLE $\hat{\theta}_{MLE} = Y_{\max}$ of θ in the uniform distribution $\mathcal{U}(0, \theta)$ (see Example 2.2) is uniformly better (in terms of MSE) than the MME $\hat{\theta}_{MME} = 2\bar{Y}$ (see Example 2.10).
(*Hint*: you first need to derive the distribution of $Y_{\max}$ to calculate the $MSE(\hat{\theta}, \theta)$.)

2. Is $\hat{\theta}$ unbiased? If not, find an unbiased estimator based on $\hat{\theta}$.

3. Consider a family of estimators $aY_{\max}$, where a may depend on n but not on the data. Which of the estimators considered in the previous paragraphs belong to this family? Find the best (minimal MSE) estimator within the family.

Exercise 2.13 Consider a series of Bernoulli trials $Y_1, ..., Y_n \sim \mathcal{B}(1, p)$. Show that the MLE $\hat{p} = \bar{Y}$ (see Example 2.1) is the UMVUE that achieves the Cramer–Rao lower bound.

Exercise 2.14 Let $Y_1, ..., Y_n \sim \mathcal{N}(\mu, \sigma^2)$, where μ is known. Show that $\hat{\sigma}^2 = \frac{1}{n} \sum_{i=1}^n (Y_i - \mu)^2$ is an unbiased estimator of σ^2, achieves the Cramer–Rao lower bound and is, therefore, the UMVUE.

Exercise 2.15 Continue Exercise 1.4.

1. Find the MLE of p.

2. Is the MLE unbiased?

3. *Is it UMVUE?

Exercise 2.16 An annual number of earthquakes in a certain seismically active region is distributed Poisson with an unknown mean λ. Let $Y_1, ..., Y_n$ be the number of earthquakes in the area during the last n years (it is possible to assume that the Y_i are independent).

1. Find the MLE $\hat{\lambda}$ for λ.

2. Is $\hat{\lambda}$ unbiased?

3. Does $\hat{\lambda}$ achieve the Cramer–Rao lower bound?

Exercise 2.17 An aircraft launched a missile towards a ground target. Let X and Y be south–north and east–west random deviations of the missile from the target, respectively. It may be assumed that X and Y are i.i.d $\mathcal{N}(0, \sigma^2)$. Define $R = \sqrt{X^2 + Y^2}$ to be the distance between the target and the actual location of the missile's explosion.

1. Show that the distribution of R has the density

$$f_\sigma(r) = \frac{r}{\sigma^2} e^{-\frac{r^2}{2\sigma^2}}, \; r \geq 0$$

(Rayleigh distribution). Does it belong to the exponential family?
(*Hint*: first find the distribution of $R^2 = X^2 + Y^2$.)

2. Find the MLE of σ^2.

3. Is the MLE unbiased? Is it UMVUE?

Exercise 2.18 A large telephone company wants to estimate the average number of daily telephone calls made by its private clients. It is known that on the average women make a times more calls than men. It is reasonable to assume that the number of daily calls has a Poisson distribution. During a certain day, the company registered the numbers of daily calls made by n female and m male randomly-chosen clients.

1. What is the statistical model for the data? Does the distribution of the data belong to the exponential family?

2. Find the MLE for the average daily number of calls made by female and male clients.

3. Are the MLEs from the previous paragraph unbiased?

Exercise 2.19 Consider a simple linear regression model without an intercept:

$$Y_i = \beta x_i + \varepsilon_i, \quad i = 1,...,n,$$

where ε_i are i.i.d $\mathcal{N}(0,\sigma^2)$.

1. Find the MLE for β. Is it unbiased? Find its MSE.

2. Consider an estimator $\tilde{\beta} = \frac{\sum_{i=1}^n Y_i}{\sum_{i=1}^n x_i}$. Is it unbiased? Find its MSE and compare it with that of the MLE.

3. Find the MLE for σ.

4. Find the MLE for β/σ.

5. *Derive the Fisher information matrix $I(\beta,\sigma)$.

6. *Find the Cramer–Rao lower bound for an unbiased estimator of β/σ.

Exercise 2.20 Let $Y_1,...,Y_n \sim Geom(p)$.

1. Find the MLE of p.

2. Find an unbiased estimator of p.
   (*Hint*: first find a trivial unbiased estimator of p and then improve it by "Rao–Blackwellization"; recall that $\sum_{i=1}^n Y_i \sim NB(n,p)$.)

3. Is the unbiased estimator from the previous paragraph UMVUE?

Exercise 2.21 A quality control laboratory randomly selected n batches of k components produced by a factory and inspected them for defectives. It is assumed that the produced components are independent, each with the same probability of being defective. The numbers of detected defectives were respectively $Y_1,...,Y_n$. One is interested in the probability p that a batch of k components contains no defectives.

1. Find the MLE of p.

2. Find an unbiased estimator of p.
   (*Hint*: first find a trivial unbiased estimator of p and then improve it by "Rao–Blackwellization.")

Exercise 2.22 *Let $Y_1,...,Y_n \sim f_\theta(y)$. Show that the M-estimator corresponding to $\rho(\theta,y) = \alpha(y-\theta)_+ + (1-\alpha)(\theta-y)_+$ for a given $0 < \alpha < 1$ is the sample $\alpha \cdot 100\%$-quantile $Y_{(\alpha)}$.

Exercise 2.23 *Prove Proposition 2.2 for the Fisher information matrix $I(\theta)$.

Chapter 3

Confidence Intervals, Bounds, and Regions

3.1 Introduction

When reporting on the results of polls, the *New York Times* on February 10, 2010 stated:

"In theory, in 19 cases out of 20, overall results based on such samples will differ by no more than three percentage points in either direction from what would have been obtained by seeking to interview all American adults. For smaller subgroups, the margin of sampling error is larger. Shifts in results between polls over time also have a larger sampling error. In addition to sampling error, the practical difficulties of conducting any survey of public opinion may introduce other sources of error into the poll. Variation in the wording and order of questions, for example, may lead to somewhat different results."

Understanding the paragraph quoted above is the subject of this chapter.

In Chapter 2, we discussed the estimation problem. When we estimate a parameter we essentially "guess" its value. This should be a "well educated guess," hopefully the best of its kind (in whatever sense). However, statistical estimation is made with error. Presenting just the estimator, no matter how good it is, is usually not enough—one should give an idea about the estimation error. The words "estimation" and "estimate" (in contrast to "measure" and "value") point to the fact that an estimate is an inexact appraisal of a value. Great efforts of statistics are invested in trying to quantify the estimation error and find ways to express it in a well-defined way.

To quantify the estimation error seems like some contradiction in terms. On the one hand, there is an unknown parameter. On the other hand, there is an estimate—a realization of a random variable (estimator) calculated exactly from the data but which is still only our best guess for the parameter of interest. Moreover, in most cases, the probability that the estimate is indeed the true value is ... zero! For example, estimating the unknown mean $\mu = EY$ of a continuous random variable by a sample mean $\bar{Y}$, we have $P(\bar{Y} = \mu) = 0$ (why?). We cannot estimate the estimation error, because if we could, we would use it to improve the estimate. Nevertheless, we can probably estimate its order of magnitude. In this chapter, we discuss the standard statistical methods to quantify the estimation error, and how we can make precise statements about it.

3.2 Quoting the estimation error

As we have argued above, any estimator has error, that is, if θ is estimated by $\hat{\theta}$, then $\hat{\theta} - \theta$ is usually different from 0. The standard error measure is the mean squared error $MSE = E(\hat{\theta} - \theta)^2$ (see Section 2.6). When together with the value of the estimator we are given its MSE, we get a feeling of how precise the estimate is.

It is more common to quote the standard error (SE) defined by

$$SE(\hat{\theta}, \theta) = \sqrt{MSE(\hat{\theta}, \theta)} = \sqrt{E(\hat{\theta} - \theta)^2}.$$

Unlike the MSE, the SE is measured in the same units as the estimator. If we estimate the length of an object, then the SE can be expressed in meters. If we estimate the mean income, the SE may be expressed in US Dollars.

DOI: 10.1201/9781003175407-3

However, the MSE expresses an *average* squared estimation error but tells nothing about its distribution, which generally might be complicated. One would be interested, in particular, in the probability $P(|\hat{\theta} - \theta| > \varepsilon)$ that the estimation error exceeds a certain accuracy level $\varepsilon > 0$. Markov's inequality enables us to translate the MSE into an upper bound on this probability: $P(|\hat{\theta} - \theta| \geq \varepsilon) \leq \frac{MSE}{\varepsilon^2}$. However, this bound is usually very conservative and the actual probability may be much smaller. Typically, a quite close approximation of the error distribution is obtained by the central limit theorem (see Section 5.5). At this stage, we will quote the "empirical law" that states that in many practical situations, the probability that a statistic will differ from its expectation by more than two standard deviations is less than 0.05, and it is very unlikely that the difference will be more than three times the standard deviation. Note that it is not a rigorous probabilistic statement and hence should be applied with care!

According to the above empirical law, it is very likely (approximately 95%) that the estimate's error is not larger than 2SE, and it will rarely be larger than 3SE. Many technical articles often quote the SE or 2SE. Thus, one can read that the estimate was $13,543.3 \pm 23.5$. The 23.5 is either the SE or the 2SE.... When a newspaper reporting on the results of a poll writes that "margin of error: +/- 4 percentage points," it probably means that the SE is 2%.

Example 3.1 (SE of the binomial mean) A political poll was taken in which 1024 randomly selected people were asked whether they support Candidate A over Candidate B. Out of those who were surveyed, 495 supported Candidate A. What was the percentage of support he has in the entire population? The simple statistical model for the data is binomial: the number of respondents Y that voted for Candidate A in the poll is assumed to be a binomial random variable $\mathscr{B}(1024, p)$. The natural estimate of p is the proportion of supporters Candidate A received in the poll: $\hat{p} = 495/1024 \approx 0.48$. Now we face a difficulty: $SE(\hat{p}, p) = \sqrt{\frac{p(1-p)}{n}}$ (see Example 2.15) but it depends on the unknown parameter p we are trying to estimate! Unfortunately, this is typical for MSE/SE. For the problem at hand, a possible option is to replace p in $SE(\hat{p}, p)$ by 0.5 and use $SE = \frac{1}{2\sqrt{n}}$. There are two reasons to do that. First, the crucial question for this survey is whether Candidate A will get the majority of votes and $p = 0.5$ becomes then the critical point. Second, probably more important, the maximum of $\sqrt{\frac{p(1-p)}{n}}$ is achieved at $p = 0.5$, and in case of doubt, we would prefer to be conservative. We should remark that $SE(\hat{p}, p)$ is not very sensitive to the value of p. For example, if p is 0.4 then $SE(\hat{p}, p)$ is only 2% shorter. Thus, our SE upper bound is $\frac{1}{2\sqrt{n}} = \frac{1}{64}$ and $2SE = \frac{1}{32} \approx 0.03$. We can therefore summarize that 48% support Candidate A with a 3% margin of error. Based on these results, for such a tough fight we cannot essentially predict with high confidence which of the two candidates will get the majority of votes among the entire population of voters. A remedy in this case could be a larger sample size to reduce the SE.

We would like to complete the considered poll example with several important remarks that may be relevant in other cases as well. In practice, a public opinion survey is often not conducted by random sampling, and the analysis is likely to be more complex than by just calculating the sample proportion. The main reason for this is that the population is mixed (for example, different genders, origins, and socio-economic classes). The ease of sampling of different groups within the society is not uniform, it might be hard to get to some of them, and different statistical means are needed to deal with this problem. In addition, not all citizens like to participate in polls, and the level of participation may be different within different demographic groups. We do not really know how different the population that actually participates in the sample is from the population of actual voters. Moreover, not all answers given to the polling company are honest.

Furthermore, the standard error of the estimate considers only the statistical error, while there may be more serious *systematic* errors. For example, the poll is taken at a given time, and it is not clear by how much public opinion shifted from the polling time to the publication time, and certainly how relevant it is to the election day. Finally, as the *New York Times* mentioned, small changes in the wording or order of questions may cause a big difference in the final result. The situation may not be much different even in exact laboratory experiments, where an experimenter often measures the quantity of interest only indirectly. The translation of actual measurements to what he really is interested in may introduce a systematic error, which he has no direct way to gauge.

It is also important to note that most statistical results reported today are not calculated by hand. Too often one finds results saying that the estimate was 12.3764 with the standard error of 0.3749. It is good practice not to quote an estimate with more digits than the statistical evidence can support. Thus 12.3 and 12.300 should mean two somewhat different things: 12.3 means that we cannot say much about the second digit after the decimal point, while 12.300 means that the second and third digits are exactly 0.

3.3 Confidence intervals

In Section 3.2, we suggested quoting the SE besides the estimator itself. However, standard statistical practice is different, and it is not very intuitive. There are a few conceptual difficulties in being precise when talking about the error. The parameter of interest is an unknown constant. After observing the data and calculating the estimate, we think about two numbers: the known estimate, which is a realization of a random variable (estimator), and the unknown non-random parameter. How can we talk about their difference?

Suppose $Y_1, \ldots, Y_n$ is a random normal sample from $\mathcal{N}(\mu, \sigma^2)$, where μ is the parameter of interest and σ is known. The standard estimator for μ is the sample mean $\bar{Y} = \frac{1}{n}\sum_{i=1}^{n} Y_i$. In particular, it is the MLE, MME, and UMVUE of μ (see Chapter 2). Its standard error (SE) is known and is equal to $\sigma/\sqrt{n}$. To give a concrete example, assume the sample mean of 100 observations was 10 and $\sigma^2 = 4$. What can we say about μ? In fact, just that it should be "not too far" from 10, but it is still an unknown quantity. Since a normal variable is rarely further away from the mean than twice the standard deviation, we are inclined to say that since the estimation error *was* unlikely to be more than 0.4 (twice the standard error), we are quite confident that μ is between 9.6 and 10.4. However, this statement makes no *probabilistic* sense. The true μ is either in the interval $(9.6, 10.4)$ or it is not. The unknown μ is not a random variable! It has a single fixed value, whether or not it is known to a statistician.

The "trick" that has been devised is to move from a probabilistic statement about the unknown (but with a fixed value) parameter to a probabilistic statement about the method. We say something like: "The interval $(9.6, 10.4)$ for the value of μ was constructed by a method which is 95% successful. The method ensures that only for 5% of the time, the true μ is not in the interval constructed." We stress again that the probabilistic statement here is about the method, not about the actual calculated interval. Since the method is usually successful, we have *confidence* in its output. Such arguments lead us to the following definition:

Definition 3.1 (confidence interval) Let $Y_1, \ldots, Y_n \sim f_\theta(y)$, $\theta \in \Theta$. A $(1-\alpha)100\%$ confidence interval for θ is the pair of scalar-valued statistics $L = L(Y_1, \ldots, Y_n)$ and $U = U(Y_1, \ldots, Y_n)$ such that

$$P(L \leq \theta \text{ and } \theta \leq U) = P(L \leq \theta \leq U) \geq 1 - \alpha$$

for all $\theta \in \Theta$, the inequality is as close to an equality as possible.

Again, the idea is to shift the point of view from the two actual endpoints of the interval (two numbers) to the procedure seen before the data were observed (two random variables). The statement is that before the data were sampled, it was known that the confidence interval that would be constructed will cover the true (unknown) value of the parameter of interest with probability $1 - \alpha$. It means that among *all* (hypothetical!) $(1-\alpha)100\%$ confidence intervals constructed for *all* possible samples from the sample space, $(1-\alpha)100\%$ of them would cover it. However, we observe just a *single* sample, and we can only feel a certain level of confidence (depending on α) that it comes from those $(1-\alpha)100\%$ "good" samples. Figure 3.1 shows plots for 90% confidence intervals for the unknown μ from 100 independent normal samples $\mathcal{N}(\mu, 1)$ of size 100. In fact, approximately 10% do not include the true value. Talking about confidence intervals, a statistician, in a way, is performing a thought experiment (*Gedankenexperiment*) similar to the one presented in this figure.

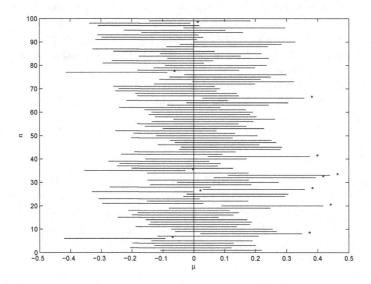

Figure 3.1: 90% confidence intervals for the normal mean μ computed from 100 independent samples of size 100 from the normal distribution $\mathcal{N}(\mu, 1)$. The true value of the parameter was $\mu = 0$. Intervals that do not include the true value are marked by asterisks.

What is the right value of α? Nothing in the statistical theory dictates a particular choice. However, as statisticians, we say that the standard is 0.1, 0.05, or 0.01, in particular 0.05. One rarely finds any other value used. The only real justification for this practice is *cosi fan tutte*, thus do they all. Theoretically, the tradeoff is between a small α, which gives a less informative but very likely to be true statement, and a large α implying a more precise statement but with a higher probability of being false. Suppose we want to estimate the increase in the rate of cancer caused by a specific agent. We may have to decide between saying something like "We have 90% confidence that the rate increase is in the interval $(1.07, 1.63)$," "We have 95% confidence that the rate increase is in the interval $(1.01, 1.69)$," or "We have 99% confidence that the rate increase is in the interval $(0.88, 1.81)$." As α decreases, we are more confident that the claim is sound, but the claim itself becomes weaker and weaker. In fact, in this example, we may say that either it is "quite likely" that the agent is helpful ($\alpha = 0.1$), or that "most likely" it does something, either good or bad ($\alpha = 0.01$), since the 99% confidence interval includes values larger and smaller than 1.

Example 3.2 (*z-**confidence interval for the normal mean when the variance is known**) Let Y_1, ..., Y_n be a simple random sample from $\mathcal{N}(\mu, \sigma^2)$ distribution with known variance σ^2. Let z_α be the $(1 - \alpha)100\%$-quantile of the standard normal distribution, that is, $\Phi(z_\alpha) = 1 - \alpha$. Since $\bar{Y} - \mu \sim \mathcal{N}(0, \frac{\sigma^2}{n})$ we know that $P(|\bar{Y} - \mu| \leq z_{\alpha/2} \frac{\sigma}{\sqrt{n}}) = 1 - \alpha$. But $|\bar{Y} - \mu| \leq z_{\alpha/2} \frac{\sigma}{\sqrt{n}}$ iff $\bar{Y} - z_{\alpha/2} \frac{\sigma}{\sqrt{n}} \leq \mu \leq \bar{Y} + z_{\alpha/2} \frac{\sigma}{\sqrt{n}}$ and, therefore,

$$P\left(\bar{Y} - z_{\alpha/2} \frac{\sigma}{\sqrt{n}} \leq \mu \leq \bar{Y} + z_{\alpha/2} \frac{\sigma}{\sqrt{n}}\right) = 1 - \alpha. \tag{3.1}$$

Hence,

$$\left(\bar{Y} - z_{\alpha/2} \frac{\sigma}{\sqrt{n}}, \ \bar{Y} + z_{\alpha/2} \frac{\sigma}{\sqrt{n}}\right)$$

is a $(1 - \alpha)100\%$ confidence interval for the unknown normal mean μ. We stress again that the equation (3.1) should be interpreted with care. It is tempting to read it as "After the mean of the sample is calculated we know that the probability that μ is between $\bar{Y} - z_{\alpha/2}\frac{\sigma}{\sqrt{n}}$ and $\bar{Y} + z_{\alpha/2}\frac{\sigma}{\sqrt{n}}$ is $1 - \alpha$." However, as we have argued, this is not true: μ is not a random variable, it is an unknown constant, and it is either within the quoted interval or not. The proper way to read the equations is: "The probability that a sample to be taken will be such that $\bar{Y} - z_{\alpha/2}\frac{\sigma}{\sqrt{n}} \leq \mu$ and $\bar{Y} + z_{\alpha/2}\frac{\sigma}{\sqrt{n}} \geq \mu$ at the same time is $1 - \alpha$."

What was the "trick" we used in constructing the confidence interval in Example 3.2? In fact, it was based on a *pivot*:

Definition 3.2 (pivot) A pivot is a function $\psi(Y_1, \ldots, Y_n; \theta)$ of the data and the parameters, whose distribution does not depend on unknown parameters.

The pivot we used in the confidence interval for the normal mean in Example 3.2 was $\bar{Y} - \mu$, whose distribution is $\mathcal{N}(0, \frac{\sigma^2}{n})$. Note, the pivot is not a statistic and cannot be calculated from the data, exactly because it depends on the unknown parameters. On the other hand, since its *distribution* does not depend on the unknown parameters, we can find an interval A_α such that $P(\psi(Y_1, \ldots, Y_n; \theta) \in A_\alpha) = 1 - \alpha$. Then we can invert the inclusion and define the interval (or region in general) $C_\alpha = \{\theta : \psi(Y_1, \ldots, Y_n; \theta) \in A_\alpha\}$. The set C_α is then a $(1 - \alpha)100\%$ confidence set. Note that C_α is a *random* set because it depends on the random sample.

Does a pivot always exist? For a random sample $Y_1, \ldots, Y_n$ from any continuous distribution with a cdf $F_\theta(y)$, $-\sum_{i=1}^{n} \ln F_\theta(Y_i) \sim \frac{1}{2}\chi_{2n}^2$ and, therefore, is a pivot (see Exercise 3.2). However, in the general case, it might be difficult (if possible) to invert the corresponding confidence interval for this pivot to a confidence interval for the original parameter of interest θ.

More convenient pivots are easy to find when the distribution belongs to a scale-location family, where the density of an observation is parameterized by two parameters μ and $\sigma > 0$, and $f_{\mu,\sigma}(x) = \sigma^{-1}f_{(0;1)}(\sigma^{-1}(x - \mu))$. Standard estimators $\hat{\mu}$ of the shift μ (the sample mean, sample median, and the 75% sample quantile, for example) and $\hat{\sigma}$ of the scale (such as the sample standard deviation, the sample inter-quantile range) ensure that the distribution of

$$\psi = \frac{\hat{\mu} - \mu}{\hat{\sigma}}$$

is a pivot. A possible way for constructing a confidence interval is then

1. Find a value c_α such that

$$P\left(\left|\frac{\hat{\mu} - \mu}{\hat{\sigma}}\right| \leq c_\alpha\right) \geq 1 - \alpha.$$

2. Construct the confidence interval

$$(\hat{\mu} - c_\alpha\hat{\sigma}, \, \hat{\mu} + c_\alpha\hat{\sigma}).$$

Example 3.3 (*t*-confidence interval for the normal mean with unknown variance) Let $Y_1, \ldots, Y_n$ be a random sample from $\mathcal{N}(\mu, \sigma^2)$, where both μ and σ^2 are unknown. The standard example of a pivot in this case is the *t*-pivot:

$$t = \frac{\bar{Y} - \mu}{s/\sqrt{n}} \sim t_{n-1}, \tag{3.2}$$

where $\bar{Y} = \frac{1}{n}\sum_{i=1}^{n} Y_i$, $s^2 = \frac{1}{n-1}\sum_{i=1}^{n}(Y_i - \bar{Y})^2$ and t_{n-1} is the *t*-distribution with $n - 1$ degrees of freedom (see Appendix A.4.14). Hence, an $(1 - \alpha)100\%$ confidence interval for μ is given by

$$\left(\bar{Y} - t_{n-1,\alpha/2}\frac{s}{\sqrt{n}}, \, \bar{Y} + t_{n-1,\alpha/2}\frac{s}{\sqrt{n}}\right), \tag{3.3}$$

where $t_{n-1,\alpha}$ is the $(1 - \alpha)100\%$ quantile of the t_{n-1} distribution.

Return to Mr. Skeptic's data from Example 1.2 and construct a 95% confidence interval for μ. Substituting $\bar{y} = 196.6, s = 6.13, n = 5$ and $t_{4,.025} = 2.78$ into (3.3) yields $(188.98, 204.22)$.

Question 3.1 Can Mr. Skeptic claim that with probability 95% the true μ lies between 188.98 and 204.22?

Is a confidence interval for θ necessarily unique? Definitely not! First, different choices of pivots lead to different forms of confidence intervals. Moreover, even for a given pivot one can typically construct an infinite set of confidence intervals at the same confidence level. For example, consider again Example 3.2. In fact, for *any* $\alpha_1 < \alpha$,

$$P\left(\bar{Y} - z_{\alpha-\alpha_1}\frac{\sigma}{\sqrt{n}} \leq \mu \leq \bar{Y} + z_{\alpha_1}\frac{\sigma}{\sqrt{n}}\right) = 1 - \alpha$$

as in (3.1) and the corresponding interval

$$\left(\bar{Y} - z_{\alpha-\alpha_1}\frac{\sigma}{\sqrt{n}}, \; \bar{Y} + z_{\alpha_2}\frac{\sigma}{\sqrt{n}}\right) \tag{3.4}$$

is a $(1-\alpha)100\%$-confidence interval for μ.

What is the "best" choice for a confidence interval? A conventional, somewhat *ad hoc* approach is based on error symmetry: a confidence interval may fail when the true parameter is either below its lower or above its upper bound. The two errors have the same probability if the confidence interval (L, U) at level $(1-\alpha)100\%$ satisfies $P(L > \theta) = P(U < \theta) = \frac{\alpha}{2}$. Such an approach is natural, for example, when the distribution of a pivot is symmetric. In fact, this was the implicit motivation to choose a symmetric confidence interval with $\alpha_1 = \frac{\alpha}{2}$ in (3.4) in Example 3.2.

A more appealing approach would be to seek a confidence interval of a minimal expected length. For example, the expected length of a confidence interval (3.4) is $(z_{\alpha_1} + z_{\alpha-\alpha_1})\frac{\sigma}{\sqrt{n}}$. One can easily show (see Exercise 3.7) that it is minimized when $\alpha_1 = \frac{\alpha}{2}$—another argument in favor of a symmetric confidence interval for this example.

Finding the confidence interval with the minimal expected length, however, in general, leads to a nonlinear minimization problem that might not have a solution in closed form (see, e.g., Exercise 3.8).

Example 3.4 (**confidence interval for the normal variance**) We consider again the random sample from the normal distribution, as studied in Example 3.3, except that this time we are interested in constructing confidence interval for σ^2. The natural pivot now is based on $s^2 = \frac{1}{n-1}\sum_{i=1}^n (Y_i - \bar{Y})^2$:

$$\frac{(n-1)s^2}{\sigma^2} \sim \chi^2_{n-1},$$

where χ^2_{n-1} is the χ^2 distribution with $n-1$ degrees of freedom. The confidence interval of the minimal expected length cannot be obtained in closed form in this case (see Exercise 3.8), while based on error symmetry, the corresponding $(1-\alpha)100\%$ confidence interval for σ^2 is given by

$$\left(\frac{(n-1)s^2}{\chi^2_{n-1,\alpha/2}}, \; \frac{(n-1)s^2}{\chi^2_{n-1,1-\alpha/2}}\right),$$

where $\chi^2_{n-1,\alpha}$ is $(1-\alpha)100\%$-quantile of the χ^2_{n-1} distribution.

A parameter of interest may be not the original parameter θ but its function $g(\theta)$. We then have:

Proposition 3.1 Let (L, U) be a $(1-\alpha)100\%$ confidence interval for θ, and $g(\cdot)$ is a strictly increasing function. Then, $\big(g(L), g(U)\big)$ is a $(1-\alpha)100\%$ confidence interval for $g(\theta)$. Similarly, if $g(\cdot)$ is strictly decreasing, the corresponding $(1-\alpha)100\%$ confidence interval for $g(\theta)$ is $\big(g(U), g(L)\big)$.

We leave the proof of Proposition 3.1 to readers as a simple exercise (see Exercise 3.9).

Example 3.5 (**confidence interval for the standard deviation**) In Example 3.4, we constructed a $(1-\alpha)100\%$ confidence interval for the variance σ^2 of the normal distribution. Proposition 3.1 immediately implies that

$$\left(\frac{\sqrt{n-1}\,s}{\sqrt{\chi^2_{n-1,\alpha/2}}}, \frac{\sqrt{n-1}\,s}{\sqrt{\chi^2_{n-1,1-\alpha/2}}}\right)$$

is a $(1-\alpha)100\%$ confidence interval for the standard deviation σ.

Example 3.6 (**continuation of Examples 2.4 and 2.5**) Consider a random sample $t_1, ..., t_n \sim \exp(\theta)$. In addition to MLEs for θ, the expected waiting time $\mu = Et$ and the probability p of waiting for more 5 minutes for a bus, John wants to construct 95% confidence intervals for all three parameters.

Each $t_i \sim \exp(\theta)$ and, therefore, $\theta t_i \sim \exp(1) = \frac{1}{2}\chi^2_2$. Thus, $2\theta \sum_{i=1}^n t_i \sim \chi^2_{2n}$ is a pivot,

$$P(\chi^2_{2n,0.975} \le 2\theta n\bar{t} \le \chi^2_{2n,0.025}) = 0.95$$

and the 95% confidence interval for θ is

$$\left(\frac{\chi^2_{2n,0.975}}{2n\bar{t}}, \frac{\chi^2_{2n,0.025}}{2n\bar{t}}\right) = (0.084, 0.173)$$

(recall that $n = 30$ and $\bar{t} = 8$ *min*). From Proposition 3.1 the 95% confidence intervals for $\mu = \theta^{-1}$ and $p = e^{-5\theta}$ are then respectively $(0.173^{-1}, 0.084^{-1}) = (5.780, 11.905)$ and $(e^{-5\cdot 0.173}, e^{-5\cdot 0.084}) = (0.421, 0.657)$.

The normal confidence intervals are undoubtedly the most important. We do not claim that most data sets are sampled from a normal distribution—this is definitely far from being true. However, what really matters is whether the distribution of an estimator $\hat{\theta} = \hat{\theta}(Y_1, ... , Y_n)$ is close to the normal rather than the distribution of $Y_1, ..., Y_n$ themselves. In Chapter 5, we argue that many estimators based on large or even medium size samples are indeed approximately normal and, therefore, the normal confidence intervals can (at least approximately) be used.

3.4   Confidence bounds

The confidence intervals considered in the previous section were, in a way, symmetric around the estimator. The idea was to find a short range of values of the parameter which is in some sense consistent with the data (we guess it would be more precise to say that the data is consistent with these values). Sometimes, however, we are more interested in an upper (or lower) bound on the parameter. Thus, for example, a meaningful statement would be that with confidence level of 95%, no more than 7% of the student population are drug abusers. In this case, we will talk about confidence bound for the unknown parameter:

Definition 3.3 (**confidence bounds**) Let $Y_1, ..., Y_n \sim f_\theta(y)$, $\theta \in \Theta$. The scalar-valued statistic $U = U(Y_1, ..., Y_n)$ is a $(1-\alpha)100\%$ confidence upper bound for θ, if $P(U \ge \theta) \ge 1-\alpha$ for all $\theta \in \Theta$. Similarly, the scalar-valued statistic $L = L(Y_1, ..., Y_n)$ is a $(1-\alpha)100\%$ confidence lower bound for θ, if for all $\theta \in \Theta$, $P(L \le \theta) \ge 1-\alpha$.

Similar to confidence intervals, the construction of confidence bound is easy when there is a pivot.

Example 3.7 (**confidence bounds for the normal mean with known variance**) Let $Y_1, ..., Y_n$ be a random sample from $\mathcal{N}(\mu, \sigma^2)$ and the variance σ^2 is known. We have already seen in Example 3.2 that $\bar{Y} - \mu \sim \mathcal{N}(0, \frac{\sigma^2}{n})$ is a pivot. One immediately gets then that $\bar{Y} - z_\alpha \frac{\sigma}{\sqrt{n}}$ and $\bar{Y} + z_\alpha \frac{\sigma}{\sqrt{n}}$ are respectively the upper and lower $(1-\alpha)100\%$ confidence bounds for the unknown μ.

Example 3.8 (**confidence bound for θ in the exponential distribution**) Suppose that $t_1, \ldots, t_n$ is a random sample from the exponential distribution $\exp(\theta)$. As we have shown in Example 3.6, $2\theta \sum_{i=1}^{n} t_i \sim \chi_{2n}^2$ and is a pivot. Hence, $P(2\theta \sum_{i=1}^{n} t_i < \chi_{2n;\alpha}^2) = 1 - \alpha$ and $\chi_{2n;\alpha}^2 / \sum_{i=1}^{n} t_i$ is a $(1 - \alpha)100\%$ confidence upper bound for θ. Similarly, the corresponding $(1 - \alpha)100\%$ confidence lower bound is $\chi_{2n;1-\alpha}^2 / \sum_{i=1}^{n} t_i$.

3.5 *Confidence regions

Confidence regions are a generalization of confidence intervals for the multiparameter cases.

In a standard public opinion poll tens of different questions are asked. Accordingly, tens of parameters are estimated, and the corresponding tens of confidence intervals are constructed simultaneously. This happens in many other situations as well. For example, in a typical fMRI experiment, the reaction of hundreds or thousands of voxels (3D pixels, or units of an image) are examined. In survey of the medical implications on a given community that suffers from some environmental hazard, many possible diseases are checked. Finally, in a typical genetic experiment, many genes are tested in parallel.

We could construct a confidence interval for a given parameter, as if it is considered on its own. For example, in the poll in Example 3.1, the confidence interval for the support of Candidate A could be based only on the answers to the relevant question, ignoring the existence of the other answers. However, there is a difficulty here. Suppose that the answers to the 72 questions in the questionnaire are all statistically independent. Even a very good 95% confidence interval is doomed to fail 1 out 20 times. Hence, on average, $72/20 = 3.6$ of confidence intervals will not cover the true values. The problem is more severe in the environmental hazard example given above, where confidence intervals for different diseases are constructed. If some interval does not include the general population rate, it may indicate that an action is required. However, if many diseases are checked simultaneously, this is likely to happen by chance alone and will cause false alarms.

The goal then is to construct a p-dimensional confidence region that will *simultaneously* cover *all* p parameters at a given confidence level:

Definition 3.4 (**confidence region**) Let $Y_1, \ldots, Y_n$ be i.i.d. $f_\theta(y)$, where $\theta = (\theta_1, \ldots, \theta_p) \in \Theta$ is a p-dimensional vector of parameters. A $(1 - \alpha)100\%$ confidence region for θ is a random set $S(\mathbf{Y})$ such that it includes θ, $P(S(\mathbf{Y}) \ni \theta) \geq 1 - \alpha$ for all $\theta \in \Theta$.

How does one construct a multidimensional confidence region? Generally, it might not be easy. However, there exists a simple *Bonferroni's method* that is very general, does not require any extra assumptions, and nevertheless, typically yields reasonable results. It is based on constructing *individual* confidence intervals I_j for each θ_j, $j = 1, \ldots, p$, that have the overall *family-wise* confidence level *at least* $(1 - \alpha)100\%$, that is, $P(\bigcap_1^p \{\theta_j \in I_j\}) \geq 1 - \alpha$. Ignoring multiplicity effect and constructing each I_j for θ_j at level $(1 - \alpha)100\%$ evidently cannot ensure the same family-wise confidence level. However, it turns out that the adjustment for multiplicity can be achieved remarkably easily: it is sufficient that each individual confidence interval I_j be at the confidence level $(1 - \frac{\alpha}{p})100\%$:

Proposition 3.2 Suppose that I_j is an individual $(1 - \frac{\alpha}{p})100\%$-level confidence interval for the parameters θ_j, $j = 1, \ldots, p$. Then, $P(\bigcap_1^p \{\theta_j \in I_j\}) \geq 1 - \alpha$.

Proof. Since the complement of an intersection is a union of the complements, and the probability of a union is less than or equal to the sum of the probabilities, we have

$$P(\bigcup_{j=1}^{p} \{\theta_j \notin I_j\}) \leq \sum_{j=1}^{p} P(\theta_j \notin I_j) = p\frac{\alpha}{p} = \alpha.$$

$\square$

The Bonferroni's confidence regions are evidently always p-dimensional rectangles.

Example 3.9 Suppose we have m normal samples $\mathcal{N}(\mu_i, \sigma_i^2)$ of sizes n_i, $i = 1, \ldots, m$. The $100(1 - \alpha)\%$ individual t-confidence interval for each μ_i is $\bar{Y}_i \pm t_{n-1,\alpha} \frac{s_i}{\sqrt{n_i}}$, where s_i^2 is an unbiased estimate of σ_i^2 (see Example 3.3). However, if we want to keep the *family-wise confidence level* $100(1 - \alpha)\%$, the corresponding Bonferroni's intervals are $\bar{Y} \pm t_{n-1,\alpha/m} \frac{s_i}{\sqrt{n_i}}$. The latter will be evidently larger. For example, if all sample sizes $n = 20$ and $m = 100$, for $\alpha = 0.05$, the corresponding intervals' length relative ratio is

$$100 \left(\frac{t_{n-1,\alpha/m}}{t_{n-1,\alpha}} - 1 \right)\% = 100 \left(\frac{4.19}{2.09} - 1 \right)\% = 100\%$$

and, therefore, Bonferroni's intervals will be twice as large.

The disadvantage of Bonferroni's method is its conservativeness since the resulting family-wise confidence level is only bounded from below by $(1 - \alpha)100\%$ while its true (unknown) value is higher. The corresponding Bonferroni's confidence region could then be possibly reduced in size without losing the required confidence level. Sometimes one can utilize specific features of the model at hand to construct confidence regions in other ways.

Example 3.10 Let $Y_1, \ldots, Y_n$ be i.i.d., $\mathcal{N}(\mu, \sigma^2)$. We want to construct a $(1 - \alpha)100\%$ confidence region for both μ and σ. Following Bonferroni's method, we construct the two individual confidence intervals for μ and σ, each one at level $(1 - \frac{\alpha}{2})100\%$. The resulting Bonferroni's confidence region is then

$$CR_B = \left\{ (\mu, \sigma) : \bar{Y} - t_{n-1,\alpha/4} \frac{s}{\sqrt{n}} \leq \mu \leq \bar{Y} + t_{n-1,\alpha/4} \frac{s}{\sqrt{n}} \right.$$
$$\left. \& \quad \frac{\sqrt{n-1}\, s}{\sqrt{\chi_{n-1,\alpha/4}^2}} \leq \sigma \leq \frac{\sqrt{n-1}\, s}{\sqrt{\chi_{n-1,1-\alpha/4}^2}} \right\} \tag{3.5}$$

(see Examples 3.3 and 3.5).

However, for this particular case, there exist alternative approaches to construct a confidence region for μ and σ at *exact* confidence level $(1 - \alpha)100\%$. Define two events

$$\mathscr{A} = \left\{ -z_{\alpha_1/2} \leq \frac{\bar{Y} - \mu}{\sigma/\sqrt{n}} \leq z_{\alpha_1/2} \right\}$$
$$\mathscr{B} = \left\{ \sqrt{\chi_{n-1,1-\alpha_1/2}^2} \leq \frac{\sqrt{n-1}\, s}{\sigma} \leq \sqrt{\chi_{n-1,\alpha_1/2}^2} \right\},$$

where $\alpha_1 = 1 - \sqrt{1 - \alpha}$. Note that the only random variable in the definition of $\mathscr{A}$ is $\bar{Y}$, and s is the only random variable in the definition of $\mathscr{B}$. For the normal sample, $\bar{Y}$ and s^2 are independent (see Example 1.7), and hence $\mathscr{A}$ and $\mathscr{B}$ are independent events. Then,

$$P(\mathscr{A} \bigcap \mathscr{B}) = P(\mathscr{A})P(\mathscr{B}) = (1 - \alpha_1)^2 = 1 - \alpha.$$

Inverting $\mathscr{A} \bigcap \mathscr{B}$ we obtain a joint confidence region for (μ, σ):

$$CR_E = \left\{ (\mu, \sigma) : \frac{\sqrt{n-1}\, s}{\sqrt{\chi_{n-1,\alpha_1/2}^2}} \leq \sigma \leq \frac{\sqrt{n-1}\, s}{\sqrt{\chi_{n-1,1-\alpha_1/2}^2}} \right.$$
$$\left. \& \quad \bar{Y} - z_{\alpha_1/2} \frac{\sigma}{\sqrt{n}} \leq \mu \leq \bar{Y} + z_{\alpha_1/2} \frac{\sigma}{\sqrt{n}} \right\}. \tag{3.6}$$

which is a trapezoid (see Figure 3.2).

There is also another possible way to construct an exact joint confidence region for μ and σ at level $(1-\alpha)100\%$. Note that $\frac{\sum_{i=1}^{n}(Y_i-\mu)^2}{\sigma^2} = \frac{(n-1)s^2+n(\bar{Y}-\mu)^2}{\sigma^2} \sim \chi_n^2$. Hence, a $(1-\alpha)100\%$ confidence region for (μ,σ) is

$$CR_{\chi^2} = \left\{ (\mu,\sigma) : \chi_{n,1-\alpha/2}^2 < \frac{(n-1)s^2+n(\bar{Y}-\mu)^2}{\sigma^2} < \chi_{n,\alpha/2}^2 \right\}$$

$$= \left\{ (\mu,\sigma) : \sqrt{\frac{(n-1)s^2+n(\bar{Y}-\mu)^2}{\chi_{n,\alpha/2}^2}} < \sigma < \sqrt{\frac{(n-1)s^2+n(\bar{Y}-\mu)^2}{\chi_{n,1-\alpha/2}^2}} \right\} \qquad (3.7)$$

However, CR_{χ^2} is an unbounded strip (see Figure 3.2) and, therefore, not so useful.

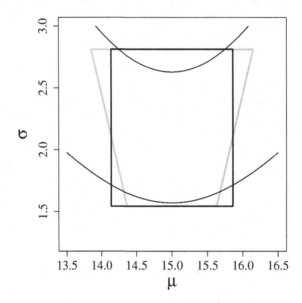

Figure 3.2: Three confidence regions for a $\mathcal{N}(\mu,\sigma^2)$ sample, where $n = 30$, $\bar{Y}_{30} = 15$, $s^2 = 4$, and $\alpha = 0.05$: the rectangle Bonferroni confidence region CR_B from (3.5), the trapezoid CR_E from (3.6) and the unbounded strip CR_{χ^2} from (3.7).

3.6   Exercises

Exercise 3.1 Let $Y_1,\ldots,Y_n$ be a random sample from the log-normal distribution $f_{\mu,\sigma}(y) = \frac{1}{\sqrt{2\pi}\sigma y} e^{-\frac{(\ln y - \mu)^2}{2\sigma^2}}$, where both μ and σ are unknown. Find $100(1-\alpha)\%$ confidence intervals for μ and σ.
(*Hint*: the distribution of $X = \ln Y$ may be useful.)

Exercise 3.2

1. Let Y be a continuous random variable with a cdf $F(y)$. Define a random variable $Z = F(Y)$. Show that $Z \sim \mathcal{U}(0,1)$ for any $F(\cdot)$.

2. Let $Y_1,\ldots,Y_n$ be a random sample from a distribution with a cdf $F_\theta(y)$, where θ is an unknown parameter. Show that $\Psi(\mathbf{Y};\theta) = -\sum_{i=1}^{n} \ln F_\theta(Y_i) \sim \frac{1}{2}\chi_{2n}^2$ and, therefore, is a pivot.
   (*Hint*: using the results of the previous paragraph, find first the distribution of $-\ln F_\theta(Y_i)$, relate it to the χ^2-distribution, and use the properties of the latter.)

Exercise 3.3 Let $Y_1, \ldots, Y_n$ be i.i.d. $\mathcal{U}(0, \theta)$.

1. Verify that $Y_{\max}/\theta$ is a pivot.
   (*Hint*: the distribution of $Y_{\max}$ was derived in Exercise 2.12.)

2. Show that $(Y_{\max}, \alpha^{-\frac{1}{n}} Y_{\max})$ is a $(1 - \alpha)100\%$ confidence interval for θ.

Exercise 3.4 Let $Y_1, \ldots, Y_n \sim \mathcal{N}(\mu, a^2\mu^2)$, where a is known. Find a pivot and use it to construct a $100(1 - \alpha)\%$ confidence interval for μ.

Exercise 3.5 Continue Example 2.6. Find $100(1 - \alpha)\%$ confidence intervals for the expected service times per clients of Alice and Kate.
(*Hint*: recall that $exp(\theta) = \frac{1}{2\theta}\chi_2^2$ to find the distribution of a sum of exponential random variables.)

Exercise 3.6 A reaction time for a certain stimulus for both cats and dogs is known to be exponentially distributed. In addition, biologists believe that an average reaction time for cats is ρ times faster than that for dogs. A biologist performed a series of experiments and measured the reaction times $X_1, \ldots, X_n$ of n dogs and $Y_1, \ldots, Y_m$ of m cats.

1. Find the MLE for average reaction times of dogs and cats.

2. Construct $100(1 - \alpha)\%$ confidence intervals for average reaction times of dogs and cats.
   (*Hint*: the distributions of $\sum_{i=1}^n X_i$ and $\sum_{j=1}^m Y_j$ may be useful.)

3. Assume now that ρ is unknown. Find its MLE and a $100(1 - \alpha)\%$ confidence interval for it.
   (*Hint*: recall the relationships between χ^2 and the F-distribution.)

Exercise 3.7 Show that the $(1 - \alpha)100\%$ confidence interval (3.4) for the unknown normal mean μ has a minimal expected length when it is symmetric, that is, $\alpha_1 = \alpha/2$.

Exercise 3.8 Continue Example 3.4. Show that the χ^2-confidence interval for the unknown variance σ^2 with a minimal expected length is $\left(\frac{(n-1)s^2}{b}, \frac{(n-1)s^2}{a} \right)$, where a and b satisfy

$$a^2 \chi_{n-1}^2(a) = b^2 \chi_{n-1}^2(b) \quad \text{and} \quad \int_a^b \chi_{n-1}^2(t)dt = 1 - \alpha,$$

where $\chi_{n-1}^2(\cdot)$ is the density of χ_{n-1}^2 distribution.

Exercise 3.9 (Proposition 3.1) Let (L, U) be a $(1 - \alpha)100\%$ confidence interval for θ, and $g(\cdot)$ a strictly increasing function. Show that $(g(L), g(U))$ is a $(1 - \alpha)100\%$ confidence interval for $g(\theta)$.

Exercise 3.10 Continue Exercise 2.3. Find $100(1 - \alpha)\%$ confidence intervals for θ, the expected lifetime of a device μ and for the probability p that a device will still be operating after a hours.

Exercise 3.11 Let $t_1, \ldots, t_n \sim exp(\theta)$. In Example 3.6, we constructed 95% confidence intervals for θ, $\mu = Et = 1/\theta$ and $p = P(t > 5) = e^{-5\theta}$ based on the pivot $2\theta \sum_{i=1}^n t_i \sim \chi_{2n}^2$. Show that another pivot is $\theta t_{\min} \sim exp(n)$. Use it to construct 95% confidence intervals for θ, μ, and p, and compare them with those from Example 3.6 for the same data.

Exercise 3.12 If $L_{\alpha/2}$ and $U_{\alpha/2}$ are the lower and upper $(1 - \alpha/2)100\%$ confidence bounds, then $(L_{\alpha/2}, U_{\alpha/2})$ is a $(1 - \alpha)100\%$-level confidence interval for θ.

Exercise 3.13 *Given a sample of vectors $\mathbf{Y}_1, \ldots, \mathbf{Y}_n$ from a p-dimensional multinormal distribution with an unknown mean vector μ and a known variance-covariance matrix V, find Bonferroni and exact $100(1 - \alpha)\%$ confidence regions for μ_j, $j = 1, \ldots, p$.
(*Hint*: recall that $(\mathbf{Y}_i - \mu)^\top V^{-1} (\mathbf{Y}_i - \mu) \sim \chi_p^2$.)

Chapter 4

Hypothesis Testing

4.1 Introduction

In many situations the research question yields one of two possible answers *"yes"* or *"no"* and a statistician is required to choose the "correct" one based on the data. For example, the researcher in Example 1.1 is probably less interested in estimating the proportion of newborn girls but really wants to know whether the proportion is 0.5 ("yes") or not ("no"). Similarly, the main concern of Mr. Skeptic from Example 1.2 is whether the expected weight of a cheese pack is indeed 200g as is claimed, or the supermarket is "cheating"; the FDA must decide whether a new drug is indeed more effective than its existing counterparts; a quality control engineer monitors a technological process to check whether all its parameters are at the required levels or it went out of control; an air traffic controller at the airport must decide whether a signal appearing on a radar comes from an aircraft or is just a bird. The two possible answers can be viewed as two *hypotheses*—the null hypothesis (H_0) and the alternative (H_1). The process of choosing between them based on the data is known as *hypothesis testing*.

More formally, the problem of hypothesis testing can be formulated as follows. Suppose that the observed data $\mathbf{Y} \sim f_\theta(\mathbf{y})$, where the distribution $f_\theta(\mathbf{y})$ belongs to a parametric family of distributions $\mathscr{F}_\theta$, $\theta \in \Theta$. Under the null hypothesis $\theta \in \Theta_0 \subset \Theta$, while under the alternative $\theta \in \Theta_1 \subset \Theta$, where $\Theta_0 \cap \Theta_1 = \emptyset$. Although generally it is not necessary that $\Theta_0 \cup \Theta_1 = \Theta$, nevertheless, since we assume only two possibilities (hypotheses) for θ, one can essentially think that the entire parameter space Θ is reduced to $\Theta_0 \cup \Theta_1$. If Θ_0 consists of a single point θ, the null hypothesis is called *simple*, otherwise it is *composite*. The same notations apply to the alternative hypothesis and Θ_1. Given the data $\mathbf{Y}$ we need to find a rule (*test*) to decide whether we accept or reject the null hypothesis H_0 based on a *test statistic* $T(\mathbf{Y})$ [1]. Typically $T(\mathbf{Y})$ is chosen in such a way that it tends to be small under H_0: the larger is it, the stronger is the evidence against H_0 in favor of H_1 [2]. The null hypothesis is then rejected if $T(\mathbf{Y})$ is "too large" for it, that is, if $T(\mathbf{Y}) \geq C$ for some *critical value C*.

For the newborn babies data in Example 1.1, where $Y_1, ..., Y_{20} \sim \mathscr{B}(1, p)$, the hypotheses are $H_0 : p = 0.5$ (simple null) vs. $H_1 : p \neq 0.5$ (composite alternative), and a reasonable test would be to reject the null (that the proportions of girls and boys are equal) if the proportion of girls in the sample $\hat{p} = \frac{1}{20} \sum_{i=1}^{20} Y_i$ is too different from 0.5, that is, if $T(\mathbf{Y}) = |\hat{p} - 0.5| \geq C$, say, for $C = 0.2$.

Mr. Skeptic in Example 1.2 wants to test $H_0 : \mu = 200$ (simple null) vs. $H_1 : \mu \neq 200$ (composite alternative) in (1.1) or perhaps, $H_0 : \mu \geq 200$ (composite null) vs. $H_1 : \mu < 200$ (composite alternative) depending on what he means by "the supermarket is cheating"—that the true μ is different from 200g or strictly less than 200g. In both cases it seems natural to look at the average sample weight $\bar{Y}$. In the former case, we reject H_0 if $T(\mathbf{Y}) = |\bar{Y} - 200| \geq C$, while for the latter, if $T(\mathbf{Y}) = 200 - \bar{Y} \geq C$, where, for example, $C = 2$.

---

[1] Acceptance/rejection of H_0 is evidently equivalent to rejection/acceptance of H_1, but traditionally the decision is specified in terms of H_0.

[2] Such a convention is just for convenience of further exposition. If there is a test statistic $T(\mathbf{Y})$ tending to be small under H_1, one can always use $-T(\mathbf{Y})$ instead.

DOI: 10.1201/9781003175407-4

It is essential to understand that since the inference on hypotheses is based on *random* data, there is always a possibility for a wrong decision: we might either erroneously reject H_0 (Type I error) or accept it (Type II error). For example, suppose there were 17 girls among the 20 newborn babies. It looks as if there is clear evidence in favor of H_1. However, there is still a nonzero probability that there will be 17 (or, in fact, even all 20!) girls among 20 babies under the null hypothesis $p = 0.5$, which might lead to its wrong rejection (Type I error). The key question in hypothesis testing is a proper choice of a test statistic and the corresponding critical value that would minimize the probability of an erroneous decision. Unfortunately, as we shall see, it is generally impossible to minimize the probabilities of errors of both types: decreasing one of them comes at the expense of increasing the other. As an extreme case, consider a situation where one decides to always reject the null hypothesis regardless of the data. Evidently, in this case, there is zero probability of a Type II error, but on the other hand, the probability of a Type I error is one. Similarly, if one always accepts the null hypothesis, the probability of a Type I error is zero, but the probability of a Type II error is one. We have to think therefore about some way to balance between the two types of errors.

Question 4.1 Consider the following trivial test for testing two hypotheses: we ignore the data, toss a (fair) coin and reject the null hypothesis if the outcome is tails. What are the probabilities of Type I and Type II errors for this coin tossing test?

We start from the case with two simple hypotheses, define main notations and derive an optimal test (in a sense to be determined). Then, we will try to extend those ideas to more general setups.

4.2   Simple hypotheses

4.2.1   Type I and Type II errors

Let the data $\mathbf{Y} \sim f_\theta(\mathbf{y})$ and we wish to test two simple hypotheses:

$$H_0 : \theta = \theta_0 \quad vs. \quad H_1 : \theta = \theta_1 \tag{4.1}$$

Suppose that we choose a test statistic $T(\mathbf{Y})$ and reject the null hypothesis if $T(\mathbf{Y}) \geq C$ for some critical value C. This induces the partition of the sample space Ω into two disjoint regions: the *rejection region* $\Omega_1 = \{\mathbf{y} : T(\mathbf{y}) \geq C\}$ and its complement, the *acceptance region* $\Omega_0 = \bar{\Omega}_1 = \{\mathbf{y} : T(\mathbf{y}) < C\}$.

As we have argued in Section 4.1, in hypotheses testing there is always a risk of making an erroneous decision. The probability of a Type I error (erroneous rejection of the null hypothesis), also known as the *(significance) level* of the test is

$$\alpha = P_{\theta_0}(\text{reject } H_0) = P_{\theta_0}(T(\mathbf{Y}) \geq C) = P_{\theta_0}(\mathbf{Y} \in \Omega_1)$$

Similarly, the probability of Type II error (erroneous acceptance of the null hypothesis) is

$$\beta = P_{\theta_1}(\text{accept } H_0) = P_{\theta_1}(T(\mathbf{Y}) < C) = P_{\theta_1}(\mathbf{Y} \in \Omega_0)$$

The *power* of a test $\pi = 1 - \beta = P_{\theta_1}(\text{reject } H_0) = P_{\theta_1}(T(\mathbf{Y}) \geq C) = P_{\theta_1}(\mathbf{Y} \in \Omega_1)$. See Table 4.1.

Example 4.1 A car company introduces a new car model and advertises that it is more economical and consumes less fuel than other cars of the same class. In particular, the company claims that the new car on average will run 15 kilometers per liter on the highway, while its existing competitors run only 12. "Auto-World Journal" wants to check the credibility of the claim by testing a random sample of 5 new cars produced by the company.

As usual, we start from building a corresponding statistical model. Let $Y_1, ..., Y_5$ be the resulting fuel consumptions. Assume that $Y_i \sim \mathcal{N}(\mu, \sigma^2)$, where $\sigma^2 = 4$ is known. The goal then is to test two simple hypotheses $H_0 : \mu = 12$ (claim is not true) vs. $H_1 : \mu = 15$ (claim is true) [3]. An intuitively

---

[3]Note that for the *unknown* σ, the above hypotheses are not simple anymore: $\Theta_0 = \{(12, \sigma), \ \sigma \geq 0\}$ and $\Theta_1 = \{(15, \sigma), \ \sigma \geq 0\}$.

Table 4.1: Two types of errors in hypothesis testing.

Decision	Truth	
	H_0	H_1
accept H_0	correct decision $(1-\alpha)$	Type II error (β)
reject H_0	Type I error (α)	correct decision $(\pi = 1 - \beta)$

reasonable test would be to believe the advertisement (reject H_0) if $\bar{Y}$ is "too large." "Auto-World Journal" decided to reject H_0 if $\bar{Y} \geq 14$. Note that $\bar{Y} \overset{H_0}{\sim} \mathcal{N}(12, .8)$ and $\bar{Y} \overset{H_1}{\sim} \mathcal{N}(15, .8)$. The corresponding error probabilities of both types are then

$$\alpha = P_{H_0}(\bar{Y} \geq 14) = 1 - \Phi\left(\frac{14 - 12}{\sqrt{.8}}\right) = 0.0127$$

and

$$\beta = P_{H_1}(\bar{Y} < 14) = \Phi\left(\frac{14 - 15}{\sqrt{.8}}\right) = 0.1318$$

(see Figure 4.1) meaning that if the new car is indeed more economical, the test will detect it with quite a high probability $\pi = 1 - \beta = 0.8682$, while if it is not true, we will erroneously believe in the company's advertisement with the (low) probability $\alpha = 0.0127$.

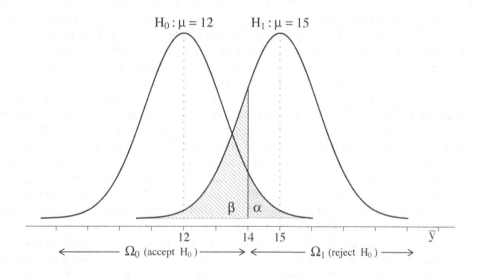

Figure 4.1: Error probabilities of both types for Example 4.1.

What would happen if "Auto-World Journal" would swap the two hypotheses around choosing $H_0' : \mu = 15$ (claim is true) vs. $H_1' : \mu = 12$ (claim is not true) and using the same test (believing the claim and accepting H_0' if $\bar{Y} \geq 14$)? In this case,

$$\alpha' = P_{H_0'}(\bar{Y} < 14) = \Phi\left(\frac{14 - 15}{\sqrt{.8}}\right) = \beta = 0.1318$$

while

$$\beta' = P_{H_1'}(\bar{Y} \geq 14) = 1 - \Phi\left(\frac{12 - 15}{\sqrt{.8}}\right) = \alpha = 0.0127$$

Hence, for a *fixed* C, there is a symmetry between the two hypotheses: by swapping them around, α becomes the new β' and vice versa.

Question 4.2 Show that the conclusion about the symmetry between the two hypotheses for a fixed critical value established in Example 4.1 for normal data is true in general as well.

Example 4.2 We toss a coin 10 times to test the null hypothesis that the coin is fair with $P(Tail) = P(Head) = 0.5$, against the alternative $P(Tail) = 0.8$. It is decided to reject the null hypothesis if there will be at least 7 tails among the 10 tossings. We define indicator variables $Y_i = 1$ if the i-th toss results in tail and $Y_i = 0$ otherwise, $i = 1, ..., 10$. Then, $Y_i \sim \mathscr{B}(1, p)$, $S = \sum_{i=1}^{10} Y_i \sim \mathscr{B}(10, p)$, where $H_0 : p = 0.5$ vs. $H_1 : p = 0.8$, and the null hypothesis is rejected if $S \geq 7$. The sample size 10 is still too small to use the normal approximation so we calculate the error probabilities directly from the binomial distribution:

$$\alpha = P_{p=0.5}(S \geq 7) = 0.5^{10} \sum_{j=7}^{10} \binom{10}{j} = 0.1719$$

and

$$\beta = P_{p=0.8}(S < 7) = \sum_{j=0}^{6} \binom{10}{j} 0.8^j \cdot 0.2^{10-j} = 0.1209$$

4.2.2 Choice of a critical value

So far, for a given test statistic $T(\mathbf{Y})$, we have rejected the null hypothesis if $T(\mathbf{Y}) \geq C$, where a critical value C was *fixed* in advance, and calculated the corresponding error probabilities. However, the choice of C is not usually so obvious, and it is not always clear how large it should be in order to reject H_0. Thus, in Example 4.1, for $\bar{y} = 17$, there would be almost no doubt about the credibility of the advertisement. But is, for example, $\bar{y} = 13$ sufficiently large to believe that a new car is indeed more economical, or can it still be explained by the randomness of the data? What about $\bar{y} = 12.1$?

Recall that $\alpha = P_{\theta_0}(T(\mathbf{Y}) \geq C)$ and $\beta = P_{\theta_1}(T(\mathbf{Y}) < C)$. Increasing C leads then to the reduction of the rejection region $\Omega_1 = \{\mathbf{y} : T(\mathbf{y}) \geq C\}$ and, therefore, to a smaller α but larger β and vice versa. Table 4.2 gives the probabilities of both error types for various values of C in Example 4.1. Hence, it is impossible to find an optimal C that would minimize both α and β. One possible way is to look for a C that minimizes the sum of error probabilities, $\alpha + \beta$ (see, e.g., Exercise 4.1). It might indeed seem a reasonable approach, but as we shall see later, its extension to composite hypotheses is problematic, and the standard criterion is different. Instead, typically, one controls the probability of the Type I error at some conventional low level (say, 0.01 or 0.05) and finds the corresponding C from the equation $\alpha = P_{\theta_0}(T(\mathbf{Y}) \geq C)$. In this case, there no longer is symmetry between the two hypotheses and the choice of H_0 becomes crucial. Two researchers analyzing the same data might come to different conclusions depending on their choices for the null and alternative. Although there are no formal rules, the null hypothesis usually represents the current or conventional state and a "strong evidence" should be provided to reject it in favor of the alternative corresponding to a change. In this sense, we set $H_0 : \mu = 12$ in Example 4.1. In many practical situations, the relative importance of two possible errors in testing is not the same. For example, in quality control, the consequences of not detecting defective parts might be much more serious than those of false alarms. In this case, one will choose the hypothesis that the product is defective as the null to control the probability of missing it at a required low level α.

Question 4.3 Show that if with a given test staistic a null hypothesis is not rejected at level α, it will not be rejected at any smaller significance level as well.

Table 4.2: Critical values C and the corresponding error probabilities α and β of both types for Example 4.1.

C	α	β
12.0	0.5	0.0004
12.5	0.2881	0.0026
13.0	0.1318	0.0127
13.5	0.0468	0.0468
14.0	0.0127	0.1318
14.5	0.0026	0.2881
15.0	0.0004	0.5

Example 4.3 (continuation of Example 4.1) Instead of fixing $C = 14$ as in Example 4.1, "Auto-World Journal" finds a critical value corresponding to $\alpha = 0.05$. The resulting C is the solution of the equation $\alpha = P_{\mu=12}(\bar{Y} \geq C) = 1 - \Phi(\frac{C-12}{.8})$ and, therefore, $C = 12 + z_{0.05}\sqrt{.8} = 13.47$. Similarly, for $\alpha = 0.01$, $C = 12 + z_{0.01}\sqrt{.8} = 14.08$.

Example 4.4 (continuation of Example 4.2) Fix $\alpha = 0.1$ and find the corresponding critical value C for the number of tails S in Example 4.2. We need to solve

$$0.1 = P_{p=0.5}(S \geq C) = 0.5^{10} \sum_{j=C}^{10} \binom{10}{j} \tag{4.2}$$

However, unlike the normal data in Example 4.3, S is a *discrete* random variable that takes integer values $0, ..., 10$. Therefore, there is no meaning in C being non-integer and the RHS of (4.2) can take also only 11 different values. Thus, there is no solution of (4.2) for all α. In this case, we find the minimal C such that the RHS of (4.2) does not exceed the required significance level 0.1. The solution cannot be obtained in the closed form. We start from the maximal possible $C = 10$ corresponding to the minimal possible α and reduce C increasing α until for the first time it exceeds 0.1. The C from the previous step will then be the resulting critical value. Table 4.3 illustrates the process:

Table 4.3: Critical values C and the corresponding error probabilities α and β of both types for Example 4.4.

C	α	β
10	0.0010	0.8326
9	0.0107	0.6242
8	0.0547	0.3222
7	0.1719	0.1209

We see that for $C = 7$, the corresponding α is already $0.1719 > 0.1$ and, hence, we should take the previous $C = 8$: the null hypothesis that the coin is fair is rejected at level 0.1 if the number of tails S is at least 8. The level of this test is 0.0547. Similarly, for $\alpha = 0.05$ and $\alpha = 0.01$ the resulting critical values are respectively 9 and 10.

4.2.3   The p-value

Consider again a general hypothesis testing (4.1) with two simple hypotheses. Following our previous arguments, for a given test statistic $T(\mathbf{Y})$, we calculate its value $t_{obs} = T(\mathbf{y})$ for the observed data

y. Until now we compared t_{obs} with a critical value C corresponding to a given significance level α: $\alpha = P_{\theta_0}(T(\mathbf{Y}) \geq C)$, and rejected the null hypothesis if $t_{obs} \geq C$. We would like now to standardize t_{obs} to a scale which will tell us directly its "unlikeness" under the null. Namely, consider the probability under H_0 that the test statistic $T(\mathbf{Y})$ has a value of at least t_{obs}. Such a probability is called a p-value or *significance of a result*, that is,

$$p - \text{value} = P_{\theta_0}(T(\mathbf{Y}) \geq t_{obs}) \qquad (4.3)$$

The smaller the p-value, the more "unlike" or "extreme" the observed value t_{obs} is under the null hypothesis and the stronger the evidence against the null. How small should the p-value be in order to reject H_0? Recall that $P_{\theta_0}(T(\mathbf{Y}) \geq C) = \alpha$, while $P_{\theta_0}(T(\mathbf{Y}) \geq t_{obs}) = p - \text{value}$. But then, $t_{obs} \geq C$ iff p-value $\leq \alpha$! In other words, in terms of the p-value, a general decision rule is to reject the null hypothesis if the p-value is less than α. A p-value can be viewed as the minimal significance level for which one would reject the null hypothesis for a given value of a test statistic.

Question 4.4 Show that for any $0 \leq \alpha \leq 1$, $P_{H_0}(p - \text{value} \leq \alpha) \leq \alpha$.

Proposition 4.1 Assume that the test statistic $T(\mathbf{Y})$ has a continuous distribution. Then, under the null hypothesis, the corresponding p-value has a uniform $\mathscr{U}(0, 1)$ distribution.

The proof is left as an exercise to the reader (see Exercise 4.4).

One can either calculate a critical value C w.r.t. a chosen α and compare it with t_{obs} or calculate the p-value and compare it with α: the decision rule is exactly the same, they are two sides of the same coin. It may be harder to evaluate the "significance" of the difference between t_{obs} and a critical value C (in particular, it should be viewed relatively to the standard deviation of a test statistic), while a p-value has a clear (absolute) probabilistic meaning. In addition, note that if α is changed, the p-value is still the same, while the corresponding C must be re-calculated w.r.t. a new α. That is why the output of statistical packages usually provides p-value and leaves the choice of α to a user.

Example 4.5 (continuation of Examples 4.1 and 4.3) As a result of the "Auto-World Journal" experiment, five tested cars ran 13.8 kilometers per liter on average on the highway. What will the decision be here? Is the new model indeed more economical? Let's calculate the significance of such a result:

$$p - \text{value} = P_{\mu=12}(\bar{Y} \geq 13.8) = 1 - \Phi\left(\frac{13.8 - 12}{\sqrt{.8}}\right) = 0.022.$$

How can we interpret it? If the new model, in fact, is not more economical than the existing competitors, the probability that the average run of five of its cars was larger than the observed 13.8 kilometers per liter would be 0.022. Is such a chance too small? On the one hand, we would accept the company's claim at a significance level of 0.05, but on the other hand, the evidence in its favor would still not be enough if we want to be very conservative and choose $\alpha = 0.01$. In this situation it might be hard to make a final conclusion based on such a small sample size. The conclusions obviously remain the same if instead we would compare $\bar{y}_{obs} = 13.8$ with the corresponding critical values 13.47 ($\alpha = 0.05$) and 14.08 ($\alpha = 0.01$). See Figure 4.2.

Example 4.6 (continuation of Examples 4.2 and 4.4) Suppose we observed 6 tails among 10 tosses. The corresponding p-value is

$$p\text{-value} = P_{p=.5}(S \geq 6) = 0.5^{10} \sum_{j=6}^{10} \binom{10}{j} = 0.377$$

which is large enough to indicate that such a result is not so extreme under the null hypothesis and we will not reject it even for a quite "liberal" $\alpha = 0.1$.

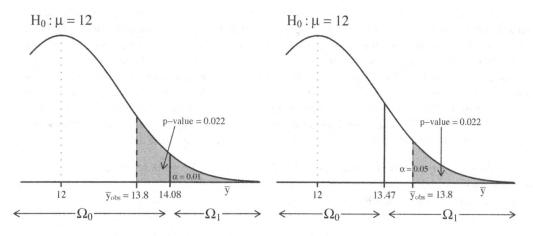

Figure 4.2: *p*-value and α for Example 4.5: $\alpha = 0.05$ (left figure) and $\alpha = 0.01$ (right figure).

4.2.4 *Maximal power tests. Neyman–Pearson lemma*

So far, a test statistic $T(\mathbf{Y})$ was chosen in advance and we discussed the corresponding probabilities of both error types, the power of the test, p-value, critical value, etc. The core question in hypothesis testing, however, is the proper choice of $T(\mathbf{Y})$. What is the "optimal" test?

Ideally, one would like to find a test with minimal error probabilities of both types. However, no test can "kill two birds with one stone": as we have already mentioned, a trivial test that never rejects the null hypothesis regardless of the data evidently has a zero probability of a Type I error but, on the other hand, the probability of a Type II error is one. We have argued in previous sections that one usually wants to control the probability of a Type I error at a certain fixed level. The aim then would be to find a test (test statistic) that achieves the minimal probability of a Type II error (or, equivalently, the maximal power) among all tests at a given level. Does such a test exist? What is it?

Let $\mathbf{Y} \sim \mathbf{f}_\theta(\mathbf{y})$ and test two simple hypotheses $H_0 : \theta = \theta_0$ vs. $H_1 : \theta = \theta_1$. Define the *likelihood ratio*

$$\lambda(\mathbf{y}) = \frac{L(\theta_1; \mathbf{y})}{L(\theta_0, \mathbf{y})} = \frac{f_{\theta_1}(\mathbf{y})}{f_{\theta_0}(\mathbf{y})}$$

The value of $\lambda(\mathbf{y})$, in fact, shows how much θ_1 is more likely for the observed data than θ_0. Obviously, the larger $\lambda(\mathbf{y})$, the stronger the evidence against the null. The corresponding *likelihood ratio test* (LRT) at a given level α is of the form: reject $H_0 : \theta = \theta_0$ in favor of $H_1 : \theta = \theta_1$ if $\lambda(\mathbf{Y}) \geq C$, where the critical value C satisfies $\alpha \leq P_{\theta_0}(\lambda(\mathbf{Y}) \geq C)$ with the inequality as close to equality as possible. As it happens, the LRT is exactly the optimal test we are looking for, that is, the one with the maximal power (MP) among all tests at a significance level α:

Lemma 4.1 (Neyman–Pearson lemma) Assume that the observed data $\mathbf{Y} \sim f_\theta(\mathbf{y})$ and consider the two simple hypotheses:

$$H_0 : \theta = \theta_0 \quad vs. \quad H_1 : \theta = \theta_1$$

The likelihood ratio test with the rejection region

$$\Omega_1 = \left\{ \mathbf{y} : \lambda(\mathbf{y}) = \frac{f_{\theta_1}(\mathbf{y})}{f_{\theta_0}(\mathbf{y})} \geq C \right\} \tag{4.4}$$

is the MP test among all tests at significance levels not larger than α, where

$$\alpha = P_{\theta_0}(\lambda(\mathbf{Y}) \geq C). \tag{4.5}$$

An attentive reader noticed that the Neyman–Pearson lemma was formulated for a fixed critical value C rather than α, and the corresponding significance level was calculated from (4.5). For a continuous distribution $f_\theta(\mathbf{y})$, it is typically the same since α is a one-to-one function of C, where for every α there exists a unique C such that $\alpha = P_{\theta_0}(\lambda(\mathbf{Y}) \geq C)$, and Lemma 4.1 can be equivalently reformulated for a fixed α. One should be, however, somewhat more careful for discrete distributions, where $\lambda(\mathbf{y})$ can only take discrete values and the solution of (4.5) exists only for the corresponding discrete set of possible value of α. For other values of α, a critical value is the minimal possible C such that $P_{\theta_0}(\lambda(\mathbf{Y}) \geq C) \leq \alpha$ and the resulting test will be, in fact, overconservative in the sense that its true significance level $\alpha' = P_{\theta_0}(\lambda(\mathbf{Y}) \geq C) < \alpha$. See also Example 4.4.

Proof. We consider the case of a continuous distribution $f_\theta(\mathbf{y})$. The proof for the discrete case will be the same but with integrals replaced by the corresponding sums.

Let π be the power of the LRT (4.4), that is,

$$\pi = P_{\theta_1}(\lambda(\mathbf{Y}) \geq C) = P_{\theta_1}(\mathbf{Y} \in \Omega_1) = \int_{\Omega_1} f_{\theta_1}(\mathbf{y})d\mathbf{y}.$$

Consider any other test at level $\alpha' \leq \alpha$ and let Ω_1' and π' be respectively its rejection region and power:

$$\alpha' = P_{\theta_0}(\mathbf{Y} \in \Omega') = \int_{\Omega'} f_{\theta_0}(\mathbf{y})d\mathbf{y}, \quad \pi' = P_{\theta_1}(\mathbf{Y} \in \Omega') = \int_{\Omega'} f_{\theta_1}(\mathbf{y})d\mathbf{y}.$$

We want to show that $\pi \geq \pi'$.

$$
\begin{aligned}
\pi - \pi' &= \int_{\Omega_1} f_{\theta_1}(\mathbf{y})d\mathbf{y} - \int_{\Omega_1'} f_{\theta_1}(\mathbf{y})d\mathbf{y} \\
&= \int_{\Omega_1 \cap \Omega_0'} f_{\theta_1}(\mathbf{y})d\mathbf{y} + \int_{\Omega_1 \cap \Omega_1'} f_{\theta_1}(\mathbf{y})d\mathbf{y} - \int_{\Omega_1' \cap \Omega_1} f_{\theta_1}(\mathbf{y})d\mathbf{y} - \int_{\Omega_1' \cap \Omega_0} f_{\theta_1}(\mathbf{y})d\mathbf{y} \\
&= \int_{\Omega_1 \cap \Omega_0'} f_{\theta_1}(\mathbf{y})d\mathbf{y} - \int_{\Omega_1' \cap \Omega_0} f_{\theta_1}(\mathbf{y})d\mathbf{y}. \tag{4.6}
\end{aligned}
$$

Note that from (4.4), $f_{\theta_1}(\mathbf{y}) \geq C f_{\theta_0}(\mathbf{y})$ for all $\mathbf{y} \in \Omega_1$ and $f_{\theta_1}(\mathbf{y}) < C f_{\theta_0}(\mathbf{y})$ for all $\mathbf{y} \in \Omega_0$. Hence, (4.6) implies

$$
\begin{aligned}
\pi - \pi' &\geq C\left(\int_{\Omega_1 \cap \Omega_0'} f_{\theta_0}(\mathbf{y})d\mathbf{y} - \int_{\Omega_1' \cap \Omega_0} f_{\theta_0}(\mathbf{y})d\mathbf{y} \right) \\
&= C\left(\int_{\Omega_1 \cap \Omega_0'} f_{\theta_0}(\mathbf{y})d\mathbf{y} + \int_{\Omega_1 \cap \Omega_1'} f_{\theta_0}(\mathbf{y})d\mathbf{y} - \int_{\Omega_1' \cap \Omega_1} f_{\theta_0}(\mathbf{y})d\mathbf{y} - \int_{\Omega_1' \cap \Omega_0} f_{\theta_0}(\mathbf{y})d\mathbf{y} \right) \\
&= C\left(\int_{\Omega_1} f_{\theta_0}(\mathbf{y})d\mathbf{y} - \int_{\Omega_1'} f_{\theta_0}(\mathbf{y})d\mathbf{y} \right) \\
&= C(\alpha - \alpha') \geq 0.
\end{aligned}
$$

□

Example 4.7 To understand better the main idea of the likelihood ratio test, consider first the following example. Suppose that you meet a man who you know to be either from the UK or France. He has no accent and you feel uncomfortable asking him directly about his origin. Instead you offer him a drink: beer, brandy/cognac [4], whisky or wine, and try to guess his origin according to his choice of drink. You want to find the MP test that will identify the man's origin, where H_0 : *France* and H_1 : *UK* at the significance level of $\alpha = 0.25$. Suppose it is known that the distributions of alcohol consumption in the UK and France are as given in Table 4.4.

---

[4]Cognac is essentially brandy produced in the wine-growing region surrounding the town Cognac in France.

Table 4.4: Distribution of alcohol consumption in the UK and France in percentages.

State	Beer	Brandy	Whisky	Wine
France	10%	20%	10%	60%
UK	50%	10%	20%	20%

This example might look somewhat different from the hypothesis testing framework considered previously since we do not test here two simple hypotheses on a *parameter(s)* of a distribution but on a distribution itself. However, in fact, we can consider all the four probabilities of the distribution as its *parameters*: $\theta_1 = P(beer)$, $\theta_2 = P(brandy)$, $\theta_3 = P(whisky)$, $\theta_4 = P(wine)$. Thus, we essentially test two simple hypotheses

$$H_0 : \theta_1 = 0.1, \theta_2 = 0.2, \theta_3 = 0.1, \theta_4 = 0.6 \quad vs. \quad H_1 : \theta_1 = 0.5, \theta_2 = 0.1, \theta_3 = 0.2, \theta_4 = 0.2$$

and can apply then all the developed theory for these kinds of hypotheses as well.

For this simple example, we can easily identify all possible tests (rejection regions) at a level not larger than 0.25 and choose the one which has the maximal power. One can immediately see that there are just two such tests which have the rejection regions $\{beer, whisky\}$ and $\{brandy\}$. The corresponding levels are $\alpha_1 = P_{France}(beer, whisky) = 0.1 + 0.1 = 0.2$ and $\alpha_2 = P_{France}(brandy) = 0.2$, and powers $\pi_1 = P_{UK}(beer, whisky) = 0.5 + 0.2 = 0.7$ and $\alpha_2 = P_{UK}(brandy) = 0.2$. Any other rejection region will result in $\alpha > 0.25$ [5]. Thus, the MP test will be to reject H_0 that the man is French if he prefers beer or whisky.

We now show that the same MP test would be obtained by the Neyman–Pearson lemma. The possible values of the likelihood ratio test statistic $\lambda(y) = P_{UK}(y)/P_{France}(y)$ are respectively 5 (for *beer*—meaning that if the man was British, the probability that he would prefer beer is five times more than that if he was French), $\frac{1}{2}$ (for *brandy*), 2 (for *whisky*) and $\frac{1}{3}$ (for *wine*). Since we are dealing with discrete data, we find the critical value corresponding to $\alpha = 0.25$ and proceed as in Example 4.4. We start from the largest possible likelihood ratio value $\lambda = 5$ corresponding to the smallest rejection region $\{beer\}$. The resulting significance level is $P_{France}(beer) = 0.1$ and we can keep trying to expand the rejection region by adding *whisky* ($\lambda(whisky) = 2$ is the second largest). The expanded rejection region is now $\{beer, whisky\}$ and the significance level is $P_{France}(beer, whisky) = 0.1 + 0.1 = 0.2$. The next candidate would be *brandy* ($\lambda(brandy) = \frac{1}{2}$) but this test would increase the significance level to $0.2 + 0.2 = 0.4$. Hence, we would stop at the previous step that results in the MP test with the rejection region $\{beer, whisky\}$ obtained previously. The entire process is illustrated by Table 4.5:

Table 4.5: Construction of the LRT at level $\alpha = 0.25$.

Critical value	Rejection region	α	π
5	beer	0.1	0.5
2	**beer, whisky**	**0.2**	**0.7**
1/2	beer, whisky, brandy	0.4	0.8
1/3	beer, whisky, brandy, wine	1	1

We continue the example. Actually, the choice of $H_0 : France$ and $H_1 : UK$ was quite arbitrary. What would happen if we instead choose $H_0 : UK$ and $H_1 : France$, and test them at the same significance level of 0.25? We leave it as an exercise to the reader (see Exercise 4.2) to verify

---

[5]One can point out two other tests with rejection regions $\{beer\}$ or $\{whisky\}$ at the significance level of $0.1 < \alpha = 0.25$ but they can be merged together to increase the power without exceeding the allowed significance level. In this sense, we are essentially talking about tests with rejection regions that cannot be expanded further.

that following the same algorithm, the Neyman–Pearson lemma will lead to the MP test with the rejection region {*wine*}: decide that the man is British if he prefers beer, whisky or brandy, and French if he prefers wine. One can easily see that the decisions for the two possible choices of the null hypothesis are different if the preferable drink is brandy. It should not be surprising since there is no symmetry between the two hypotheses for a fixed α as we have discussed above.

Example 4.8 (**MP test for the normal mean**) Let $Y_1, ..., Y_n \sim \mathcal{N}(\mu, \sigma^2)$, where the variance σ^2 is assumed to be known. We want to test two simple hypotheses about the mean μ, namely, $H_0 : \mu = \mu_0$ against $H_1 : \mu = \mu_1$, where $\mu_1 > \mu_0$. Consider the corresponding likelihood ratio:

$$
\begin{aligned}
\lambda(\mathbf{y}) &= \frac{L(\mu_1; \mathbf{y})}{L(\mu_0; \mathbf{y})} = \exp\left\{ -\frac{1}{2\sigma^2}\left(\sum_{i=1}^{n}(y_i - \mu_1)^2 - \sum_{i=1}^{n}(y_i - \mu_0)^2 \right) \right\} \\
&= \exp\left\{ -\frac{1}{2\sigma^2}\sum_{i=1}^{n}(\mu_0 - \mu_1)(2y_i - \mu_0 - \mu_1) \right\} = \exp\left\{ \frac{1}{\sigma^2}(\mu_1 - \mu_0)n\left(\bar{y} - \frac{\mu_1 + \mu_0}{2} \right) \right\}.
\end{aligned}
\tag{4.7}
$$

According to the Neyman–Pearson lemma, the MP test at level α rejects the null hypothesis if $\lambda(\mathbf{Y})$ in (4.7) is larger than the critical value C satisfying $P_{\mu_0}(\lambda(\mathbf{Y}) \geq C) = \alpha$. To solve this equation and find C one should first derive the distribution of the RHS of (4.7) under the null hypothesis which requires tedious calculus. However, one can note instead that for $\mu_1 > \mu_0$ the likelihood ratio $\lambda(\mathbf{y})$ in (4.7) is an increasing function of the sufficient statistic $\bar{y}$ and $\lambda(\mathbf{y}) \geq C$ iff $\bar{y} \geq C^*$, where

$$
C = \exp\left\{ -\frac{1}{\sigma^2}(\mu_1 - \mu_0)n\left(C^* - \frac{\mu_1 + \mu_0}{2} \right) \right\}
$$

(see Figure 4.3).

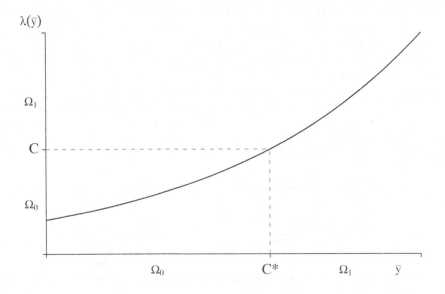

Figure 4.3: The likelihood ratio $\lambda(\mathbf{y})$ in (4.7) as a function of $\bar{y}$.

The MP test can then be re-written in terms of $\bar{Y}$: reject the null hypothesis if $\bar{Y} \geq C^*$, where $\alpha = P_{\mu_0}(\lambda(\mathbf{Y}) \geq C) = P_{\mu_0}(\bar{Y} \geq C^*)$. Unlike the distribution of the original likelihood ratio $\lambda(\mathbf{Y})$, the distribution of $\bar{Y}$ under the null hypothesis is well-known: $\bar{Y} \overset{H_0}{\sim} \mathcal{N}(\mu_0, \frac{\sigma^2}{n})$ and the corresponding critical value C^* can be easily calculated: $C^* = \mu_0 + z_\alpha \frac{\sigma}{\sqrt{n}}$. Thus, the MP test for testing $H_0 : \mu = \mu_0$

against $H_1 : \mu = \mu_1$, where $\mu_1 > \mu_0$, rejects the null hypothesis if

$$\bar{Y} \geq \mu_0 + z_\alpha \frac{\sigma}{\sqrt{n}} \tag{4.8}$$

Note that this is exactly the same test we have used in Examples 4.1 and 4.3 based on "common sense considerations."

Following the same arguments for $\mu_0 > \mu_1$, the likelihood ratio $\lambda(\mathbf{y})$ (4.7) is now a decreasing function of $\bar{y}$ and the resulting MP test rejects the null hypothesis if

$$\bar{Y} \leq \mu_0 - z_\alpha \frac{\sigma}{\sqrt{n}} \tag{4.9}$$

(see Exercise 4.3).

Example 4.9 (**MP test for the proportion**) Let $Y_1, ..., Y_n$ be a series of Bernoulli trials and consider testing two simple hypotheses on the probability of success p: $H_0 : p = p_0$ against $H_1 : p = p_1$, where $p_1 > p_0$. The likelihood ratio $\lambda(\mathbf{y})$ is

$$\lambda(\mathbf{y}) = \frac{L(p_1; \mathbf{y})}{L(p_0; \mathbf{y})} = \left(\frac{p_1}{p_0} \right)^{\sum_{i=1}^n y_i} \left(\frac{1 - p_1}{1 - p_0} \right)^{n - \sum_{i=1}^n y_i}$$

and is an increasing function of $\sum_{i=1}^n y_i$. Thus, the LRT statistic is $\sum_{i=1}^n Y_i$ and the corresponding MP test is of the form: reject the null hypothesis if $\sum_{i=1}^n Y_i \geq C$. This is actually the test we have used in Examples 4.2 and 4.4 for testing the fairness of a coin. Under the null hypothesis, $\sum_{i=1}^n Y_i \overset{H_0}{\sim} \mathcal{B}(n, p_0)$ which is a discrete distribution, and to find the critical value C corresponding to a given level α one should proceed as in Example 4.4. For a "sufficiently large" sample size n, it is possible to use the normal approximation $\sum_{i=1}^n Y_i \overset{H_0}{\sim} \mathcal{N}(np_0, np_0(1 - p_0))$ yielding $C = np_0 + z_\alpha \sqrt{np_0(1 - p_0)}$.

The Neyman–Pearson lemma says that the likelihood ratio $\lambda(\mathbf{y}) = \frac{f_{\theta_1}(\mathbf{y})}{f_{\theta_0}(\mathbf{y})}$ is the optimal (MP) test statistic for testing two simple hypotheses $H_0 : \theta = \theta_0$ against $H_1 : \theta = \theta_1$ and one should reject H_0 if $\lambda(\mathbf{Y}) \geq C$. The main problem in its practical implementation is finding the corresponding critical value C. For a fixed significance level α, C is the solution of $\alpha = P_{\theta_0}(\lambda(\mathbf{Y}) \geq C)$, where the distribution of $\lambda(\mathbf{Y})$ might be quite complicated. As we have seen in Example 4.8 and Example 4.9, sometimes monotonicity of $\lambda(\mathbf{y})$ in some other statistic $T(\mathbf{y})$ with a simpler distribution allows one to re-formulate the MP test in terms of $T(\mathbf{y})$ but generally this is not the case. For large samples various asymptotic approximations for distribution of $\lambda(\mathbf{Y})$ can be used (see Section 5.12).

4.3 Composite hypotheses

4.3.1 Power function

The situation in which both the null and the alternative hypotheses are simple is quite rare in practical applications — it is usual that at least the alternative hypothesis is composite. Testing the fairness of a coin, the probability of a tail under the alternative is actually unknown, it just differs from 0.5. Mr. Skeptic in Example 1.2 cannot specify μ if a supermarket is indeed "cheating." As we have discussed in Section 4.1, he tests either a simple $H_0 : \mu = 200$ against a composite $H_1 : \mu \neq 200$ or two composite hypotheses $H_0 : \mu \geq 200$ against $H_1 : \mu < 200$ depending on his definition of "cheating."

Consider the general case, where the data $\mathbf{Y} \sim f_\theta(\mathbf{y})$, $\theta \in \Theta$, and we test

$$H_0 : \theta \in \Theta_0 \subset \Theta \quad vs. \quad H_1 : \theta \in \Theta_1 \subset \Theta, \tag{4.10}$$

$\Theta_0 \cap \Theta_1 = \emptyset$ and $\Theta_0 \cup \Theta_1 = \Theta$.

We should first extend the definitions of significance level, power, p-value, etc. introduced for simple hypotheses to this general case. Recall that for simple hypotheses, where $\Theta_0 = \{\theta_0\}$ and $\Theta_1 = \{\theta_1\}$, for a given test with a rejection region Ω_1, the significance level $\alpha = P_{\theta_0}(reject\ H_0) = P_{\theta_0}(\mathbf{Y} \in \Omega_1)$ and power $\pi = P_{\theta_1}(reject\ H_0) = P_{\theta_1}(\mathbf{Y} \in \Omega_1)$. We now extend their definitions for the case, where $\theta_0 \in \Theta_0$ and $\theta_1 \in \Theta_1$ are not explicitly specified:

Definition 4.1 (power function) The *power function* $\pi(\theta)$ of a test is the probability of rejecting H_0 as a function of θ, that is,

$$\pi(\theta) = P_\theta(reject\ H_0) = P_\theta(\mathbf{Y} \in \Omega_1).$$

Evidently, for simple hypotheses, $\alpha = \pi(\theta_0)$ and $\pi = \pi(\theta_1)$. In a way, the power function is the analogue of significance level for $\theta \in \Theta_0$ (erroneous rejection of H_0) and of power for $\theta \in \Theta_1$ (correct rejection of H_0). An ideal test would have a power function equal to zero for $\theta \in \Theta_0$ and equal to one for $\theta \in \Theta_1$. In reality, a power function of a "good" test should be low for $\theta \in \Theta_0$ and high for $\theta \in \Theta_1$ (see Figure 4.4).

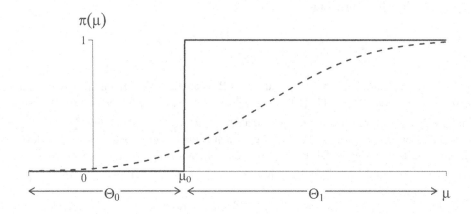

Figure 4.4: The power functions of an ideal test (solid) and a "good" test (dashed).

Example 4.10 Let $Y_1, ..., Y_n \sim \mathcal{N}(\mu, \sigma^2)$, where σ^2 is known, and consider testing two composite (one-sided) hypotheses $H_0 : \mu \leq \mu_0$ against $H_1 : \mu > \mu_0$. Intuitively, one would reject H_0 for "too large" values of $\bar{Y}$. Consider then a test, where H_0 is rejected if $\bar{Y} \geq C$ for a given critical value C. In this case,

$$\pi(\mu) = P_\mu(\bar{Y} \geq C) = 1 - \Phi\left(\frac{C - \mu}{\sigma/\sqrt{n}}\right)$$

which is an increasing function of μ with lower values for $\mu \leq \mu_0$ and higher for $\mu > \mu_0$ (see Figure 4.5). In this sense, the proposed test looks "reasonable" though we cannot yet verify its optimality.

Example 4.11 We continue the previous Example 4.10 for a normal sample but test now a simple null hypothesis $H_0 : \mu = \mu_0$ against a composite (two-sided) alternative $H_1 : \mu \neq \mu_0$. We reject H_0 if $\bar{Y}$ is "too far" from μ_0, that is, if $|\bar{Y} - \mu_0| \geq \varepsilon$ for some fixed $\varepsilon > 0$. In this case,

$$\pi(\mu) = P_\mu(|\bar{Y} - \mu_0| \geq \varepsilon) = P_\mu(\bar{Y} \leq \mu_0 - \varepsilon) + P_\mu(\bar{Y} \geq \mu_0 + \varepsilon) = \Phi\left(\frac{\mu_0 - \varepsilon - \mu}{\sigma/\sqrt{n}}\right) + 1 - \Phi\left(\frac{\mu_0 + \varepsilon - \mu}{\sigma/\sqrt{n}}\right)$$

(see Figure 4.5).

Definition 4.2 (significance level for composite hypotheses) A (significance) level of a given test with a power function $\pi(\theta)$ is $\alpha = \sup_{\theta \in \Theta_0} \pi(\theta)$.

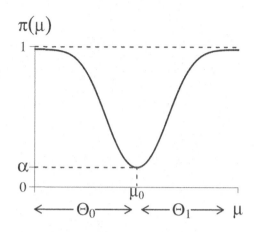

Figure 4.5: The power functions for a one-sided test of Example 4.10 (left figure) and a two-sided test of Example 4.11 (right figure).

Such an extension of a definition of a significance level for the case of composite hypotheses can be naturally interpreted as the maximal probability of a Type I error over all possible $\theta \in \Theta_0$ (the worst case scenario).

Similarly, for the observed value t_{obs} of a given test statistic $T(\mathbf{Y})$, the p-value$= \sup_{\theta \in \Theta_0} P_\theta(T(\mathbf{Y}) \geq t_{obs})$.

Example 4.12 (continuation of Examples 4.10 and 4.11) We have seen that the power function $\pi(\mu)$ in Example 4.10 is an increasing function of μ. Therefore,

$$\alpha = \sup_{\mu \leq \mu_0} \pi(\mu) = \pi(\mu_0) = 1 - \Phi\left(\frac{C - \mu_0}{\sigma/\sqrt{n}}\right) \qquad (4.11)$$

In particular, to ensure a given significance level α, the corresponding $C = \mu_0 + z_\alpha \sigma/\sqrt{n}$. For the observed $\bar{y}_{obs}$ the corresponding p-value$= \sup_{\mu \leq \mu_0} P_\mu(\bar{Y} \geq \bar{y}_{obs}) = P_{\mu_0}(\bar{Y} \geq \bar{y}_{obs}) = 1 - \Phi\left(\frac{\bar{y}_{obs} - \mu_0}{\sigma/\sqrt{n}}\right)$. In Example 4.11, the null hypothesis is simple, so trivially $\alpha = \pi(\mu_0) = 2\Phi\left(-\frac{\varepsilon}{\sigma/\sqrt{n}}\right)$ and the p-value$= 2\Phi\left(-\frac{|\bar{y}_{obs}|}{\sigma/\sqrt{n}}\right)$. See Figure 4.5.

4.3.2 Uniformly most powerful tests

As we have argued, the main problem in hypothesis testing is the proper choice of a test. In Section 4.2.4 we discussed that for simple hypotheses one compares the performance of two different tests with the same significance level by their powers and looks for the maximal power test. What then happens with composite hypotheses, where there is no a single power *value* but a power *function*?

Consider two tests with power functions $\pi_1(\theta)$ and $\pi_2(\theta)$ with the same significance level α. Let us look at two possible scenarios in Figure 4.6. In the first case $\pi_1(\theta) > \pi_2(\theta)$ for *all* $\theta \in \Theta_1$ and, thus, we can say that the first test *uniformly* outperforms its counterpart. In the second case, however, neither of the two is uniformly better and we cannot prefer either of them. Unfortunately, this is quite common when comparing tests. In a way, it is reminiscent of the situation of comparing estimators by MSE discussed in Section 2.6.

The question is whether we can find a test that achieves a maximal power *uniformly* for all $\theta \in \Theta_1$ among all tests at the given significance level:

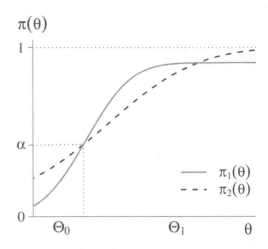

Figure 4.6: Power functions for two pairs of tests at level α.

Definition 4.3 (uniformly most powerful test) A test is called a *uniformly most powerful* (UMP) among all tests at level α if its power function $\pi(\theta)$ satisfies:

1. $\sup_{\theta \in \Theta_0} \pi(\theta) = \alpha$

2. $\pi(\theta) \geq \pi_1(\theta)$ for all $\theta \in \Theta_1$, where $\pi_1(\theta)$ is a power function of any other test at a level not larger than α

The requirement of uniformly most powerfulness for *all* $\theta \in \Theta_1$ among *all* possible tests at a given level is very strong and the first natural question is whether it is not *too* strong, that is, whether a UMP test ever exists? The following example provides an evidence of its existence at least for one particular case:

Example 4.13 (one-sided test for the normal mean) Consider testing a simple hypothesis $H_0 :$ $\mu = \mu_0$ on a normal mean against the one-sided alternative $H_1 : \mu > \mu_0$ with a known variance σ^2. Fix an arbitrary $\mu_1 > \mu_0$ and test two *simple* hypotheses $H_0 : \mu = \mu_0$ against $H_1 : \mu = \mu_1$ at level α. We already know that the MP (LRT) test at level α rejects H_0 if $\bar{Y} \geq \mu_0 + z_\alpha \frac{\sigma}{\sqrt{n}}$ (see (4.8) in Example 4.8). But this test does not depend on a particular μ_1 under the alternative and thus is the MP test for *any* $\mu_1 > \mu_0$. Hence, it is the UMP test.

Moreover, intuitively it is clear (though it can be rigorously proved—see the proof of Proposition 4.2) that to distinguish between $\mu = \mu_0$ against $\mu > \mu_0$ is more difficult than for any other $\mu \leq \mu_0$, so the above test (aka one-sided z-test) is also UMP for testing a composite one-sided hypotheses $H_0 : \mu \leq \mu_0$ against $H_1 : \mu > \mu_0$.

Similarly, testing $H_0 : \mu = \mu_0$ or $H_0 : \mu \geq \mu_0$ against $H_1 : \mu < \mu_0$, the UMP test at level α is to reject H_0 if $\bar{Y} \leq \mu_0 - z_\alpha \frac{\sigma}{\sqrt{n}}$ (see (4.9)).

Encouraged by the above example, one may now hope that a UMP test *always* exists. Unfortunately, it is not true as is demonstrated by the following simple counterexample:

Example 4.14 (two-sided test for the normal mean) We continue testing a normal mean μ with a known variance σ^2 but this time consider $H_0 : \mu = \mu_0$ against a two-sided alternative $H_1 : \mu \neq \mu_0$. In the previous Example 4.13 we derived separately two *different* UMP z-tests for one-sided alternatives $\mu < \mu_0$ and $\mu > \mu_0$. Any other α-level test would be outperformed either on the right with the first of these tests, or on the left by the second. Thus, there is no *single* test which is UMP for all $\mu \neq \mu_0$.

For this example, however, it is still possible to find a UMP test among a *restricted* set of tests. As we have argued in Example 4.11, for a two-sided alternative it would be natural to reject the null hypothesis if $|\bar{Y} - \mu_0|$ is "too large." We now show that if we consider only tests depending on the statistic $T(\mathbf{Y}) = |\bar{Y} - \mu_0|$, rejecting $H_0 : \mu = \mu_0$ if $|\bar{Y} - \mu_0| \geq z_{\frac{\alpha}{2}} \frac{\sigma}{\sqrt{n}}$ is a UMP test at level α among them.

Re-write the tested hypotheses as $H_0 : \Delta = 0$ *vs.* $H_1 : \Delta \neq 0$, where $\Delta = \mu - \mu_0$. We first derive the density $f_\Delta(t)$ of $T(\mathbf{Y})$. Since $\bar{Y} \sim \mathcal{N}(\mu, \frac{\sigma^2}{n})$, the cdf $F_\Delta(t)$ is

$$F_\Delta(t) = P_\Delta(|\bar{Y} - \mu_0| < t) = \Phi\left(\frac{t - \Delta}{\sigma^2/n}\right) - \Phi\left(\frac{-t - \Delta}{\sigma^2/n}\right),$$

and, therefore,

$$f_\Delta(t) = \frac{1}{\sqrt{2\pi}\sigma/\sqrt{n}} e^{-\frac{(t-\Delta)^2}{2\sigma^2/n}} + \frac{1}{\sqrt{2\pi}\sigma/\sqrt{n}} e^{-\frac{(t+\Delta)^2}{2\sigma^2/n}} = \frac{1}{\sqrt{2\pi}\sigma/\sqrt{n}} e^{-\frac{t^2}{2\sigma^2/n}} e^{-\frac{\Delta^2}{2\sigma^2/n}} \left(e^{\frac{t\Delta}{\sigma^2/n}} + e^{-\frac{t\Delta}{\sigma^2/n}}\right).$$

Fix any $\Delta_1 \neq 0$. The likelihood ratio

$$\lambda(T(\mathbf{y})) = \frac{f_{\Delta_1}(T(\mathbf{y}))}{f_0(T(\mathbf{y}))} = e^{-\frac{\Delta_1^2}{2\sigma^2/n}} \left(e^{\frac{T(\mathbf{y})\Delta_1}{\sigma^2/n}} + e^{-\frac{T(\mathbf{y})\Delta_1}{\sigma^2/n}}\right)$$

is an increasing function of $T(\mathbf{y})$. Thus, by Neyman–Pearson lemma, among all tests based on $T(\mathbf{y})$, the α-level MP test for testing two simple hypotheses $H_0 : \Delta = 0$ *vs.* $\Delta = \Delta_1$ rejects H_0 iff $T(\mathbf{Y}) \geq C$, where $\alpha = P_0(T(\mathbf{Y}) \geq C)$. Under the null hypothesis $T(\mathbf{Y}) = \frac{\sigma}{\sqrt{n}}|z|$, where z is a standard normal random variable $\mathcal{N}(0, 1)$, and, therefore, $C = z_{\frac{\alpha}{2}} \frac{\sigma}{\sqrt{n}}$. This two-sided z-test does not depend on the particular $\Delta_1 \neq 0$ and, hence, is the UMP test.

In addition, it can be shown that this test is also the so-called generalized likelihood ratio test discussed later in Section 4.3.3.

Example 4.14 above demonstrates that a UMP test does not exist even for such a simple hypothesis testing setup. In fact, this is typically the case. Moreover, for multiparameter cases, UMP tests do not exist at all except for some singular cases. As an illustration, consider again Example 4.8 for a normal sample, testing $H_0 : \mu = \mu_0$ against $H_1 : \mu = \mu_1$, where $\mu_1 > \mu_0$, but for a more realistic scenario with an *unknown* variance σ^2. Note that in this case the considered hypotheses are not simple any more: $\Theta_0 = \{(\mu_0, \sigma), \sigma \geq 0\}$ and $\Theta_1 = \{(\mu_1, \sigma), \sigma \geq 0\}$. The UMP test, if it exists, should be then the most powerful test uniformly for all $\sigma \geq 0$. However, for any given σ, the corresponding MP test rejects H_0 if $\bar{Y} \geq \mu_0 + z_\alpha \frac{\sigma}{\sqrt{n}}$ (see Example 4.8) and does depend on σ. Thus, although our primary interest is in μ while σ is, in fact, a nuisance parameter, it is nevertheless accounted for in the most powerful test.

There is still one important class of hypothesis testing problems for a one-dimensional parameter θ, where a UMP test does exist. We start from the following definition:

Definition 4.4 (monotone likelihood ratio) A family of distributions $\mathscr{F}_\theta$, $\theta \in \Theta$ with a one-dimensional parameter θ has a monotone likelihood ratio in a statistic $T(\mathbf{Y})$ if for any $\theta_1 < \theta_2$, the likelihood ratio $f_{\theta_2}(\mathbf{y})/f_{\theta_1}(\mathbf{y})$ is a nondecreasing function of $T(\mathbf{y})$.

Obviously, if the likelihood ratio is a *nonincreasing* function of $T(\mathbf{Y})$, $\mathscr{F}_\theta$ has a monotone likelihood ratio in $-T(\mathbf{Y})$.

Question 4.5 Show that the one-parameter exponential family (see Definition 1.5 in Section 1.6) has a monotone likelihood ratio in a sufficient statistic $T(\mathbf{Y})$ if $c(\theta)$ is a nondecreasing function of θ.

Proposition 4.2 Let $\mathbf{Y} \sim f_\theta(\mathbf{y})$, where $f_\theta \in \mathscr{F}_\theta$ has a monotone likelihood ratio in a statistic $T(\mathbf{Y})$. Then there exists a UMP test at level α for testing the one-sided hypothesis $H_0 : \theta \leq \theta_0$ vs. the one-sided hypothesis $H_1 : \theta > \theta_0$, where H_0 is rejected if

$$T(\mathbf{Y}) \geq C \quad \text{and} \quad P_{\theta_0}(T(\mathbf{Y}) \geq C) = \alpha \qquad (4.12)$$

(with the usual obvious modifications for discrete distributions).

Proposition 4.2 states that if the family $\mathscr{F}_\theta$ has a monotone likelihood ratio in a statistic $T(\mathbf{Y})$ and the tested hypotheses are one-sided, then

a) A UMP test does exist.

b) $T(\mathbf{Y})$ can be used as test statistic.

c) To calculate the corresponding critical value one should use the distribution of $T(\mathbf{Y})$ for $\theta = \theta_0$.

Proof. Take any $\theta_1 > \theta_0$ and consider testing two *simple* hypotheses $H_0 : \theta = \theta_0$ against $H_1 : \theta = \theta_1$ at level α. According to the Neyman–Pearson lemma, the MP test is a LRT and rejects the null if $\lambda(\mathbf{Y}) \geq \tilde{C}$, where $P_{\theta_0}(\lambda(\mathbf{Y}) \geq \tilde{C}) = \alpha$. However, due to monotonicity of $\lambda(\mathbf{Y})$ in $T(\mathbf{Y})$ it can be equivalently re-formulated in terms of $T(\mathbf{Y})$ and the null is rejected if $T(\mathbf{Y}) \geq C$, where $P_{\theta_0}(T(\mathbf{Y}) \geq C) = P_{\theta_0}(\lambda(\mathbf{Y}) \geq \tilde{C}) = \alpha$, which is exactly the test (4.12). Moreover, since it does not depend on a particular $\theta_1 > \theta_0$, it is UMP for testing $H_0 : \theta = \theta_0$ against $H_1 : \theta > \theta_0$.

To complete the proof we need to verify that the test (4.12) has significance level α for a composite $H_0 : \theta \leq \theta_0$, that is, that $\sup_{\theta \leq \theta_0} P_\theta(T(\mathbf{Y}) \geq C) = P_{\theta_0}(P_{\theta_0}(T(\mathbf{Y}) \geq C) = \alpha$. Consider any $\theta_2 < \theta_0$ and test two simple hypotheses $H_0 : \theta = \theta_2$ vs. $H_1 : \theta = \theta_0$, where the null hypothesis is rejected if $T(\mathbf{Y}) \geq C$ for a fixed critical value C defined in (4.12). Repeating the previous arguments, this is a LRT (MP) test at level $\alpha' = P_{\theta_2}(T(\mathbf{Y}) \geq C)$ with a power $\pi' = P_{\theta_0}(T(\mathbf{Y}) \geq C) = \alpha$. Thus, all we need to show is that $\alpha' \leq \alpha$. Consider the following another trivial test for testing $H_0 : \theta = \theta_2$ vs. $H_1 : \theta = \theta_0$: regardless of the data, generate a uniformly random variable $u \sim U[0,1]$ and reject the null if $u \leq \alpha'$. Obviously, such a test is of level α' and with power α' as well (why?). Since the test based on $T(\mathbf{Y})$ is a MP test among all tests of level α', its power $\pi' = \alpha$ cannot be less than that of any other test at the same level and, therefore, $\alpha \geq \alpha'$. $\square$

Proposition 4.2 can be immediately modified for testing $H_0 : \theta \geq \theta_0$ against $H_1 : \theta < \theta_0$. Repeating all the arguments in the proof with the corresponding obvious modifications, one can easily verify that the UMP test in this case is to reject H_0 if $T(\mathbf{Y}) \leq C$, where $\alpha = P_{\theta_0}(T(\mathbf{Y}) \leq C)$.

We now consider several examples from the one-parameter exponential family with the monotone likelihood ratio (see Question 4.5).

Example 4.15 (normal data with a known variance) A normal sample from $\mathcal{N}(\mu, \sigma^2)$ with the known σ^2 has a monotone likelihood ratio in $\bar{Y}$. Hence, the UMP test at level α for testing $H_0 : \mu \leq \mu_0$ against $H_1 : \mu > \mu_0$ is to reject the null if $\bar{Y} \geq \mu_0 + z_\alpha \frac{\sigma}{\sqrt{n}}$. We have already proved this result in Example 4.13 directly.

Example 4.16 (Bernoulli trials) Let $Y_1, ..., Y_n \sim \mathscr{B}(1, p)$ and test $H_0 : p \leq p_0$ against $H_1 : p > p_0$. A distribution of a series of Bernoulli trials has a monotone likelihood ratio in $\sum_{i=1}^n Y_i \sim \mathscr{B}(n, p)$. For simplicity of exposition, assume that n is "sufficiently large" and we can use the normal approximation to the distribution of the test statistic. The resulting UMP test rejects the null if $\sum_{i=1}^n Y_i \geq np_0 + z_\alpha \sqrt{np_0(1 - p_0)}$. This is exactly the MP test from Example 4.9 for testing two simple hypotheses $H_0 : p = p_0$ against $H_1 : p = p_1$, where $p_1 > p_0$, but since it does not depend on a specific p_1, it is also the UMP test for testing two composite one-sided hypotheses. In particular, it can be used for testing the fairness of a coin $H_0 : p = 0.5$ against a general composite alternative that has a tendency for tails $H_1 : p > 0.5$ (unlike a specific simple $H_1 : p = 0.8$ considered in Example 4.2).

Example 4.17 (exponential data) Let $Y_1, ..., Y_n \sim \exp(\theta)$. The corresponding likelihood ratio is monotone in $-\sum_{i=1}^n Y_i$ and the UMP test at level α rejects $H_0 : \theta \leq \theta_0$ in favor of $H_1 : \theta > \theta_0$

if $-\sum_{i=1}^{n} Y_i \geq -C$ or, equivalently, $\sum_{i=1}^{n} Y_i \leq C$, where $P_{\theta_0}(\sum_{i=1}^{n} Y_i \leq C) = \alpha$. Recall that $\theta Y_i \sim$ exp$(1) = \frac{1}{2}\chi_2^2$ and therefore $2\theta_0 \sum_{i=1}^{n} Y_i \overset{\theta_0}{\sim} \chi_{2n}^2$ (see, e.g., Example 3.6). The critical value is then $C = \frac{1}{2\theta_0} \chi_{2n;1-\alpha}^2$.

4.3.3 Generalized likelihood ratio tests

We have seen in the Section 4.3.2 that a UMP test for testing composite hypotheses usually does not exist—the requirement of being the most powerful test uniformly for all possible $\theta \in \Theta_1$ and for all possible tests of the same level is too ambitious. Sometimes a UMP test can be found over a restricted set of tests (see Example 4.14) but in the general case even this is either impossible or the resulting restricted set is very limited. On the other hand, it seems tempting to generalize the likelihood ratio test of Neyman–Pearson that has a clear intuitive meaning and optimality properties when testing simple hypotheses to composite alternatives.

Let $\mathbf{Y} \sim f_\theta(\mathbf{Y})$, $\theta \in \Theta$ and test

$$H_0 : \theta \in \Theta_0 \quad vs. \quad H_1 : \theta \in \Theta_1,$$

where $\Theta_0 \cap \Theta_1 = \emptyset$ and $\Theta_0 \cup \Theta_1 = \Theta$. Recall that for simple hypotheses, where $\Theta_0 = \{\theta_0\}$ and $\Theta_1 = \{\theta_1\}$, the likelihood ratio $\lambda(\mathbf{y}) = L(\theta_1, \mathbf{y})/L(\theta_0, \mathbf{y}) = f_{\theta_1}(\mathbf{y})/f_{\theta_0}(\mathbf{y})$. A natural generalization of the likelihood ratio for composite hypotheses would then be

$$\lambda^*(\mathbf{y}) = \frac{\sup_{\theta \in \Theta_1} L(\theta, \mathbf{y})}{\sup_{\theta \in \Theta_0} L(\theta, \mathbf{y})} = \frac{\sup_{\theta \in \Theta_1} f_\theta(\mathbf{y})}{\sup_{\theta \in \Theta_0} f_\theta(\mathbf{y})}. \tag{4.13}$$

The $\lambda^*(\mathbf{y})$ in (4.13) evidently coincides with the likelihood ratio for the case of simple hypotheses and indicates how much more likely the alternative hypothesis is for the observed data than the null. The larger a value of $\lambda^*(\mathbf{y})$, the stronger the evidence against H_0 and in this sense, it is a "reasonable" test statistic.

Usually, it is more convenient to use the equivalent statistic

$$\lambda(\mathbf{y}) = \frac{\sup_{\theta \in \Theta} L(\theta, \mathbf{y})}{\sup_{\theta \in \Theta_0} L(\theta, \mathbf{y})} = \frac{\sup_{\theta \in \Theta} f_\theta(\mathbf{y})}{\sup_{\theta \in \Theta_0} f_\theta(\mathbf{y})}, \tag{4.14}$$

where obviously $\lambda(\mathbf{y}) = \max(\lambda^*(\mathbf{y}), 1)$ and is a nondecreasing function of $\lambda^*(\mathbf{y})$. The $\lambda(\mathbf{y})$ from (4.14) is called a *generalized likelihood ratio* (GLR).

The corresponding *generalized likelihood ratio test* (GLRT) at level α rejects the null if

$$\lambda(\mathbf{y}) = \frac{\sup_{\theta \in \Theta} L(\theta, \mathbf{y})}{\sup_{\theta \in \Theta_0} L(\theta, \mathbf{y})} \geq C, \tag{4.15}$$

where the critical value C satisfies $\sup_{\theta \in \Theta_0} P_\theta(\lambda(\mathbf{Y}) \geq C) = \alpha$ (with the usual modifications for a discrete distribution).

Question 4.6 Show that the critical value C in (4.15) has to be greater than or equal to one. Moreover, C is strictly greater than one for any $\alpha < 1$.

To calculate the GLRT statistic $\lambda(\mathbf{y})$ in (4.14) one should:

1. Find the MLE $\hat{\theta}$ for θ to calculate the numerator $\sup_{\theta \in \Theta} L(\theta, \mathbf{y}) = L(\hat{\theta}, \mathbf{y})$.

2. Find the MLE $\hat{\theta}_0$ for θ under the restrictions $\theta \in \Theta_0$ to calculate the denumerator $\sup_{\theta \in \Theta_0} L(\theta, \mathbf{y}) = L(\hat{\theta}_0, \mathbf{y})$.

3. Form the generalized likelihood ratio $\lambda(\mathbf{y}) = L(\hat{\theta}, \mathbf{y})/L(\hat{\theta}_0, \mathbf{y})$ and find an equivalent simpler test statistic $T(\mathbf{y})$ (if possible) such that $\lambda(\mathbf{y})$ is its increasing function.

4. Find the corresponding critical value for $T(\mathbf{y})$ solving $\alpha = \sup_{\theta \in \Theta_0} P_\theta(T(\mathbf{Y}) \geq C)$.

As it happens, a variety of well-known and widely used statistical tests (e.g., one- and two-sample t-tests for normal means, χ^2-test for the normal variance, F-test for normal variances, F-test for comparing nested models in regression, Pearson's χ^2-test for goodness-of-fit, etc.) are, in fact, examples of GLRTs. Several examples below serve as an illustration.

Example 4.18 (**one-sample** t-**test**) Let $Y_1, ..., Y_n \sim \mathcal{N}(\mu, \sigma^2)$, where σ^2 is unknown, and we want to test $H_0 : \mu = \mu_0$ against the two-sided alternative $\mu \neq \mu_0$. Recall that the normal likelihood is given by

$$L(\mu, \sigma; \mathbf{y}) = \left(\frac{1}{\sqrt{2\pi \sigma^2}} \right)^n e^{-\frac{\sum_{i=1}^n (y_i - \mu)^2}{2\sigma^2}} \tag{4.16}$$

Following the general scheme for calculating the GLR $\lambda(\mathbf{y})$, we plug in the MLEs $\hat{\mu} = \bar{y}$ and $\hat{\sigma}^2 = \frac{1}{n} \sum_{i=1}^n (y_i - \bar{y})^2$ (see Example 2.7) in (4.16) to get

$$L(\hat{\mu}, \hat{\sigma}; \mathbf{y}) = \left(\frac{1}{\sqrt{2\pi \hat{\sigma}^2}} \right)^n e^{-\frac{\sum_{i=1}^n (y_i - \bar{y})^2}{2\hat{\sigma}^2}} = \left(\frac{1}{\sqrt{2\pi \hat{\sigma}^2}} \right)^n e^{-\frac{n}{2}}$$

Under the null hypothesis, obviously, $\hat{\mu}_0 = \mu_0$ and using a straightforward calculus that we leave to the reader yields $\hat{\sigma}_0^2 = \frac{1}{n} \sum_{i=1}^n (y_i - \mu_0)^2$. Thus,

$$L(\hat{\mu}_0, \hat{\sigma}_0; \mathbf{y}) = \left(\frac{1}{\sqrt{2\pi \hat{\sigma}_0^2}} \right)^n e^{-\frac{\sum_{i=1}^n (y_i - \mu_0)^2}{2\hat{\sigma}_0^2}} = \left(\frac{1}{\sqrt{2\pi \hat{\sigma}_0^2}} \right)^n e^{-\frac{n}{2}}$$

and the GLR is

$$
\begin{aligned}
\lambda(\mathbf{y}) &= \frac{L(\hat{\mu}, \hat{\sigma}; \mathbf{y})}{L(\hat{\mu}_0, \hat{\sigma}_0; \mathbf{y})} = \left(\frac{\hat{\sigma}_0^2}{\hat{\sigma}^2} \right)^{\frac{n}{2}} = \left(\frac{\sum_{i=1}^n (y_i - \mu_0)^2}{\sum_{i=1}^n (y_i - \bar{y})^2} \right)^{\frac{n}{2}} = \left(\frac{\sum_{i=1}^n (y_i - \bar{y})^2 + n(\bar{y} - \mu_0)^2}{\sum_{i=1}^n (y_i - \bar{y})^2} \right)^{\frac{n}{2}} \\
&= \left(1 + \frac{n(\bar{y} - \mu_0)^2}{\sum_{i=1}^n (y_i - \bar{y})^2} \right)^{\frac{n}{2}} = \left(1 + \frac{1}{n-1} \left(\frac{\bar{y} - \mu_0}{s/\sqrt{n}} \right)^2 \right)^{\frac{n}{2}},
\end{aligned}
$$

where $s^2 = \frac{1}{n-1} \sum_{i=1}^n (y_i - \bar{y})^2$. The GLR $\lambda(\mathbf{y})$ is an increasing function of a statistic $|T(\mathbf{y})|$, where $T(\mathbf{y}) = \frac{\bar{y} - \mu_0}{s/\sqrt{n}}$, and we reject $H_0 : \mu = \mu_0$ if $|T(\mathbf{y})| \geq C$, where $P_{\mu_0}(|T(\mathbf{Y})| \geq C) = \alpha$. Note that for $\mu = \mu_0$, $T(\mathbf{Y})$ has a Student's t-distribution t_{n-1} with $n-1$ degrees of freedom (see (3.2)) and, therefore, $C = t_{n-1;\alpha/2}$. The resulting GLRT rejects then $H_0 : \mu = \mu_0$ in favor of $H_1 : \mu \neq \mu_0$ if

$$\frac{|\bar{Y} - \mu_0|}{s/\sqrt{n}} \geq t_{n-1;\alpha/2},$$

which is the celebrated one-sample two-sided t-test.

We continue the example but now test the one-sided hypotheses $H_0 : \mu \leq \mu_0$ against $H_1 : \mu > \mu_0$. The MLEs $\hat{\mu}$, $\hat{\sigma}^2$ and the maximum likelihood $L(\hat{\mu}, \hat{\sigma}; \mathbf{y})$ in the numerator of $\lambda(\mathbf{y})$ obviously are not changed. To obtain $\hat{\mu}_0$ and $\hat{\sigma}_0$ we need to maximize the likelihood $L(\mu, \sigma; \mathbf{y})$ (or, equivalently, the log-likelihood $l(\mu, \sigma; \mathbf{y})$) w.r.t. μ and σ under the constraint $\mu \leq \mu_0$. Thus, we should minimize

$$n \ln \sigma + \frac{\sum_{i=1}^n (y_i - \mu)^2}{2\sigma^2} = n \ln \sigma + \frac{\sum_{i=1}^n (y_i - \bar{y})^2}{2\sigma^2} + \frac{n(\bar{y} - \mu)^2}{2\sigma^2}$$

over the set $\{(\mu, \sigma), \mu \leq \mu_0, \sigma \geq 0\}$. Note that $\hat{\sigma}_0^2 = \frac{1}{n} \sum_{i=1}^n (y_i - \hat{\mu}_0)^2$. To find $\hat{\mu}_0$, we need to minimize $(\bar{y} - \mu)^2$ w.r.t. μ subject to constraint $\mu \leq \mu_0$.

One can easily verify that for $\bar{y} \leq \mu_0$, $\hat{\mu}_0 = \bar{y} = \hat{\mu}$ (see Figure 4.7), the GLR $\lambda(\mathbf{y}) = 1$ and the null hypothesis is therefore never rejected (see Question 4.6) . For $\bar{y} > \mu_0$, the minimum is achieved at the boundary point $\hat{\mu}_0 = \mu_0$ (see Figure 4.7) and $\lambda(\mathbf{y})$ is identical to that for the two-sided

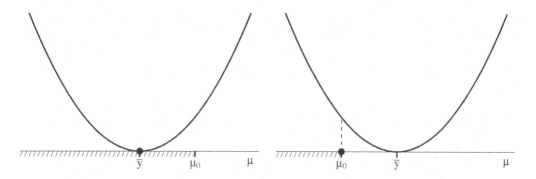

Figure 4.7: Minimizing $(\bar{y} - \mu)^2$ subject to constraint $\mu \leq \mu_0$: $\bar{y} \leq \mu_0$ (left figure) and $\bar{y} > \mu_0$ (right figure).

alternative considered above, being an increasing function of $T(\mathbf{y}) = \frac{\bar{y} - \mu_0}{s/\sqrt{n}}$. The null hypothesis is rejected in this case if $T(\mathbf{y}) \geq C$, where $\alpha = \sup_{\mu \leq \mu_0} P_\mu (T(\mathbf{Y}) \geq C)$. Similar to the known variance case, one can easily verify that $P_\mu (T(\mathbf{Y}) \geq C)$ is an increasing function of μ and, therefore, $\alpha = P_{\mu_0} (T(\mathbf{Y}) \geq C)$ implying $C = t_{n-1;\alpha}$.

Summarizing both cases leads to the one-sample one-sided t-test that rejects $H_0 : \mu \leq \mu_0$ in favor of $H_1 : \mu > \mu_0$ if

$$\frac{\bar{Y} - \mu_0}{s/\sqrt{n}} \geq t_{n-1;\alpha}$$

We leave it as a simple exercise to the reader to show that the GLRT for testing $H_0 : \mu \geq \mu_0$ vs. $H_1 : \mu < \mu_0$ rejects H_0 if $T(\mathbf{Y}) = \frac{\bar{Y} - \mu_0}{s/\sqrt{n}} \leq -t_{n-1;\alpha}$.

We can now return to Example 1.2 in Section 1.1 and help Mr. Skeptic to test whether the supermarket was cheating claiming that the mean weight of a cheese pack is 200g. Recall that the average weight of 5 sampled cheese packs was 196.6g and s was 6.13g (see Example 3.3). The resulting t-statistic $T(\mathbf{y}) = \frac{\bar{y} - \mu_0}{s/\sqrt{n}} = -1.24$. As we have argued, the inference depends on the definition of "cheating." If by "cheating" Mr. Skeptic means that the true mean weight is *different* from 200g, that is, $H_0 : \mu = 200$ against $H_1 : \mu \neq 200$, he should use the two-sided t-test. The value of the test statistic $|T(\mathbf{y})| = 1.24 < t_{4;0.025} = 2.78$ and thus the null hypothesis is not rejected at level $\alpha = 0.05$. If "cheating" means that μ is less than 200g, then $H_0 : \mu \geq 200$ vs. $H_1 : \mu < 200$ and Mr. Skeptic should apply the one-sided t-test at level 0.05 comparing $T(\mathbf{y}) = -1.24$ with $-t_{4;0.05} = -2.13$ and still not reject H_0. Thus, in both cases Mr. Skeptic cannot sue a supermarket for cheating consumers. However, such a small sample size ($n = 5$) obviously does not allow one to make any final conclusions.

Example 4.19 (χ^2-**test for the variance**) Consider again $Y_1, ..., Y_n \sim \mathcal{N}(\mu, \sigma^2)$, where both parameters are unknown, but now we are interested in making inference about the variance σ^2. In particular, test $H_0 : \sigma^2 = \sigma_0^2$ against $H_1 : \sigma^2 \neq \sigma_0^2$ at level α.

From the previous Example 4.18 we have $\hat{\mu} = \bar{y}$, $\hat{\sigma}^2 = \frac{1}{n} \sum_{i=1}^n (y_i - \bar{y})^2$ and

$$L(\hat{\mu}, \hat{\sigma}; \mathbf{y}) = \left(\frac{1}{\sqrt{2\pi \hat{\sigma}^2}} \right)^n e^{-\frac{n}{2}}$$

The MLE $\hat{\mu} = \bar{y}$ does not depend on σ^2 and remains the same under the null hypothesis. Thus,

$$L(\hat{\mu}_0, \hat{\sigma}_0^2; \mathbf{y}) = L(\hat{\mu}, \sigma_0^2; \mathbf{y}) = \left(\frac{1}{\sqrt{2\pi \sigma_0^2}} \right)^n e^{-\frac{\sum_{i=1}^n (y_i - \bar{y})^2}{2\sigma_0^2}}$$

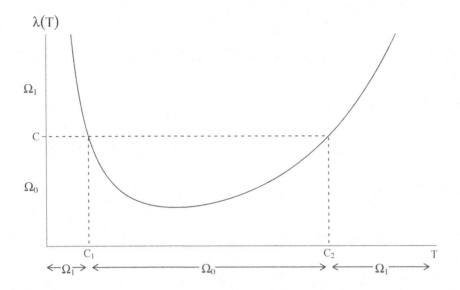

Figure 4.8: The generalized likelihood ratio $\lambda(\mathbf{y})$ as a function of $T(\mathbf{y})$ in Example 4.19.

and the GLR

$$\lambda(\mathbf{y}) = \left(\frac{\sigma_0^2}{\hat{\sigma}^2}\right)^{n/2} e^{(n/2)(\hat{\sigma}^2/\sigma_0^2 - 1)} = n^{n/2} T(\mathbf{y})^{-n/2} e^{(T(\mathbf{y}) - n)/2},$$

where $T(\mathbf{Y}) = \frac{n\hat{\sigma}^2}{\sigma_0^2} = \frac{\sum_{i=1}^{n}(Y_i - \bar{Y})^2}{\sigma_0^2} \overset{H_0}{\sim} \chi_{n-1}^2$ (see Example 3.4). Hence, $\lambda(\mathbf{y}) \geq C$ iff $T(\mathbf{y}) \leq C_1$ or $T(\mathbf{y}) \geq C_2$, where

$$X_{n-1}^2(C_2) - X_{n-1}^2(C_1) = 1 - \alpha, \quad C_2 - C_1 = n \ln\left(\frac{C_2}{C_1}\right) \tag{4.17}$$

and X_{n-1}^2 is the cdf of the χ_{n-1}^2 distribution (see Figure 4.8). There are no closed solutions for C_1 and C_2 but it can be shown that the commonly used quantiles $\chi_{n-1;1-\alpha/2}^2$ and $\chi_{n-1;\alpha/2}^2$ approximately satisfy (4.17), and the resulting two-sided χ^2-test for the normal variance does not reject the null if

$$\chi_{n-1;1-\alpha/2}^2 \leq \frac{\sum_{i=1}^{n}(Y_i - \bar{Y})^2}{\sigma_0^2} \leq \chi_{n-1;\alpha/2}^2$$

The corresponding one-sided tests are left as exercises (see Exercise 4.11).

4.4 Duality between hypothesis testing and confidence intervals (regions)

Example 4.20 Let $Y_1, ..., Y_n \sim \mathcal{N}(\mu, \sigma^2)$, where both parameters are unknown, and we want to test $H_0 : \mu = \mu_0$ against a two-sided alternative $H_1 : \mu \neq \mu_0$ at level α. As we already know from Example 4.18, the GLRT is the one-sample two-sided t-test that does not reject H_0 if $\frac{|\bar{Y} - \mu_0|}{s/\sqrt{n}} < t_{n-1;\alpha/2}$. Let us now look at this problem from a slightly different point of view and ask: What is the set of values of μ_0 for which we would not reject H_0 for a given $\bar{Y}$? The immediate answer is that it is the set $\{\mu : \frac{|\bar{Y} - \mu|}{s/\sqrt{n}} < t_{n-1;\alpha/2}\} = \{\mu : \bar{Y} - t_{n-1;\alpha/2} < \mu < \bar{Y} + t_{n-1;\alpha/2}\}$, which is exactly the $(1 - \alpha)100\%$ confidence interval for μ (see (3.3) in Example 3.3). Is this just a coincidence or an indication of a general phenomenon?

Proposition 4.3 Consider a general one-parameter case, where the observed sample $\mathbf{Y} \sim f_\theta(\mathbf{y})$, and for any $\theta_0 \in \Theta$ test $H_0 : \theta = \theta_0$ against the two-sided alternative $\theta \neq \theta_0$ at level α by some test depending on θ_0. Let $S(\mathbf{Y})$ be the set of all values θ_0 for which the tests would not reject H_0 for a given $\mathbf{Y}$. Then, $S(\mathbf{Y})$ forms a $(1 - \alpha)100\%$ confidence region for θ.

Conversely, let $S(\mathbf{Y})$ be a $(1 - \alpha)100\%$ confidence interval (region) for θ and consider testing $H_0 : \theta = \theta_0$ against $H_1 : \theta \neq \theta_0$ at level α by the following test: H_0 is not rejected if $\theta_0 \in S(\mathbf{Y})$. Such a test is of level α.

Proof. The proof is quite trivial. Let $\Omega_0(\theta_0)$ be the acceptance region of a test. Evidently, $\mathbf{y} \in \Omega_0(\theta_0)$ iff $S(\mathbf{Y}) \ni \theta_0$. Thus,

$$P_{\theta_0}\left(S(\mathbf{Y}) \ni \theta_0\right) = P_{\theta_0}\left(\mathbf{Y} \in \Omega_0(\theta_0)\right) = 1 - \alpha.$$

$\square$

In particular, it follows from Proposition 4.3 that if one already has derived a $(1 - \alpha)100\%$ confidence interval for an unknown parameter θ and then wants to test $H_0 : \theta = \theta_0$ against a two-sided alternative $H_1 : \theta \neq \theta_0$ at level α, he should simply check whether θ_0 lies within the confidence interval and then accept H_0 or reject it otherwise.

Example 4.21 (continuation of Example 3.6) An announcement at the bus station states that the expected waiting time for a bus is 7 minutes. Waiting for a bus on the average 8 minutes during 30 days, John wants to test whether it is indeed true at $\alpha = 0.05$. In Example 3.6, John already derived the 95% confidence interval for $\mu = Et$: $(5.780, 11.905)$. Hence, to test $H_0 : \mu = 7$ vs. $H_1 : \mu \neq 7$ he should just check that 7 is within this confidence interval and to accept the null hypothesis.

Finally, it must be mentioned that duality between hypothesis testing and confidence intervals (or, more generally, confidence sets) established in Proposition 4.3 can be straightforwardly extended to the multiparameter case and also to establish a similar duality between *one*-sided hypotheses testing and the corresponding confidence *bounds* (see Section 3.4).

4.5   Sequential testing

So far we considered hypothesis testing after observing the entire data. Alternatively, in various practical experiments, the data is analyzed *sequentially* as they are collected and further sampling is stopped once there is already enough evidence for making a conclusion. Sequential testing may save experimental costs. Furthermore, in some situations the decision must be made in "real time." For example, a quality control engineer controls a certain chemical technological process that requires keeping the temperature at a fixed level. The temperature is measured every 5 minutes and the engineer must decide whether the process is still operating properly or the temperature has been changed and the process should be terminated. This is also an example, where sequential hypothesis testing is essential.

Suppose that we observe a series of successive observations $Y_1, Y_2, \ldots \sim f_\theta(y)$ and want to test two simple hypotheses $H_0 : \theta = \theta_0$ vs. $H_1 : \theta = \theta_1$. At each "time" t we can either stop sampling and accept H_0, stop sampling and reject H_0 or continue sampling. Consider the likelihood ratio $\lambda_t(\mathbf{y})$ at time t:

$$\lambda_t(\mathbf{y}) = \frac{L(\theta_1; y_1, \ldots, y_t)}{L(\theta_0; y_1, \ldots, y_t)}$$

According to the Neyman–Pearson lemma, $\lambda_t(\mathbf{Y})$ is the optimal test statistic for testing H_0 vs. H_1 at given t. It is natural then to use $\lambda_t(\mathbf{Y})$ as a test statistic for sequential testing as well and to define a sequential likelihood ratio test as follows. At any given time t we decide:

reject H_0,	if $\lambda_t(\mathbf{Y}) \geq B$
accept H_0,	if $\lambda_t(\mathbf{Y}) \leq A$
sample the next observation Y_{t+1},	if $A < \lambda_t(\mathbf{Y}) < B$

for some critical values A and B. This is known as a *sequential Wald test*. The procedure evidently terminates at the stopping time $T = \min\{t : \lambda_t(\mathbf{Y}) \leq A \text{ or } \lambda_t(\mathbf{Y}) \geq B\}$. Note that unlike the "standard" hypotheses testing setup, where there are two possible decisions: to accept or to reject the null, in sequential testing at any given time there is an additional "neutral" option: not to decide yet and wait for the next observation. It can be shown that the stopping time T is finite with probability one, that is, $P(T < \infty) = 1$.

We now discuss the resulting error probabilities of both types for the proposed sequential Wald test. Define the events

$$\mathscr{A}_t = \{H_0 \text{ is accepted at time } t\} = \{A < \lambda_\tau(\mathbf{Y}) < B, \ \tau = 1, ..., t-1; \ \lambda_t(\mathbf{Y}) \leq A\}$$

and

$$\mathscr{B}_t = \{H_0 \text{ is rejected at time } t\} = \{A < \lambda_\tau(\mathbf{Y}) < B, \ \tau = 1, ..., t-1; \ \lambda_t(\mathbf{Y}) \geq B\}$$

Then,

$$
\begin{aligned}
\alpha = P_{\theta_0}(\text{reject } H_0) &= \sum_{t=1}^{\infty} P_{\theta_0}(\mathscr{B}_t) = \sum_{t=1}^{\infty} \iint_{\mathscr{B}_t} f_{\theta_0}(y_1, ..., y_t) dy_1 ... dy_t \\
&= \sum_{t=1}^{\infty} \iint_{\mathscr{B}_t} \frac{f_{\theta_0}(y_1, ..., y_t)}{f_{\theta_1}(y_1, ..., y_t)} f_{\theta_1}(y_1, ..., y_t) dy_1 ... dy_t \\
&= \sum_{t=1}^{\infty} \iint_{\mathscr{B}_t} \lambda_t^{-1}(\mathbf{y}) f_{\theta_1}(y_1, ..., y_t) dy_1 ... dy_t \leq \frac{1}{B} \sum_{t=1}^{\infty} \iint_{\mathscr{B}_t} f_{\theta_1}(y_1, ..., y_t) dy_1 ... dy_t \\
&= \frac{1}{B} \sum_{t=1}^{\infty} P_{\theta_1}(\mathscr{B}_t) = \frac{1}{B}(1 - \beta)
\end{aligned}
\tag{4.18}
$$

Similarly,

$$\beta = P_{\theta_1}(\text{accept } H_0) = \sum_{t=1}^{\infty} P_{\theta_1}(\mathscr{A}_t) = \sum_{t=1}^{\infty} \iint_{\mathscr{A}_t} f_{\theta_1}(y_1, ..., y_t) dy_1 ... dy_t \leq A \sum_{t=1}^{\infty} P_{\theta_0}(\mathscr{A}_t) = A(1 - \alpha)$$

$$\tag{4.19}$$

The above inequalities yield

$$\frac{\alpha}{1 - \beta} \leq \frac{1}{B} \quad \text{and} \quad \frac{\beta}{1 - \alpha} \leq A \tag{4.20}$$

In practice, the inequalities in (4.20) are considered as if they are equalities and for the given error probabilities α and β, the critical values

$$A = \frac{\beta}{1 - \alpha} \quad \text{and} \quad B = \frac{1 - \beta}{\alpha}$$

are used. The resulting Wald sequential test then accepts H_0 if $\lambda_t(\mathbf{Y}) \leq \frac{\beta}{1-\alpha}$, rejects H_0 if $\lambda_t(\mathbf{Y}) \geq \frac{1-\beta}{\alpha}$ and waits for the next observation Y_{t+1} otherwise.

Replacing inequalities by equalities in (4.20) implies that the actual error probabilities α' and β' of the Wald sequential test are somewhat different from the required α and β. From (4.18) and (4.19) we have

$$\alpha' \leq \frac{1}{B}(1 - \beta') = \frac{\alpha}{1 - \beta}(1 - \beta')$$

and

$$\beta' \leq A(1 - \alpha') = \frac{\beta}{1 - \alpha}(1 - \alpha')$$

that after a simple calculus yields

$$\alpha' \le \frac{\alpha}{1-\beta}, \quad \beta' \le \frac{\beta}{1-\alpha}$$

and

$$\alpha' + \beta' \le \alpha + \beta$$

Thus, it could not be that both $\alpha' > \alpha$ and $\beta' > \beta$.

Example 4.22 (sequential Wald test for the normal mean) Assume that we observe a series $Y_1, Y_2, \ldots \sim \mathcal{N}(\mu, \sigma^2)$ of successive normal observations with known variance. Derive a sequential Wald test for testing two simple hypotheses $H_0 : \mu = \mu_0$ vs. $H_1 : \mu = \mu_1$, where $\mu_1 > \mu_0$, for given α and β.

For a given t the likelihood ratio is

$$\lambda_t(\mathbf{Y}) = \exp\left\{ \frac{1}{\sigma^2}(\mu_1 - \mu_0)\left(\sum_{i=1}^{t} Y_i - \frac{\mu_1 + \mu_0}{2} t \right) \right\}$$

(see (4.7)). It is more convenient to write the test in terms of the log-likelihood $\ln \lambda_t(\mathbf{Y})$. Let $a = \ln A = \ln \frac{\beta}{1-\alpha}$ and $b = \ln B = \ln \frac{1-\beta}{\alpha}$. One then accepts H_0 if

$$\frac{1}{\sigma^2}(\mu_1 - \mu_0)\left(\sum_{i=1}^{t} Y_i - \frac{\mu_1 + \mu_0}{2} t \right) \le a,$$

rejects H_0 if

$$\frac{1}{\sigma^2}(\mu_1 - \mu_0)\left(\sum_{i=1}^{t} Y_i - \frac{\mu_1 + \mu_0}{2} t \right) \ge b$$

and continues sampling otherwise.

Equivalently, in terms of $\sum_{i=1}^{t} Y_i$, the H_0 is accepted if

$$\sum_{i=1}^{t} Y_i \le \frac{a\sigma^2}{\mu_1 - \mu_0} + \frac{\mu_1 + \mu_0}{2} t,$$

rejected if

$$\sum_{i=1}^{t} Y_i \ge \frac{b\sigma^2}{\mu_1 - \mu_0} + \frac{\mu_1 + \mu_0}{2} t$$

and the decision is not made otherwise. The resulting test has a clear geometric illustration (see Figure 4.9). We can draw a sample trajectory of the test statistic $\sum_{i=1}^{t} Y_i$ as a function of a current sample size t. The sampling is continuing as long as $\sum_{i=1}^{t} Y_i$ remains within the "corridor" of width $\frac{(b-a)\sigma^2}{\mu_1 - \mu_0}$ between two parallel straight boundaries $\frac{a\sigma^2}{\mu_1 - \mu_0} + \frac{\mu_1 + \mu_0}{2} t$ and $\frac{b\sigma^2}{\mu_1 - \mu_0} + \frac{\mu_1 + \mu_0}{2} t$. The first time $\sum_{i=1}^{t} Y_i$ falls outside the corridor, the sampling is terminated and the null hypothesis is either accepted (if $\sum_{i=1}^{t} Y_i$ falls below the corridor) or rejected (if $\sum_{i=1}^{t} y_i$ falls above the corridor).

In particular, consider testing $H_0 : \mu = 0$ vs. $H_1 : \mu = 1$, where $\sigma^2 = 1$ and $\alpha = \beta = 0.05$. The corresponding $a = \ln(1/19) \approx -2.9$, $b = \ln(19) = -a \approx 2.9$ and the boundaries are $-2.9 + 0.5t$ and $2.9 + 0.5t$ (see Figure 4.9). To demonstrate the performance of the test we continued generating a random sample (on computer) from the null distribution $\mathcal{N}(0,1)$ as long as the decision could not yet be made: $y_1 = -1.539, y_2 = 0.166, y_3 = -0.007, y_4 = 2.390, y_5 = -0.043, y_6 = -2.626$. After 6 observations, their cumulative sum $\sum_{i=1}^{6} y_i$ fell outside the corridor (below the lower boundary) for the first time, the sampling was terminated and the null (true) hypothesis $H_0 : \mu = 0$ was accepted.

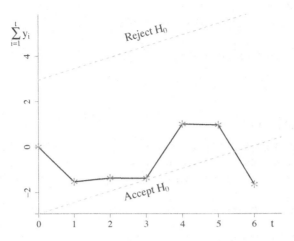

Figure 4.9: Graphical illustration of the sequential Wald test for testing $H_0 : \mu = 0$ vs. $H_1 : \mu = 1$. The boundary lines are $0.5t \pm 2.9$.

As we have mentioned, sequential testing may require a smaller sample size and, hence, save experimental costs. It becomes, however, more problematic when testing composite hypotheses. Suppose $Y_1, Y_2, \ldots \sim f_\theta(y)$ and we want to sequentially test $H_0 : \theta = \theta_0$ vs. $H_1 : \theta > \theta_0$ for a fixed α and β. Since H_1 is composite it is not clear which specific $\theta_1 > \theta_0$ should be used for controlling the Type II error. A typical remedy in this case is to set a certain threshold value $\theta_1 > \theta_0$ to control the Type II error for and to test two simple hypotheses $H_0 : \theta = \theta_0$ vs. $H_1 : \theta = \theta_1$. Such an approach, however, might strongly depend on the choice of the threshold θ_1, which is often hard to specify in practice.

We refer the interested reader to Siegmund (1985) for more detail on sequential testing.

4.6 *Multiple testing

In variety of statistical applications one faces the problem of multiple inferences when a series of hypotheses should be tested *simultaneously*. Thus, in genomic studies identification of differently expressed genes between treated and control groups implies simultaneous testing of thousands or even millions of genes. Ignoring multiplicity by an unguarded use of single-inference testing ("don't worry" approach) naturally results in a great number of falsely rejected null hypotheses. Moreover, the more null hypotheses are tested simultaneously, the larger is the chance for erroneous inferences: testing each one of m hypotheses with significance level α, about αm null hypotheses are expected to be rejected even when all null hypotheses are true. Interpreting rejection of a null as a detection or discovery, it leads, therefore, to a large number of false discoveries. To control the overall error it is then necessary to correct the significance levels of individual tests for multiplicity effect.

4.6.1 Family-wise error

The traditional concern for the multiplicity is about controlling the probability of even a single Type I error (a single false discovery)—the control of the *family-wise error* (FWER).

Consider some procedure for simultaneous testing m hypotheses with the (unknown) number of true null hypotheses $m_0 \leq m$. Let R be the number of null hypotheses rejected by the procedure and V be the number of wrong rejections (the number of Type I errors). In terms of these random variables, ignoring multiplicity and testing individually each hypothesis at level α, guarantees $E(V/m) \leq \alpha$, but $FWER = P(V \geq 1) \geq E(V/m)$ (see Question 4.7).

Question 4.7 Show that $E(V/m) \leq P(V \geq 1)$.

To control FWER at level α one should conduct individual tests at levels smaller than α. Thus, the widely known *Bonferroni procedure* (see also Section 3.5) performs each individual test at level α/m:

Proposition 4.4 Let $H_{01}, \ldots, H_{0m}$ be a family of simultaneously tested null hypotheses and $p_1, \ldots, p_m$ the corresponding p-values. Perform each individual test at level α/m, that is, reject H_{0j} if $p_j \leq \alpha/m$. Then, the resulting $FWER \leq \frac{m_0}{m}\alpha \leq \alpha$ for any configuration of true and false null hypotheses.

Proof. Let J_0 be a set of indices of true null hypotheses. Its cardinality $|J_0| = m_0$. We have

$$FWER = P(V \geq 1) = P\left(\bigcup_{J_0} \left\{p_j \leq \frac{\alpha}{m}\right\}\right) \leq \sum_{j \in J_0} P\left(p_j \leq \frac{\alpha}{m}\right) \leq m_0 \frac{\alpha}{m} \leq \alpha$$

(see Question 4.4). □

The Bonferroni procedure is very simple, does not require any assumptions on the relations between the test statistics but it may be conservative and, as a result, has a low power. For example, identifying differently expressed genes among 1000 genes with FWER control at level $\alpha = 0.05$, one selects a gene (rejects the null hypothesis) only if its $p - value < 0.05/1000 = 5 \cdot 10^{-5}$ (!) and, thus, is likely to miss some truly significant genes.

There exist other, more powerful FWER controlling procedures. The *Holm step-down procedure* orders p-values in increasing order: $p_{(1)} \leq p_{(2)} \cdots \leq p_{(m)}$ and for $j = 1, 2, \ldots$ sequentially compares them with respectively $\frac{\alpha}{m}, \frac{\alpha}{m-1}, \ldots, \alpha$ as long as $p_{(j)} \leq \frac{\alpha}{m-j+1}$. Let j^* be the first time when $p_{(j)} > \frac{\alpha}{m-j+1}$. Reject then all $j^* - 1$ null hypotheses with the smallest $j^* - 1$ p-values and accept (do not reject) others (i.e. $R = j^* - 1$). Evidently, all null hypotheses are accepted if $j^* = 1$ ($R = 0$) and rejected if no such j^* exists ($R = m$).

Note that at the first stage, the Holm procedure tests a null hypothesis $H_{0(1)}$ with the smallest p-value $p_{(1)}$ at level $\frac{\alpha}{m}$ similar to Bonferroni. However, once $H_{0(1)}$ is rejected, the next null hypothesis $H_{0(2)}$ corresponding to the second smallest p-value $p_{(2)}$ is tested at a larger level $\frac{\alpha}{m-1}$ and so on. Hence, Holm necessarily rejects all null hypotheses rejected by Bonferroni but, in addition, may reject some other null hypotheses as well and, thus, is uniformly more powerful. The following Proposition 4.5 shows that it controls FWER at level α:

Proposition 4.5 The Holm procedure controls FWER at level α for any configuration of true and false null hypotheses.

Proof. Consider again a set of indices of true null hypotheses J_0 of cardinality m_0. Let $H_{0(1)}, \ldots, H_{0(m)}$ be a family of simultaneously tested null hypotheses corresponding to the ordered p-values $p_{(1)} \leq \ldots p_{(m)}$. Let $H_{0(j)}$ be the first erroneously rejected null hypothesis following the Holm procedure. Hence, $H_{0(1)}, \ldots, H_{0(j-1)}$ were rejected correctly and, therefore, $j - 1 \leq m - m_0$, the number of false null hypotheses. We then have

$$p_{(j)} \leq \frac{\alpha}{m-j+1} \leq \frac{\alpha}{m_0},$$

that is, if at least one null hypothesis is erroneously rejected ($V \geq 1$), there has to be a null hypothesis with $p - value \leq \frac{\alpha}{m_0}$. Thus,

$$FWER = P(V \geq 1) \leq P\left(\bigcup_{J_0} \left\{p_j \leq \frac{\alpha}{m_0}\right\}\right) \leq \sum_{j \in J_0} P\left(p_j \leq \frac{\alpha}{m_0}\right) \leq \alpha.$$

□

Similarly, the *Hochberg step-up procedure* starts with the largest $p-$value $p_{(m)}$ and accepts all null hypotheses $H_{0(j)}, j = m, m-1, \ldots$ as long as $p_{(j)} > \frac{\alpha}{m-j+1}$. Hochberg is more powerful than Holm (why?) but does not control FWER in a general case without extra assumptions on relations between test statistics. In particular, it can be shown that the Hochberg procedure does control the FWER at level α if all individual test statistics (or, equivalently, p-values) are *independent*.

Even though Holm, Hochberg and some other existing FWER controlling procedures are more powerful than Bonferroni, the control of FWER in general may imply poor power properties especially when the number of tested hypotheses is large. In fact, guarding against any single false rejection by controlling the FWER is not often needed. Thus, in genomic studies that involve simultaneous testing of huge amounts of genes, it might lead to a large number of important missed findings. In this case a reasonable approach would be to relax such a tight criterion at a price of some possible false discoveries. Yet, the "don't worry" approach to ignore multiplicity is also inappropriate.

4.6.2   *False discovery rate*

The *False Discovery Rate* (FDR) criterion is a less stringent alternative to the FWER in multiple testing problems. FDR is defined as the expected proportion of falsely rejected null hypotheses (false discoveries) among those being rejected. The control of FDR assumes the control over the *proportions* of Type I errors rather than of making even one Type I error (FWER). Intuitively, when controlling the FDR, the loss caused by false rejections is inversely related to the number of rejected null hypotheses—the larger their number, the less restrictive the number of allowed false rejections is. Hence, the FDR criterion may be preferable especially when the number of tested hypotheses is large.

For a given multiple testing procedure, define the random variable $Q = V/R$—the proportion of falsely rejected hypotheses among those being rejected, when $R \geq 1$, and naturally, set $Q = 0$ when $R = 0$ since no error of false rejection can be committed in this case. FDR is defined as the expectation of Q, i.e.

$$FDR = E(Q) = E(V/R),$$

where $V/R = 0$ for $V = R = 0$ by definition.

Obviously, $E(Q) = E(V/R) \geq E(V/m)$ and, therefore, controlling FDR is more stringent than ignoring multiplicity.

On the other hand, if all null hypotheses are true ($m_0 = m$), then evidently $V = R$. Hence, $Q = 1$, if $R \geq 1$ and $Q = 0$, if $R = 0$. In this case, $E(Q) = P(V \geq 1)$ and controlling FDR is equivalent to controlling FWER. When $m_0 < m$, $E(V/R) \leq P(V \geq 1)$ and, therefore, control of FWER implies control of FDR.

Hence, $E(V/m) \leq E(V/R) \leq P(V \geq 1)$ meaning that controlling FDR is adaptively in-between ignoring multiplicity altogether and controlling FWER.

Consider the following *Benjamini–Hochberg* (step-up) multiple testing procedure. Let $p_1, \ldots, p_m$ be the p-values for the m simultaneously tested hypotheses. Order them in increasing order $p_{(1)} \leq \ldots \leq p_{(m)}$ and let $H_{0(1)}, \ldots, H_{0(m)}$ be the corresponding null hypotheses. Start with the largest $p-$value $p_{(m)}$ and for $j = m, m-1, \ldots$ continue as long as $p_{(j)} > (j/m)\alpha$. Let j^* be the first time when $p_{(j)} \leq (j/m)\alpha$. Reject then all hypotheses corresponding to the smallest j^* p-values ($R = j^*$). If no such j^* exists, reject none ($R = 0$).

The Benjamini–Hochberg procedure is similar in spirit to the FWER controlling step-up Hochberg procedure but tests $H_{0(j)}$ at a higher level $\frac{j}{m}\alpha > \frac{1}{m-j+1}\alpha$. The ratio $\frac{j/m}{1/(m-j+1)} = j(1 - \frac{j-1}{m}) \sim j$ for small j and, therefore, Benjamini–Hochberg is much less conservative than Hochberg. We now prove that for *independent* test statistics Benjamini–Hochberg controls the (less stringent) FDR at the (unknown) level $(m_0/m)\alpha \leq \alpha$:

Proposition 4.6 If p-values are independent, the Benjamini–Hochberg procedure controls FDR at level $\frac{m_0}{m}\alpha \leq \alpha$ for any configuration of true and false null hypotheses.

Proof. Let $R = j^*$ be the number of rejected null hypotheses by Benjamini–Hochberg, that is, $j^* = \max\{j : p_{(j)} \leq (j/m)q, 1 \leq j \leq m\}$, and J_0 of cardinality m_0 be again a set of indices of true null hypotheses.

If all null hypotheses are false ($J_0 = \emptyset$), then trivially $V = Q = 0$ and $FDR = 0 < \alpha$. Otherwise, without loss of generality, assume that H_{01} is true, i.e. $1 \in J_0$, and consider the remaining $m - 1$ null hypotheses $H_{02}, \ldots, H_{0m}$. With some abuse of notation, let $p_{(2)} \leq \ldots \leq p_{(m)}$ be their ordered p-values and define $j^*_{(-1)} = \max\{j : p_{(j)} \leq (j/m)\alpha, 2 \leq j \leq m\}$. If no such $j^*_{(-1)}$ exists, set $j^*_{(-1)} = 1$.

We first prove that $j^* \leq j^*_{(-1)}$. We have

$$\#\left\{j : p_j \leq \frac{j^*}{m}\alpha, j \geq 2\right\} = \left\{\begin{array}{ll} j^* & , p_1 > \frac{j^*}{m}\alpha \\ j^* - 1 & , p_1 \leq \frac{j^*}{m}\alpha \end{array}\right. \geq j^* - 1 \tag{4.21}$$

Assume that $j^* > j^*_{(-1)}$. Then, $j^*_{(-1)} \leq j^* - 1$ and by definition of $j^*_{(-1)}$, $\#\left\{j : p_j \leq \frac{j^*}{m}\alpha, j \geq 2\right\} \leq j^* - 2$ that contradicts (4.21).

Hence, in particular, if $p_1 > \frac{j^*_{(-1)}}{m}\alpha$, then $p_1 > \frac{j^*}{m}\alpha$. We now show that if $p_1 \leq \frac{j^*_{(-1)}}{m}\alpha$, then the equality $j^* = j^*_{(-1)}$ holds. Indeed, there are $j^*_{(-1)} - 1$ p-values $p_j \leq \frac{j^*_{(-1)}}{m}\alpha$, $j \geq 2$. So, under the additional condition, there are overall (including p_1) $j^*_{(-1)}$ p-values $p_j \leq \frac{j^*_{(-1)}}{m}\alpha$. Thus, by definition of j^*, $j^* \geq j^*_{(-1)}$ that implies $j^* = j^*_{(-1)}$.

We then have conditionally on $p_2, \ldots, p_m$

$$E\left(\frac{1\{P_1 \leq \frac{j^*}{m}\alpha\}}{\max(j^*, 1)} \mid P_2 = p_2, \ldots, P_m = p_m\right) = E\left(\frac{1\{P_1 \leq \frac{j^*_{(-1)}}{m}\alpha\}}{j^*_{(-1)}} \mid P_2 = p_2, \ldots, P_m = p_m\right)$$

$$= \frac{1}{j^*_{(-1)}} E\left(1\{P_1 \leq (j^*_{(-1)}/m)\alpha\} \mid P_2 = p_2, \ldots, P_m = p_m\right)$$

$$\leq \frac{\alpha}{m},$$

where the last inequality holds due to independence of p-values. Hence, also unconditionally, $E\left(\frac{1\{P_1 \leq \frac{j^*}{m}\alpha\}}{\max(j^*, 1)}\right) \leq \alpha/m$ and, therefore,

$$FDR = E(V/R) = E\left(\sum_{j \in J_0} \frac{1\{P_j \leq \frac{j^*}{m}\alpha\}}{\max(j^*, 1)}\right) = \sum_{j \in J_0} E\left(\frac{1\{P_j \leq \frac{j^*}{m}\alpha\}}{\max(j^*, 1)}\right) \leq \frac{m_0}{m}\alpha \leq \alpha \tag{4.22}$$

(see Question 4.4). Moreover, if the underlying test statistics are continuous random variables, Proposition 4.1 implies that the FDR is exactly $(m_0/m)\alpha$ in (4.22). □

It follows from the proof that, in fact, only independence of p-values corresponding to true null hypotheses is required.

In the general case (without assuming independence), applying Benjamini–Hochberg procedure with smaller $\alpha^* = \alpha/(1 + 1/2 + \ldots + 1/m) \sim \alpha/\ln m$ still ensures the FDR control at level $(m_0/m)\alpha \leq \alpha$. However, such procedure can be over-conservative in many practical situations for large m.

The corresponding *step-down* FDR controlling procedure that starts from $j = 1, 2, \ldots$ and rejects $H_{0(j)}$ as long as $p_{(j)} \leq (j/m)\alpha$, tends to loose power as m increases.

4.7 Exercises

Exercise 4.1 The advertisement of the fast food chain of restaurants "FastBurger" claims that the average waiting time for food in its branches is 30 seconds unlike 50 seconds of their competitors.

Mr. Skeptic does not believe much in advertising and decided to test its truth by the following test: he will go to one of "FastBurger's" branches, measure the waiting time and if it is less than 40 seconds (the critical waiting time he fixed) he would believe in its advertisement. Otherwise, he would conclude that the service in "FastBurger" is no faster than in other fast food companies. Mr. Skeptic also assumes that waiting time is exponentially distributed.

1. What are the hypotheses Mr. Skeptic tests? Calculate the probabilities of errors of both types for his test.

2. Can you suggest a better test to Mr. Skeptic with the same significance level?

3. Mrs. Skeptic, Mr. Skeptic's wife, agrees in general with the test her husband has used but decided instead to fix a critical waiting time that will minimize the sum of the probabilities of both error types. What will be her critical waiting time?

4. Calculate the probabilities of the errors for Mrs. Skeptic's test and compare them with those of her husband.

Exercise 4.2 Continue Example 4.7 and derive the 25%-level MP test to determine the origin of the man by asking about his drink of choice, where $H_0 : UK$ vs. $H_1 : France$. Compare the results with those obtained in Example 4.7.

Exercise 4.3 Continue Example 4.8 for testing $H_0 : \mu = \mu_0$ vs. $H_1 : \mu = \mu_1$ based on a sample $Y_1, ..., Y_n \sim \mathcal{N}(\mu, \sigma^2)$, where σ^2 is known, and now $\mu_0 > \mu_1$.

1. Show that the test that rejects H_0 if $\bar{Y} \leq \mu_0 - z_\alpha \frac{\sigma}{\sqrt{n}}$ is the MP test at level α.

2. What is the power of the test?

3. What is the corresponding p-value for the given sample?

Exercise 4.4 Prove Proposition 4.1.

Exercise 4.5 Let $Y_1, ..., Y_n \sim \mathcal{U}(0, \theta)$.

1. Derive the MP test at level α for testing $H_0 : \theta = \theta_0$ vs. $H_1 : \theta = \theta_1$, where $\theta_1 > \theta_0$.

2. Calculate the power of the MP test.

Exercise 4.6 Continue Example 4.1. "Auto-World Journal" decided to extend its pilot study on the credibility of the company's advertisement based on 5 new cars by testing more cars. How large should a minimal sample size be for testing $H_0 : \mu = 12$ (advertisement is not true) vs. $H_1 : \mu = 15$ (advertisement is true) to guarantee that the probabilities of both error types will be at most 0.01? (Recall that $\sigma^2 = 4$.)

Exercise 4.7 Let $Y_1, ..., Y_n$ be a random sample from a distribution with the density function

$$f_\theta(y) = \frac{y}{\theta} e^{-\frac{y^2}{2\theta}}, \ y \geq 0, \ \theta > 0$$

1. Find the MP test at level α for testing two simple hypotheses $H_0 : \theta = \theta_0$ vs. $H_1 : \theta = \theta_1$, where $\theta_1 > \theta_0$.

2. Is there a UMP test at level α for testing the one-sided composite hypotheses $H_0 : \theta \leq \theta_0$ vs. $H_1 : \theta > \theta_0$? What is its power function?

(*Hint*: show that $Y_i^2 \sim \exp(1/(2\theta))$ and, therefore, $\sum_{i=1}^n Y_i^2 \sim \theta \chi^2_{2n}$.)

Exercise 4.8 Repeat Exercise 4.7 for testing the hypotheses about the parameter θ of a Pareto distribution with the density $f_{\theta,\gamma}(y) = \frac{\theta}{\gamma} \left(\frac{\gamma}{y}\right)^{\theta+1}$, $y \geq \gamma$; $\theta, \gamma > 0$, where γ is known.
(*Hint*: show that $\ln(Y_i/\gamma) \sim \exp(\theta)$.)

Exercise 4.9 Continue Exercise 2.21. Derive the UMP test at level α for testing $H_0 : p \leq p_0$ vs. $H_1 : p > p_0$.

Exercise 4.10 Let $Y_1, ..., Y_n$ be a random sample from a distribution with the density function

$$f_\theta(y) = \frac{2\theta^2}{y^3}, \ y \geq \theta > 0$$

Is there a UMP test at level α for testing $H_0 : \theta \leq \theta_0$ vs. $H_1 : \theta > \theta_0$? If so, what is the test?

Exercise 4.11 (χ^2-tests for the variance) Let $Y_1, ..., Y_n \sim \mathcal{N}(\mu, \sigma^2)$.

1. Assume first that μ is known.

   (a) Find a MP test at level α for testing $H_0 : \sigma^2 = \sigma_0^2$ vs. $H_1 : \sigma^2 = \sigma_1^2$, where $\sigma_1 > \sigma_0$.

   (b) Show that the MP test from the previous paragraph is a UMP test for testing $H_0 : \sigma^2 \leq \sigma_0^2$ vs. $H_1 : \sigma^2 > \sigma_0^2$.

   (*Hint*: recall that $\sum_{i=1}^n (Y_i - \mu)^2 \sim \sigma^2 \chi_n^2$.)

2. Consider now an unknown μ. See also Exercise 4.19.

   (a) Is there a MP test for testing $H_0 : \sigma^2 = \sigma_0^2$ vs. $H_1 : \sigma^2 = \sigma_1^2$, where $\sigma_1 > \sigma_0$, in this case? If not, find the corresponding GLRT test.

   (b) Is the above test also a GLRT for testing the one-sided hypotheses $H_0 : \sigma^2 \leq \sigma_0^2$ vs. $H_1 : \sigma^2 > \sigma_0^2$?

   (c) Find the GLRT test at level α to test $H_0 : \sigma^2 \geq \sigma_0^2$ vs. $H_1 : \sigma^2 < \sigma_0^2$?

   (*Hint*: recall that $\sum_{i=1}^n (Y_i - \bar{Y})^2 \sim \sigma^2 \chi_{n-1}^2$.)

Exercise 4.12 (**two-sample t-tests**) Consider two independent normal random samples with a common (unknown) variance $X_1, ..., X_n \sim \mathcal{N}(\mu_1, \sigma^2)$ and $Y_1, ..., Y_m \sim \mathcal{N}(\mu_2, \sigma^2)$.

1. Show that the GLRT for testing $H_0 : \mu_1 \leq \mu_2$ vs. $H_1 : \mu_1 \leq \mu_2$ at level α results in the well-known two-sample one-sided t-test:

   (a) Derive the generalized likelihood ratio and show that it is a monotone function of the statistic

   $$T = \frac{\bar{X} - \bar{Y}}{S_p \sqrt{\frac{1}{n} + \frac{1}{m}}},$$

   where

   $$S_p^2 = \frac{\sum_{i=1}^n (X_i - \bar{X})^2 + \sum_{j=1}^m (Y_j - \bar{Y})^2}{n + m - 2}$$

   usually called the *a pooled variance estimate* is the unbiased estimator of the common variance σ^2.

   (b) Verify that for $\mu_1 = \mu_2$, $T \sim t_{n+m-2}$ to obtain the resulting GLRT.

   (c) Is it a UMP test?

2. Find the corresponding two-sided two-sample t-test for testing $H_0 : \mu_1 = \mu_2$ vs. $H_1 : \mu_1 \neq \mu_2$ at level α. Extend it to testing $H_0 : \mu_1 - \mu_2 = \delta$ vs. $H_1 : \mu_1 - \mu_2 \neq \delta$ for a fixed δ.

3. Exploit the duality between two-sided tests and confidence intervals to derive a $100(1 - \alpha)\%$ confidence interval for the means difference $\mu_1 - \mu_2$.

4. To test the efficiency of a new drug, a group of 19 rats was infected and then randomly assigned to case and control groups. 9 rats in the control group were treated by the existing drug, while the 10 rats in the case group received a new treatment. Here are the survival times (in days) of rats in both groups:
   case group   3 7 6 7 7 13 9 4 7
   control group   7 4 8 9 11 6 7 9 12 11

   (a) What are the assumptions of your model for the data? Formulate the hypotheses for testing the efficiency of the new drug in terms of parameters of the model.

(b) Calculate the corresponding t-statistic and the resulting p-value. What are your conclusions?

Exercise 4.13 (*F-test for the equality of variances*) Consider two independent normal samples $X_1,...,X_n \sim \mathcal{N}(\mu_1, \sigma_1^2)$ and $Y_1,...,Y_m \sim \mathcal{N}(\mu_2, \sigma_2^2)$, where all the parameters are unknown.

1. Derive the generalized likelihood ratio for testing the equality of variances $H_0 : \sigma_1^2 = \sigma_2^2$ vs. $H_1 : \sigma_1^2 \neq \sigma_2^2$ and show that it is a function of the sampled variances ratio $F = s_y^2/s_x^2$, where $s_y^2 = \frac{1}{m-1}\sum_{j=1}^{m}(Y_j - \bar{Y})^2$ and $s_x^2 = \frac{1}{n-1}\sum_{i=1}^{n}(X_i - \bar{X})^2$.

2. Show that under the null hypothesis, $F \overset{H_0}{\sim} F_{m-1,n-1}$.

3. Derive the corresponding approximate GLRT at level α.

4. In particular, such a test is relevant for testing the assumption of the equality of variances required for the two-sample t-test (see Exercise 4.12). Check this assumption for the rats survival data in Exercise 4.12 at level $\alpha = 0.05$.

Exercise 4.14 Let $Y_1,...,Y_n$ be a random sample from a distribution with the density function $f_\theta(y) = \theta y^{\theta-1}$, $0 \leq y \leq 1$, $\theta > 0$.

1. Show that the uniform distribution $\mathcal{U}(0,1)$ is a particular case of $f_\theta(y)$.

2. Find the GLRT at level α for testing the hypothesis that the data comes from $\mathcal{U}(0,1)$.

(*Hint*: show that $-\ln Y_i \sim \exp(\theta)$.)

Exercise 4.15 Continue Exercise 3.6. Test the hypothesis that the reaction time for the stimulus is the same for dogs and cats at level α.

Exercise 4.16 Observing a series of Bernoulli trials $Y_1, Y_2, ... \sim \mathcal{B}(1,p)$ we want to sequentially test $H_0 : p = p_0$ vs. $H_1 : p = p_1$, where $p_1 > p_0$. Derive the corresponding sequential Wald test for the given error probabilities α and β.

Exercise 4.17 * In a genomic study researchers want to identify differently expressed genes within a set of m genes. Let $Y_j \sim \mathcal{N}(\mu_j, \sigma^2)$ be the observed expression level of the j-th gene, $j = 1,...,m$. For simplicity assume that Y_j's are independent (maybe not so realistic in practice) and σ^2 is known. Thus, we need to test simultaneously $H_{0j} : \mu_j = 0$ (non differentially expressed gene) vs. $H_{1j} : \mu_j \neq 0$ (differentially expressed gene). Evidently, H_{0j} will be rejected if $|Y_j|$ is large (GLRT). Apply various multiple testing procedures to find the corresponding thresholds for $|Y_j|$'s:

1. What is the Bonferroni FWER controlling procedure at level α?

2. Derive the Holm and Hochberg procedures to control FWER at level α.

3. Which of the above FWER controlling procedures is the most powerful?

4. Derive the Benjamini–Hochberg FDR controlling procedure at level α.

5. For which of the above procedures the independence assumption is necessary?

Chapter 5

Asymptotic Analysis

5.1 Introduction

In Chapter 2, we considered various aspects of estimation of an unknown parameter from a finite data sample. In many situations an estimator might be a complicated function of the data, and the exact calculation of its characteristics (e.g., bias, variance and MSE) might be quite difficult if possible at all. Nevertheless, rather often, good *approximate* properties can be obtained if the sample size is "sufficiently large." Intuitively, everyone would agree that a "good" estimator should improve and approach the estimated parameter as the sample size increases: e.g., the larger the poll, the more representative its results are. Conclusions based on a large sample have more ground than those based on a small sample. But what is the formal justification for this claim?

Similar issues arise with confidence intervals and hypothesis testing considered in Chapters 3 and 4. To construct a confidence interval we should know the distribution of the underlying statistic, and that may be difficult to calculate. Can we derive at least an approximate confidence interval without knowing the *exact* distribution of the statistic for a "sufficiently large" sample? How? Furthermore, statistical theory is relatively simple when the underlying distribution is normal and much of the theory has been developed for this case. For example, most of the output of standard statistical packages is based on the assumption of normality of the resulting statistics. Is this justified? Are we able to construct a confidence interval as if the statistic is normally distributed, and hope that it will have approximately the right coverage probability? In other words, is the normal distribution "normal" and if so, why?

In hypothesis testing we need the distribution of the test statistic $T(\mathbf{Y})$ under the null hypothesis in order to find the critical value for a given significance level α by solving $\alpha = P_{H_0}(T(\mathbf{Y}) \geq C)$ or to calculate the p-value. The statistic $T(\mathbf{Y})$ might be a complicated function of the data whose distribution is hard to derive, if doable at all. Is it possible then to find its simple asymptotic approximation?

These are the questions we will investigate in this chapter. Statisticians like to answer them by considering a sequence of similar problems that differs one from another only in their sample size. Let $\hat{\theta}_n = \hat{\theta}(Y_1, ..., Y_n)$ be an estimator for θ based on a sample of size n (e.g., $\bar{Y}_n = \frac{1}{n}(Y_1 + ... + Y_n)$ or $Y_{\max,n} = \max(Y_1, ..., Y_n)$). We can then study the asymptotic properties of the sequence $\{\hat{\theta}_n\}$ as n increases. In particular, whether $\{\hat{\theta}_n\}$ converges to θ (this property is known as the *consistency* of the estimator $\hat{\theta}_n$) and in what sense—recall that $\{\hat{\theta}_n\}$ is a sequence of *random variables* (estimators) and the notion of its convergence should first be properly defined. What is the asymptotic distribution of $\hat{\theta}_n$?

It is quite simple to define convergence of a sequence of real numbers $\{a_n\}$: we say that the sequence $a_1, a_2, ...$ converges to a, if for any $\varepsilon > 0$ there exists an n_0 such that $|a_n - a| < \varepsilon$ for all $n > n_0$. We write it as $\lim_{n \to \infty} a_n = a$, or simply $a_n \to a$. For example, the sequence $1, \frac{1}{2}, \frac{1}{3}, ...$ converges to 0 since for any $\varepsilon > 0$, $a_n = 1/n < \varepsilon$ starting from $n > \varepsilon^{-1}$.

Things start to be slightly more complicated when we want to consider convergence of more complex objects. Thus, if $\mathbf{a}_n \in \mathbb{R}^d$, $n = 1, 2, ...$ is a d-dimensional sequence of vectors, the notion of convergence of $\mathbf{a}_n \to \mathbf{a} \in \mathbb{R}^d$ can be easily translated to convergence of a sequence of real

DOI: 10.1201/9781003175407-5

numbers: we say that $\mathbf{a}_n \to \mathbf{a}$ if $\|\mathbf{a}_n - \mathbf{a}\| \to 0$ in some norm (e.g., in Euclidean norm $\|\mathbf{a}_n - \mathbf{a}\|_2 = ((a_{n1} - a_1)^2 + ... + (a_{nd} - a_d)^2)^{1/2}$). Consider next a sequence of functions $f_n(x) : [0,1] \to \mathbb{R}$, $n = 1, 2, \ldots$. For example $f_n(x) = x^n$. Clearly, *pointwise* $f_n(x) \to 0$ if $x \in [0,1)$ and $f_n(1) \to 1$ as $n \to \infty$. But what does the convergence of functions mean? Similarly to vectors we can translate this problem to the convergence of real numbers in terms of norms of functions. We can consider different norms on continuous real-valued functions defined on the unit interval. For example, two common norms are the L_2 norm defined by $\|f\|_2 = \left(\int_0^1 f^2(x)dx\right)^{1/2}$ and the sup-norm defined by $\|f\|_\infty = \sup_{x \in [0,1]} |f(x)|$. We say that $f_n \to f$ if $\|f_n - f\|_\alpha \to 0$, where $\alpha = 2$ or $\alpha = \infty$, though the result may depend on its choice. Thus, if $f_n(x) = x^n$ and $f(x) = 0$ on the unit interval, then $\|f_n - f\|_2^2 = \int_0^1 x^{2n}dx = (2n+1)^{-1} \to 0$ but $\|f_n - f\|_\infty = \sup_{x \in [0,1]} |x^n| = 1 \nrightarrow 0$.

How can we define the convergence of a sequence of estimators $\{\hat{\theta}_n\}$ to a parameter θ? Unlike a sequence of real numbers, $|\hat{\theta}_n - \theta|$ is a *random* number that might be larger or smaller depending on its particular realization.

In what follows we discuss several types of convergence (consistency) of estimators. For simplicity of exposition throughout this chapter, we will not distinguish between a sequence of estimators $\{\hat{\theta}_n\}$ and an estimator $\hat{\theta}_n$, and will talk about the convergence of an estimator $\hat{\theta}_n$ as n increases.

5.2   Convergence and consistency in MSE

Recall that the most common measure of goodness-of-estimation for $\hat{\theta}_n$ is its mean squared error $MSE(\hat{\theta}_n, \theta) = E(\hat{\theta}_n - \theta)^2$ (see Section 2.6). Note that the $MSE(\hat{\theta}_n, \theta)$ is a real (nonnegative) sequence and it would be natural then to say that $\hat{\theta}_n$ converges to θ if $MSE(\hat{\theta}_n, \theta) \to 0$ as the sample size n increases:

Definition 5.1 (consistency in MSE) An estimator $\hat{\theta}_n$ is a *consistent in MSE* estimator of θ if $MSE(\hat{\theta}_n, \theta) \to 0$ as $n \to \infty$. We will denote it as $\hat{\theta}_n \xrightarrow{\text{MSE}} \theta$.

If $\hat{\theta}_n \xrightarrow{\text{MSE}} \theta$, then obviously, $bias(\hat{\theta}_n, \theta) = E\hat{\theta}_n - \theta \to 0$ and $Var(\hat{\theta}_n) \to 0$.

Example 5.1 (consistency in MSE of a sample mean) Given a random sample $Y_1, ..., Y_n$ of size n from some distribution with a finite second moment, we estimate the unknown expectation $\mu = EY$ by the sample mean $\bar{Y}_n$. The estimator $\bar{Y}_n$ is always an unbiased estimator of μ and, therefore, $MSE(\bar{Y}_n, \mu) = Var(\bar{Y}_n) = \frac{Var(Y)}{n}$ (see Example 2.14) that tends to zero as n increases. Hence, regardless of a particular distribution, a sample mean is a consistent in MSE estimator of EY.

Example 5.2 (continuation of Example 2.15) Consider again the three estimators of an unknown p in a series of Bernoulli trials $Y_1, ..., Y_n \sim \mathcal{B}(1, p)$ from Example 2.15:

$$\hat{p}_{1n} = \frac{\sum_{i=1}^n Y_i}{n}, \quad \hat{p}_{2n} = \frac{\sum_{i=1}^n Y_i + 1}{n+2} \quad \text{and} \quad \hat{p}_{3n} = Y_1$$

From the calculus there we have

$$MSE(\hat{p}_{1n}, p) = \frac{p(1-p)}{n} \underset{n \to \infty}{\to} 0,$$

$$MSE(\hat{p}_{2n}, p) = \frac{(1-2p)^2 + np(1-p)}{(n+2)^2} \underset{n \to \infty}{\to} 0,$$

$$MSE(\hat{p}_{3n}, p) = p(1-p) \underset{n \to \infty}{\nrightarrow} 0$$

and, therefore, $\hat{p}_{1n}$ and $\hat{p}_{2n}$ are both consistent estimators in MSE for p, while $\hat{p}_{3n}$ is not.

Consistency in MSE, however, might be hard to prove and sometimes even too strong to ask for. A more realistic approach is then to relax it by the weaker consistency *in probability* as a minimal asymptotic requirement for a "good" estimator. Unlike the former, the latter type of convergence can often be verified by relatively simple probabilistic tools.

5.3 Convergence and consistency in probability

We recall first a general definition of convergence in probability:

Definition 5.2 (convergence in probability) The sequence of random variables $\{Y_n\}$ is said to converge to c, if for any $\varepsilon > 0$, the probability $P(|Y_n - c| > \varepsilon) \to 0$ as $n \to \infty$. We write this as $Y_n \xrightarrow{P} c$.

Note that the above definition does not assume anything about the joint distribution of $Y_1, Y_2, \ldots$. In fact, the Y_i's can even be defined on different probability spaces. Only the distributions of each Y_i matters. However, when the Y_i's are defined on the same probability space, we can extend Definition 5.2 of convergence in probability to a *random* limit Y (provided that it is also defined on the same probability space): we say that $Y_n \xrightarrow{P} Y$ if $(Y_n - Y) \xrightarrow{P} 0$.

Let $\{X_n\}$ and $\{Y_n\}$ be two sequences of random variables. Similar to sequences of real numbers, we say that $X_n = O_p(Y_n)$ if the sequence $\{X_n/Y_n\}$ is bounded in probability: for any $\varepsilon > 0$ there is a finite M, such that $P(|X_n/Y_n| > M) < \varepsilon$ for all n. We say that $X_n = o_p(Y_n)$ if $X_n/Y_n \xrightarrow{P} 0$.

The following theorem describes some basic arithmetic of the limits in probability:

Theorem 5.1 Let $X_n \xrightarrow{P} a$ and $Y_n \xrightarrow{P} b$ defined on the same probabilistic space. Let c be real and $g(\cdot)$ a continuous function. Then,

1. $cX_n \xrightarrow{P} ca$.

2. $X_n + c \xrightarrow{P} a + c$.

3. $(X_n + Y_n) \xrightarrow{P} a + b$.

4. If $b \neq 0$ then also $X_n/Y_n \xrightarrow{P} a/b$.

5. $g(X_n) \xrightarrow{P} g(a)$.

Proof. We will prove only Parts 3 and 5, leaving the other parts to the reader. For Part 3 note that

$$P(|X_n + Y_n - a - b| > \varepsilon) \leq P(|X_n - a| + |Y_n - b| > \varepsilon).$$

However, if $|X_n - a| + |Y_n - b| > \varepsilon$, then either $|X_n - a| > \varepsilon/2$ or $|Y_n - b| > \varepsilon/2$ or both. Formally, $\{|X_n + Y_n - a - b| > \varepsilon\} \subseteq \{|X_n - a| > \varepsilon/2\} \cup \{|Y_n - b| > \varepsilon/2\}$ (why?). Since $P(A \cup B) \leq P(A) + P(B)$ for any two events A and B,

$$P(|X_n + Y_n - a - b| > \varepsilon) \leq P(|X_n - a| > \varepsilon/2) + P(|Y_n - b| > \varepsilon/2) \to 0.$$

We now prove Part 5. The continuity of $g(\cdot)$ at a means that for any $\varepsilon > 0$ there exists a $\delta > 0$ such that $|x - a| < \delta$ yields $|g(x) - g(a)| < \varepsilon$. Hence,

$$P(|g(X_n) - g(a)| \leq \varepsilon) \geq P(|X_n - a| < \delta) \to 1$$

or, equivalently, $P(|g(X_n) - g(a)| > \varepsilon) \to 0$. $\qquad\square$

Question 5.1 Complete the proof of Theorem 5.1. Prove also that if $a \neq 0$ then $1/X_n \xrightarrow{P} 1/a$ and $X_n Y_n \xrightarrow{P} ab$.

The definition of consistency in probability of an estimator (usually just called consistency for brevity) is then as follows:

Definition 5.3 (consistency in probability) An estimator $\hat{\theta}_n$ is a *consistent* (in probability) estimator of θ if $\hat{\theta}_n \xrightarrow{P} \theta$ as $n \to \infty$.

As we have mentioned, the requirement of consistency in probability is weaker than that in MSE. Indeed, suppose that $\hat{\theta}_n \xrightarrow{MSE} \theta$. Then, by Markov's inequality, $P(|\hat{\theta}_n - \theta| \geq \varepsilon) \leq E(\hat{\theta}_n - \theta)^2/\varepsilon^2 \to 0$ and, therefore, $\hat{\theta}_n \xrightarrow{P} \theta$. However, the opposite is not true as we illustrate by the following counterexample. Suppose we observe a random sample $Y_1, ..., Y_n$ of size n and want to estimate the unknown $\mu = EY$. Consider the following estimator $\hat{\mu}_n$: with probability $p_n = 1 - \frac{1}{n}$ we estimate μ by a sample mean $\bar{Y}_n$ as usual, but with small probability $1 - p_n = \frac{1}{n}$ we ignore the data and estimate μ by ... n. We leave it as an exercise to the reader to verify that such a "wild guess" still does not affect the consistency in probability and $P(|\hat{\mu}_n - \mu| \geq \varepsilon) \to 0$. However, $E\hat{\mu}_n = \mu(1 - \frac{1}{n}) + n\frac{1}{n} = \mu + 1$, the bias $E\hat{\mu}_n - \mu = 1 \not\to 0$ and, therefore, $\hat{\mu}_n$ is not consistent in MSE.

On the other hand, unlike consistency in MSE, consistency in probability can often be established by standard probabilistic tools. We start from the weak law of large numbers (WLLN) that implies that a sample mean is always a consistent estimator of the expectation:

Theorem 5.2 (weak law of large numbers, WLLN) Let $Y_1, Y_2, ...$ be i.i.d. random variables with a finite second moment. Then, $\bar{Y}_n = \frac{1}{n} \sum_{i=1}^{n} Y_i \xrightarrow{p} EY$ as $n \to \infty$. [1]

Proof. The proof immediately follows from Chebyshev's inequality:

$$P(|\bar{Y}_n - \mu| \geq \varepsilon) \leq \frac{Var(\bar{Y}_n)}{\varepsilon^2} = \frac{Var(Y)/n}{\varepsilon^2} \to 0 \tag{5.1}$$

for every $\varepsilon > 0$. $\square$

In fact, the convergence in probability of $\bar{Y}_n$ to EY could also be immediately established from the corresponding convergence in MSE (see Example 5.1). On the other hand, using somewhat more delicate techniques it can be shown that finiteness of the first moment is sufficient for the WLLN.

Example 5.3 (consistency of a sample proportion) Let $Y_1, ..., Y_n$ be a series of Bernoulli trials $\mathscr{B}(1, p)$. Consider a sample proportion $\hat{p}_n = n^{-1} \sum_{i=1}^{n} Y_i$. In Chapter 2 we showed, in particular, that $\hat{p}_n$ is the MLE and UMVUE estimator of p. By the WLLN, $\hat{p}_n \xrightarrow{p} p$ (see Figure 5.1).

Example 5.4 (consistency of a sample variance) Let $Y_1, ..., Y_n$ be a random sample from a distribution with a finite fourth moment. Let $\sigma^2 = Var(Y)$ and consider the corresponding sample variance $\hat{\sigma}_n^2 = \frac{1}{n} \sum_{i=1}^{n} (Y_i - \bar{Y}_n)^2$. After simple algebra we have

$$\begin{aligned}
\hat{\sigma}_n^2 &= \frac{1}{n} \sum_{i=1}^{n} (Y_i - \bar{Y}_n)^2 \\
&= \frac{1}{n} \sum_{i=1}^{n} (Y_i - \mu + \mu - \bar{Y}_n)^2 \\
&= \frac{1}{n} \sum_{i=1}^{n} (Y_i - \mu)^2 + \frac{2}{n} (\mu - \bar{Y}_n) \sum_{i=1}^{n} (Y_i - \mu) + (\mu - \bar{Y}_n)^2 \\
&= \frac{1}{n} \sum_{i=1}^{n} (Y_i - \mu)^2 - 2(\mu - \bar{Y}_n)^2 + (\mu - \bar{Y}_n)^2 \\
&= \frac{1}{n} \sum_{i=1}^{n} (Y_i - \mu)^2 - (\mu - \bar{Y}_n)^2.
\end{aligned} \tag{5.2}$$

The first term in the RHS of (5.2) is the sample mean of random variables $(Y_i - \mu)^2$, $i = 1, ..., n$, that according to the WLLN converges in probability to $E(Y - \mu)^2 = \sigma^2$. In Example 5.1, we verified that $\bar{Y}_n \xrightarrow{MSE} \mu$ and, therefore, by Markov's inequality, $P(|\bar{Y}_n - \mu|^2 > \varepsilon) \leq E(\bar{Y}_n - \mu)^2 / \varepsilon \to 0$, which means that the second term $(\mu - \bar{Y}_n)^2 \xrightarrow{p} 0$. Hence, $\hat{\sigma}_n^2 \xrightarrow{p} \sigma^2$ as $n \to \infty$.

Question 5.2 Show that under the conditions of Example 5.4, the unbiased estimator $s_n^2 = \frac{1}{n-1} \sum_{i=1}^{n} (Y_i - \bar{Y}_n)^2$ of σ^2 (see Example 2.16) is also consistent.

---

[1] If there is a weak law of large numbers, there also should be a strong one. It indeed exists and says that the sample mean $\bar{Y}_n$ converges to EY not just in probability but also *almost surely*, which is another, stronger type of convergence for random variables but beyond the scope of this book.

Figure 5.1: Ten random paths of a sample proportion $\hat{p}_n = n^{-1} \sum_{i=1}^n Y_i$ as a function of n, for a series of Bernoulli trials $\mathscr{B}(1, 0.5)$.

Combining the WLLN and Part 5 of Theorem 5.1 immediately yields the following corollary:

Corollary 5.1 Let $Y_1, ..., Y_n$ be a random sample from a distribution with a finite second moment. Assume that $\theta = g(EY)$ is a parameter of interest, where $g(\cdot)$ is a continuous function. Then $g(\bar{Y}_n)$ is a consistent estimator of $g(\theta)$.

Corollary 5.1 allows one to verify the consistency of a variety of estimators. Thus, the sample standard deviation $\hat{\sigma}_n = \sqrt{\frac{1}{n} \sum_{i=1}^n (Y_i - \bar{Y}_n)^2}$ is a consistent estimator of σ, $\frac{\hat{p}_n}{1-\hat{p}_n}$ in Example 5.3 is a consistent estimator of the odds $\frac{p}{1-p}$, etc.

Question 5.3 (**consistency of the MME**) Let $Y_1, ..., Y_n$ be a random sample from a distribution with a finite second moment and the parameter of interest θ is a continuous function of $\mu = EY$. Show that the MME $\hat{\theta}_{MME}$ of θ is consistent.

Although consistency itself is important, the *rate* of convergence also matters. One would obviously prefer a consistent estimator with a faster convergence rate. For example, (5.1) implies that the accuracy of estimating $\mu = EY$ by the sample mean $\bar{Y}_n$ is of order $n^{-1/2}$ in the sense that if r_n is such that $P(|\bar{Y}_n - \mu| < r_n)$ tends to zero as n increases, then $r_n/n^{-1/2} \to 0$. We will not consider these issues in detail but just mention that in many situations, $n^{-1/2}$ is the best available convergence rate for estimating a parameter (e.g., for distributions from the exponential family) and such estimators are called $\sqrt{n}$-consistent:

Definition 5.4 ($\sqrt{n}$-**consistency**) An estimator $\hat{\theta}_n$ is $\sqrt{n}$-consistent of θ, if for every $\varepsilon > 0$ there exists $0 < C < \infty$ and a $n_0 < \infty$ such that $P(|\hat{\theta}_n - \theta| < Cn^{-1/2}) \geq 1 - \varepsilon$ for all $n > n_0$.

5.4 Convergence in distribution

Consistency in MSE and in probability considered in the previous sections still does not say anything about the asymptotic distribution of an estimator needed, for example, to construct an approximate confidence interval for an unknown parameter or to calculate the critical value in the hypothesis testing (see the remarks at the beginning of the chapter). For this purpose we consider another type of convergence of random variables—convergence in distribution.

Suppose there is a sequence of random variables $Y_1, Y_2, \ldots$ (not necessarily living in the same probability space). We want to formalize the notion that their distributions converge to some particular distribution (e.g., normal or Poisson). Let $F_1(\cdot), F_2(\cdot), \ldots$ be cumulative distribution functions (cdf) of $Y_1, Y_2, \ldots$, that is, $F_n(x) = P(Y_n \leq x)$. Note that if $P(Y_n = x) = 0$, $F_n(\cdot)$ is continuous at x.

Definition 5.5 (convergence in distribution) A sequence of random variables $Y_1, Y_2, \ldots$ with cdf's $F_1(\cdot), F_2(\cdot), \ldots$ is said to converge in distribution to a random variable Y with a cdf $F(\cdot)$, if $\lim_{n \to \infty} F_n(x) = F(x)$ for every continuity point x of $F(\cdot)$. We denote this by $Y_n \xrightarrow{\mathscr{D}} Y$.

The restriction to a continuity point is essential. We want to say that if $a_n > 0$, $a_n \to 0$ and $P(Y_n = a_n) = 1 - P(Y_n = 1 - a_n) = 0.5$ then $Y_n \xrightarrow{\mathscr{D}} \mathscr{B}(1, 0.5)$. However, $F(0) = 0.5$ while $F_n(0) = 0$ for all n and hence $\lim_{n \to \infty} F_n(0) \neq F(0)$.

Convergence in distribution is important because it enables us to approximate probabilities. Quite often F_n is either complex or not explicitly known, while F is simple and known. If $Y_n \xrightarrow{\mathscr{D}} F$, we can use F to calculate approximate probabilities related to Y_n. The prime examples of limiting distributions include Poisson (as an approximation for binomial random variables with large n and small p), the normal distribution (for properly scaled means), the exponential (for the minimum of many random variables), and extreme value distribution (for the distribution of the extremely large values). The fact that these distributions are obtained as a limit is the reason why they are so useful in describing various phenomena in nature. It often happens that an observed variable is a result of many different factors, where each one of them only has a tiny impact on the overall measure (e.g., IQ is a function of a variety of genetic and environmental factors the subject was exposed to). By the central limit theorem (see Section 5.5) it is reasonable to assume then that its distribution will be close to normal. Under very general assumptions, the number of citizens suffering a heart attack at night in a city of population size n can be modeled as $\mathscr{B}(n, p)$, where the probability of a heart attack p for an individual is (luckily!) small. We can then approximate the number of actual heart attacks by a Poisson distribution (see Example 5.5). Suppose there is some device consisting of a large number of components with random working times that can be very long. If the components operate independently and the device stops functioning when even one of them does not operate properly, it can be shown that the random lifetime of the device is approximately exponentially distributed (see Appendix A.4.6).

Example 5.5 (Poisson limit) Suppose that $Y_n \sim \mathscr{B}(n, p_n)$ for each n, where $p_n \to 0$ but $np_n \to \lambda$ as $n \to \infty$. Then $Y_n \xrightarrow{\mathscr{D}} Pois(\lambda)$.

Proof. Let $F_n(x)$ and $F(x)$ be the cdfs of $\mathscr{B}(n, p_n)$ and $Pois(\lambda)$, respectively. Note that for every non-integer x, $F(x) = \sum_{i < x} P(Y = i)$ and has jumps on the non-negative integers. Similarly, for every non-integer x, $F_n(x) = \sum_{i < x} P(Y_n = i)$. Since a limit of a finite sum is the sum of limits, we need only to show that $\lim_{n \to \infty} P(Y_n = i) = P(Y = i)$ for all $i = 0, 1, 2, \ldots$ in order to prove that $Y_n \xrightarrow{\mathscr{D}} Y$. By rearranging terms we have

$$P(Y_n = i) = \frac{n!}{(n-i)!i!} p_n^i (1 - p_n)^{n-i} = \frac{(np_n)^i}{i!} (1 - p_n)^n \frac{1}{(1 - p_n)^i} \prod_{j=0}^{i-1} \frac{n-j}{n}.$$

Clearly, when $p_n \to 0$ and $np_n \to \lambda$, for any fixed i

$$\lim_{n \to \infty} \prod_{j=0}^{i-1} \frac{n-j}{n} = 1, \qquad \lim_{n \to \infty} (1 - p_n)^i = 1, \qquad \text{and} \qquad \lim_{n \to \infty} (1 - p_n)^n = \lim_{n \to \infty} e^{-np_n} = e^{-\lambda}.$$

Figure 5.2: Bar plot of the probability function of $Pois(2)$, $\mathscr{B}(10,0.2)$, $\mathscr{B}(20,0.1)$. Note that all three have a mean of 2.

Since the limit of a product is the product of the limits:

$$P(Y_n = i) \to e^{-\lambda}\frac{\lambda^i}{i!} = P(Y = i).$$

$\square$

We demonstrate the quality of the Poisson approximation by comparing two binomial distributions $\mathscr{B}(10,.2)$ and $\mathscr{B}(20,.1)$ approximated by the same Poisson distribution $Pois(2)$ in Figure 5.2.

5.5 The central limit theorem

The most important and common limiting distribution is, however, the normal distribution, which plays a central role in probability and statistics. The well-known central limit theorem (CLT) is one of the most fundamental examples where it appears. Here is a simple version of its result:

Theorem 5.3 (central limit theorem) Suppose $Y_1, Y_2, \ldots$ are i.i.d. random variables with the common (finite) expectation μ and variance σ^2. Let $\bar{Y}_n = \frac{1}{n}\sum_{i=1}^{n} Y_i$. Then, as $n \to \infty$,

$$\sqrt{n}(\bar{Y}_n - \mu) \xrightarrow{\mathscr{L}} \mathscr{N}(0, \sigma^2). \tag{5.3}$$

Formulation (5.3) is essentially a formal way to say that for sufficiently large n, the distribution of a sample mean $\bar{Y}_n$ is approximately normal with the mean μ and variance $\frac{\sigma^2}{n}$. Note that it is always true that $E\bar{Y}_n = \mu$ and $\text{Var}(\bar{Y}_n) = \frac{\sigma^2}{n}$. The CLT adds that the distribution of $\bar{Y}_n$ is approximately normal. We omit here the technical proof of the CLT that can be found in any probability textbook.

The CLT is a real refinement of the WLLN. Both consider the asymptotic behavior of $\bar{Y}_n$. The WLLN only says that $\bar{Y}_n \xrightarrow{\text{p}} \mu$, or that $\bar{Y}_n$ is very close to μ when n is large but does not tell us how close. The CLT gives a precise answer. It is as if we look at the cdf of $\bar{Y}_n$ through a magnifying glass: we center it around μ and magnify by $\sqrt{n}$. As n increases, the cdf looks more and more smooth, and becomes closer and closer to a normal cdf.

Question 5.4 Show that if, in particular, $Y_i \sim \mathcal{N}(\mu, \sigma^2)$, the distribution of $\sqrt{n}(\bar{Y}_n - \mu)$ is *exactly* normal $\mathcal{N}(0, \sigma^2)$ starting from $n = 1$.

Question 5.5 Show that if $\sqrt{n}(T_n - v) \xrightarrow{\mathscr{D}} \mathcal{N}(0, \sigma^2)$ for some v and σ^2, and $S_n = a + bT_n$, then $\sqrt{n}(S_n - bv - a) \xrightarrow{\mathscr{D}} \mathcal{N}(0, b^2\sigma^2)$.

Example 5.6 (normal approximation to the binomial distribution) Let $Y_1, Y_2, \ldots$ be a series of Bernoulli trials $\mathscr{B}(1, p)$, and $S_n = \sum_{i=1}^{n} Y_i$ and $\hat{p}_n = S_n/n$ be respectively the total number and proportion of successes. Clearly, $S_n \sim \mathscr{B}(n, p)$, where $ES_n = np$ and $\text{Var}(S_n) = np(1-p)$. Then, by the CLT,

$$\frac{1}{\sqrt{n}}(S_n - np) \xrightarrow{\mathscr{D}} \mathcal{N}(0, p(1-p)) \quad \text{and} \quad \sqrt{n}(\hat{p}_n - p) \xrightarrow{\mathscr{D}} \mathcal{N}(0, p(1-p)).$$

A common question is regarding the rate of convergence to the normal distribution in the CLT. Can we use the normal approximation for $n = 20$? Is 30 enough? 100?

Let $Y_1, \ldots, Y_n$ be a random sample from some distribution $f(y)$ with a finite variance and $S_n = \sum_{i=1}^{n} Y_i$. Consider its normalized version:

$$Z_n^* = \frac{S_n - ES_n}{\sqrt{\text{Var}(S_n)}},$$

where Z_n^* has a zero mean and a unit variance regardless of the original distribution $f(y)$ of Y or the sample size n. We then ask how close is the shape of the distribution of Z_n^* to the bell of the standard normal distribution $\mathcal{N}(0, 1)$. How large should n be such that we would be able to say that it is "approximately normal"?

The answer depends obviously on the particular distribution $f(y)$ at hand and the required accuracy. For example, the normal approximation for the distribution of S_n is *exact* even for $n = 1$ when the Y_is are normal (see Question 5.4) but might require much larger samples to achieve good accuracy for other distributions.

There exists a general uniform upper bound on the error of approximating the true cdf of $\bar{Y}_n$ by normal distribution known as Berry–Esseen bound, which we present here without a proof:

Theorem 5.4 (Berry–Esseen bound) Let $Y_1, \ldots, Y_n$ be i.i.d. random sample, where $\mu = EY$, $\sigma^2 = Var(Y)$ and $\rho = E|Y - \mu|^3 < \infty$. Then,

$$\sup_z \left| P\left(\frac{\sqrt{n}(\bar{Y}_n - \mu)}{\sigma} \leq z \right) - \Phi(z) \right| \leq \frac{C\rho}{\sigma^3 \sqrt{n}} \tag{5.4}$$

for a certain universal constant $C < 0.8$, where $\Phi(\cdot)$ is the cdf of the standard normal distribution $\mathcal{N}(0, 1)$.

Theorem 5.4 ensures that for a sample mean from *any* distribution satisfying the moment conditions, the error of the normal approximation in the CLT is of maximum order $n^{-1/2}$. However, being so general, the Berry–Esseen bound (5.4) might be imprecise and overconservative for a particular distribution at hand.

For illustrating the goodness of the normal approximation, consider a random sample $Y_1, \ldots, Y_n$ from a Gamma distribution $Gamma(\alpha, 1)$ with a unit scale parameter and a shape parameter $\alpha > 0$, where $f_{\alpha,1}(y) = \Gamma^{-1}(\alpha)y^{\alpha-1}e^{-y}$, $y \geq 0$ (see Appendix A.4.8). Note that for a unit scale parameter $EY = \text{Var}(Y) = \alpha$. One of the important properties of the Gamma family is that it is closed under the sum of independent random variables and $S_n = \sum_{i=1}^{n} Y_i \sim Gamma(n\alpha, 1)$. Hence, $Z_n^* = (S_n - n\alpha)/\sqrt{n\alpha}$ and the quality of approximation to normality depends on $n\alpha$ rather than on n alone. The distribution of the mean of 50 observations from $Gamma(1, 1)$ (the exponential $\exp(1)$ distribution) is as close to normality as 100 observations from $Gamma(0.5, 1)$, or as 10 observations from $Gamma(5, 1)$. Figure 5.3 shows different gamma densities scaled to have zero mean and unit variance.

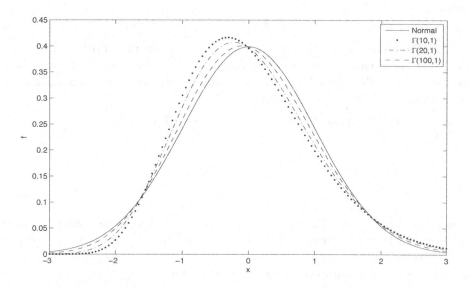

Figure 5.3: Different gamma densities, scaled to have mean 0 and variance 1, and the $\mathcal{N}(0,1)$ density.

5.6 Asymptotically normal consistency

We return now to the original question on the asymptotic distribution of estimators. In particular, we show why in many situations their limiting distribution is normal. It is indeed so "normal" that it deserves a special definition:

Definition 5.6 (asymptotically normal consistency) An estimator $\hat{\theta}_n$ is called a consistent asymptotically normal (CAN) estimator of θ if $\sqrt{n}(\hat{\theta}_n - \theta) \xrightarrow{\mathscr{L}} \mathcal{N}(0, \sigma^2(\theta))$ for some *asymptotic variance* $\sigma^2(\theta)$ that may generally depend on θ.

Less formally, Definition 5.6 implies that for sufficiently large n, the distribution of $\hat{\theta}_n$ is approximately normal $\mathcal{N}(\theta, \frac{\sigma^2(\theta)}{n})$ with the mean θ and variance of order n^{-1}.

In particular, the CLT immediately yields that the sample mean $\bar{Y}_n$ is a CAN estimator of EY with the asymptotic variance $\sigma^2 = Var(Y)$.

Asymptotically normal consistency yields $\sqrt{n}$-consistency (see Section 5.3):

Theorem 5.5 If $\hat{\theta}_n$ is a CAN estimator of θ, then it is also $\sqrt{n}$-consistent.

Proof. Due to the asymptotically normal consistency of $\hat{\theta}_n$, for any $0 < C < \infty$, $P(|\hat{\theta}_n - \theta| < Cn^{-1/2}) \to \Phi(C/\sigma) - \Phi(-C/\sigma) = 2\Phi(C/\sigma) - 1$. Hence, for any $\varepsilon > 0$ and $C > \sigma z_{\varepsilon/2}$, $P(|\hat{\theta}_n - \theta| < Cn^{-1/2}) \geq 1 - \varepsilon$ for all sufficiently large n. $\qquad\square$

The following theorem (aka *delta method*) explains why so many estimators are CAN:

Theorem 5.6 Let $\hat{\theta}_n$ be a CAN estimator of θ with an asymptotic variance $\sigma^2(\theta)$ and $g(\cdot)$ be any differentiable function. Then, if $g'(\theta) \neq 0$, $g(\hat{\theta}_n)$ is a CAN estimator of $g(\theta)$ with the asymptotic variance $g'(\theta)^2\sigma^2(\theta)$. That is, if $\sqrt{n}(\hat{\theta}_n - \theta) \xrightarrow{\mathscr{L}} \mathcal{N}(0, \sigma^2(\theta))$, then $\sqrt{n}(g(\hat{\theta}_n) - g(\theta)) \xrightarrow{\mathscr{L}} \mathcal{N}(0, g'(\theta)^2\sigma^2(\theta))$.

Proof. Note first that since $\hat{\theta}_n \xrightarrow{p} \theta$ (see Theorem 5.5), $\hat{\theta}_n - \theta = o_p(1)$. Applying Taylor's expansion one then has

$$g(\hat{\theta}_n) = g(\theta) + g'(\theta)(\hat{\theta}_n - \theta) + o_p(\hat{\theta}_n - \theta)$$

and, therefore,

$$\sqrt{n}(g(\hat{\theta}_n) - g(\theta)) = g'(\theta)\sqrt{n}(\hat{\theta}_n - \theta) + \sqrt{n}\, o_p(\hat{\theta}_n - \theta). \tag{5.5}$$

Since $\hat{\theta}_n$ is a CAN estimator of θ with asymptotic variance $\sigma^2(\theta)$, $g'(\theta)\sqrt{n}(\hat{\theta}_n - \theta) \xrightarrow{\mathscr{D}} \mathscr{N}(0, g'(\theta)^2\sigma^2(\theta))$ (see Question 5.5), while $\sqrt{n}(\hat{\theta}_n - \theta) = O_p(1)$ (see Theorem 5.5) and, hence, $\sqrt{n}\, o_p(\hat{\theta}_n - \theta) = o_p(1)$. The rest of the proof follows then directly from Slutsky's theorem that we present below and will use throughout this chapter:

Theorem 5.7 (Slutsky's theorem) Let $X_n \xrightarrow{\mathscr{D}} X$ and $U_n \xrightarrow{p} c$. Then,

1. $X_n + U_n \xrightarrow{\mathscr{D}} X + c$.

2. $X_n \cdot U_n \xrightarrow{\mathscr{D}} cX$.

3. if $c \neq 0$, $\frac{X_n}{U_n} \xrightarrow{\mathscr{D}} \frac{X}{c}$.

$\square$

In particular, let a parameter of interest $\theta = g(EY)$, where $g(\cdot)$ is any differentiable function. The CLT and Theorem 5.6 immediately imply then that $g(\bar{Y}_n)$ is a CAN estimator of θ with asymptotic variance $g'(\theta)^2 Var(Y)$.

Question 5.6 (asymptotic normal consistency of the MME) Let $Y_1, \ldots, Y_n$ be a random sample from a distribution with $EY = \mu$ and a finite variance $Var(Y)$. Show that if the parameter of interest $\theta = g(\mu)$, where $g(\cdot)$ is a differentiable function, then the MME $\hat{\theta}_{MME}$ of θ is CAN with the asymptotic variance $(g'(\mu))^2 Var(Y)$.

Example 5.7 Suppose that we have records $Y_1, \ldots, Y_n$ of annual numbers of earthquakes in a seismically active area for the last n years. Such data can typically be well modeled by a Poisson distribution with an (unknown) mean $\lambda > 0$. We are interested in $\theta = \lambda^{-1}$—the expected duration of a quiet period between two successive earthquakes. Following the above arguments for $g(\lambda) = \lambda^{-1}$, $\hat{\theta}_n = \bar{Y}_n^{-1}$ is a CAN estimator of θ, where $\sqrt{n}(\hat{\theta}_n - \theta) \xrightarrow{\mathscr{D}} \mathscr{N}(0, \lambda^{-4}\lambda) = \mathscr{N}(0, \lambda^{-3})$ or, in terms of θ itself, $\sqrt{n}(\hat{\theta}_n - \theta) \xrightarrow{\mathscr{D}} \mathscr{N}(0, \theta^3)$.

Similarly, suppose that we also want to estimate the probability p that the forthcoming year will be quiet. In this case $p = P(Y = 0) = g(\lambda) = e^{-\lambda}$. Define $\hat{p}_n = g(\bar{Y}_n) = e^{-\bar{Y}_n}$. Thus, $\sqrt{n}(\hat{p}_n - p) \xrightarrow{\mathscr{D}} \mathscr{N}(0, e^{-2\lambda}\lambda) = \mathscr{N}(0, p^2 \ln(p^{-1}))$.

Note that $\hat{\theta}_n = \bar{Y}_n^{-1}$ and $\hat{p}_n = e^{-\bar{Y}_n}$ are also MLEs of θ and p respectively.

Example 5.8 (asymptotic normality of the sample standard variance and deviation) Let $Y_1, \ldots, Y_n$ be a random sample from a distribution with, say, six finite moments. Denote its mean by μ and variance by σ^2. Consider the sample variance

$$\hat{\sigma}_n^2 = \frac{1}{n}\sum_{i=1}^n (Y_i - \bar{Y}_n)^2 = \frac{1}{n}\sum_{i=1}^n (Y_i - \mu)^2 - (\bar{Y}_n - \mu)^2$$

(see (5.2)). Hence,

$$\sqrt{n}(\hat{\sigma}_n^2 - \sigma^2) = \frac{1}{\sqrt{n}}\sum_{i=1}^n \left((Y_i - \mu)^2 - \sigma^2\right) - \sqrt{n}(\bar{Y}_n - \mu)^2.$$

However, $\left((Y_i - \mu)^2 - \sigma^2\right)$, $i = 1, \ldots, n$, are i.i.d. zero mean random variables, and by the CLT

$$\sqrt{n}\sum_{i=1}^n \left((Y_i - \mu)^2 - \sigma^2\right) \xrightarrow{\mathscr{D}} \mathscr{N}(0, \tau^2)$$

where $\tau^2 = Var((Y - \mu)^2)$. On the other hand, by Markov's inequality, for any $\varepsilon > 0$,

$$P\left(\sqrt{n}(\bar{Y}_n - \mu)^2 > \varepsilon\right) \leq \frac{E\left(\sqrt{n}(\bar{Y}_n - \mu)^2\right)}{\varepsilon} = \frac{\sqrt{n}Var(\bar{Y}_n)}{\varepsilon} = \frac{\sigma^2/\sqrt{n}}{\varepsilon} \xrightarrow[n\to\infty]{} 0$$

and therefore, $\sqrt{n}(\bar{Y}_n - \mu)^2 \xrightarrow{p} 0$.

Slutsky's Theorem 5.7 then yields $\sqrt{n}(\hat{\sigma}_n^2 - \sigma^2) \xrightarrow{\mathcal{D}} \mathcal{N}(0, \tau^2)$. Furthermore, Theorem 5.6 implies that $\hat{\sigma}_n$ is a CAN estimator for σ, where $\sqrt{n}(\hat{\sigma} - \sigma) \xrightarrow{\mathcal{D}} \mathcal{N}(0, \frac{\tau^2}{4\sigma})$.

The notations *asymptotic mean* and *asymptotic variance* for θ and $\sigma^2(\theta)$ might be somewhat misleading: although it is tempting to think that for a CAN estimator $\hat{\theta}_n$ of θ, $E\hat{\theta}_n \to \theta$ and $Var(\sqrt{n}\hat{\theta}_n) \to \sigma^2(\theta)$ as n increases, generally it is not true. The convergence in distribution does not necessarily yield convergence of moments. In fact, $E\hat{\theta}_n$ or $Var(\sqrt{n}\hat{\theta}_n)$ can even have an infinite limit:

Example 5.9 (continuation of Example 5.7) In Example 5.7, we found that for $Y_1, ..., Y_n \sim Pois(\lambda)$, $\hat{\theta} = 1/\bar{Y}_n$ is a CAN estimator for $\theta = 1/\lambda$ with the asymptotic variance λ^{-3}. However, note that for any fixed n,

$$P(\hat{\theta}_n = \infty) = P(\bar{Y}_n = 0) = \prod_{i=1}^{n} P(Y_i = 0) = e^{-n\lambda} \neq 0.$$

A (nonnegative) discrete random variable that takes an infinite value with nonzero probability necessarily has an infinite mean and variance (why?) and, therefore, $E\hat{\theta}_n$ and $Var(\sqrt{n}\hat{\theta}_n)$ are infinite for any fixed n.

5.7 Asymptotic confidence intervals

Asymptotic normality of an estimator can be used to construct an approximate normal confidence interval when its exact distribution is complicated or even not completely known.

Example 5.10 (asymptotic confidence interval for a mean with a known variance) Suppose that $Y_1, Y_2, \ldots$ are i.i.d. with unknown mean μ and known variance σ^2. Assume that the CLT holds and $\sqrt{n}(\bar{Y}_n - \mu) \xrightarrow{\mathcal{D}} \mathcal{N}(0, \sigma^2)$. Then,

$$P\left(\bar{Y}_n - z_{\alpha/2}\frac{\sigma}{\sqrt{n}} \leq \mu \leq \bar{Y}_n + z_{\alpha/2}\frac{\sigma}{\sqrt{n}}\right) = P\left(-z_{\alpha/2} \leq \sqrt{n}\frac{\bar{Y}_n - \mu}{\sigma} \leq z_{\alpha/2}\right) \to 1 - \alpha. \qquad (5.6)$$

Hence, for sufficiently large n,

$$\left(\bar{Y}_n - z_{\alpha/2}\frac{\sigma}{\sqrt{n}} \, , \, \bar{Y}_n + z_{\alpha/2}\frac{\sigma}{\sqrt{n}}\right) \qquad (5.7)$$

is a confidence interval for μ with a coverage probability of approximately $1 - \alpha$ as required.

In the general case, suppose that $\hat{\theta}_n$ is a CAN estimator of θ with an asymptotic variance $\sigma^2(\theta)$, that is, $\sqrt{n}(\hat{\theta}_n - \theta) \xrightarrow{\mathcal{D}} \mathcal{N}(0, \sigma^2(\theta))$. Then,

$$\hat{\theta}_n \pm z_{\alpha/2}\frac{\sigma(\theta)}{\sqrt{n}} \qquad (5.8)$$

is a $(1 - \alpha)100\%$ asymptotic confidence interval for θ.

The equation (5.8) is, however, not of much use since the asymptotic variance $\sigma^2(\theta)$ generally also depends on θ or is an additional unknown parameter as in a more realistic version of Example 5.10. Intuitively, a possible approach would be to replace $\sigma(\theta)$ in (5.8) by some estimator and hope that it will not strongly affect the resulting asymptotic coverage probability. Does this idea really work? Which estimator(s) $\hat{\sigma}_n$ of $\sigma(\theta)$ should be used? As it happens, the idea indeed works if $\hat{\sigma}_n$ is any *consistent* estimator of $\sigma(\theta)$:

Theorem 5.8 Let $\hat{\theta}_n$ be a CAN estimator of θ with an asymptotic variance $\sigma^2(\theta) > 0$ and $\hat{\sigma}_n$ be any consistent estimator of $\sigma(\theta)$. Then,

$$P\left(\hat{\theta}_n - z_{\alpha/2}\frac{\hat{\sigma}_n}{\sqrt{n}} \leq \theta \leq \hat{\theta}_n + z_{\alpha/2}\frac{\hat{\sigma}_n}{\sqrt{n}}\right) \xrightarrow[n\to\infty]{} 1 - \alpha.$$

Proof. The proof of this important result is remarkably simple. It is sufficient to show that $\sqrt{n}(\frac{\hat{\theta}_n-\theta}{\hat{\sigma}_n}) \xrightarrow{\mathscr{D}} \mathscr{N}(0,1)$. We have

$$\sqrt{n}\left(\frac{\hat{\theta}_n - \theta}{\hat{\sigma}_n}\right) = \sqrt{n}\left(\frac{\hat{\theta}_n - \theta}{\sigma(\theta)}\right) \cdot \frac{\sigma(\theta)}{\hat{\sigma}_n}, \tag{5.9}$$

where $\sqrt{n}\left(\frac{\hat{\theta}_n-\theta}{\sigma(\theta)}\right) \xrightarrow{\mathscr{D}} \mathscr{N}(0,1)$ and $\frac{\sigma(\theta)}{\hat{\sigma}_n} \xrightarrow{\text{p}} 1$, and Slutksy's Theorem 5.7 completes the proof. $\square$

Example 5.11 (asymptotic confidence interval for a mean with an unknown variance) Consider again Example 5.10 where we constructed an asymptotic confidence interval for $\mu = EY$ but assume now that the variance σ^2 is unknown. A sample standard deviation $\hat{\sigma}_n = \sqrt{\frac{1}{n}\sum_{i=1}^n(Y_i - \bar{Y}_n)^2}$ is a consistent estimator of σ (see Section 5.3) and we can then substitute it for σ in the asymptotic confidence interval (5.7) for μ to get

$$\bar{Y}_n \pm z_{\alpha/2}\frac{\hat{\sigma}_n}{\sqrt{n}}. \tag{5.10}$$

Alternatively, we can estimate σ by $s_n = \sqrt{\frac{1}{n-1}\sum_{i=1}^n(Y_i - \bar{Y}_n)^2}$, which is also consistent.

To illustrate the goodness of approximation for the coverage probability of the asymptotic confidence interval (5.10) we conducted a small simulation study. We generated 150,000 samples of size n from $Gamma(2,1)$ distribution, and for each sample calculated the corresponding 95% confidence interval according to (5.10). Since we generated the random variables, we have the unique privilege of knowing the truth—the real mean is 2, and could compute whether each of these 150,000 confidence intervals cover the true value. The results are reported in Table 5.1. As n increases, the goodness-of-approximation improves and the estimated coverage probability approaches to the required level of 0.95.

Table 5.1: Estimated coverage probabilities of the approximate confidence interval for the mean of a gamma distribution.

Sample size	Estimated cover probability	Mean width
20	0.92	1.20
30	0.93	0.99
40	0.93	0.86
50	0.94	0.77
60	0.94	0.71

Question 5.7 Explain why the mean width of confidence intervals in Table 5.1 decreases with n.

Example 5.12 (asymptotic confidence interval for the binomial parameter) Let $Y_1,...,Y_n$ be a series of Bernoulli trials $\mathscr{B}(1,p)$. The MLE and UMVUE point estimator of the unknown p is $\hat{p}_n = \bar{Y}_n$ (see Chapter 2). In Example 5.6, we established its asymptotically normal consistency with the asymptotic variance $p(1-p)$. Note that by WLLN $\hat{p}_n$ is a consistent estimator of p and by Theorem 5.8, the asymptotic $100(1-\alpha)\%$ confidence interval for p is then

$$\hat{p}_n \pm z_{\alpha/2}\sqrt{\frac{\hat{p}_n(1-\hat{p}_n)}{n}}. \tag{5.11}$$

In particular, for the newborn babies data from Example 1.1 ($n = 20, \hat{p} = 0.6$) the 95% approximate confidence interval for the proportion of girls is

$$0.6 \pm 1.96\sqrt{\frac{0.6 \cdot 0.4}{20}} = 0.6 \pm 0.21 = (0.39, 0.81).$$

The resulting interval is quite wide and essentially does not allow one to make a well-grounded inference about the proportion of girls in the population, in particular, whether it is 0.5 or different.

Example 5.13 (continuation of Example 5.7) We continue Example 5.7 with the earthquake Poisson data and want to construct asymptotic $100(1 - \alpha)\%$ confidence intervals for $\theta = 1/\lambda$ and $p = P(Y = 0) = e^{-\lambda}$. We can do this in at least two different ways. First, we can derive an asymptotic $100(1 - \alpha)\%$ confidence interval for $\lambda = EY$ using the general result (5.10). Since the Poisson variance is also λ and $\bar{Y}_n$ is its consistent estimator (by the WLLN), the asymptotic $100(1 - \alpha)\%$ confidence interval for λ is

$$\bar{Y}_n \pm z_{\alpha/2} \sqrt{\frac{\bar{Y}_n}{n}}$$

and using the monotonicity of θ and p as functions of λ the resulting asymptotic confidence intervals for them will respectively be

$$\left(\left(\bar{Y}_n + z_{\alpha/2} \sqrt{\frac{\bar{Y}_n}{n}} \right)^{-1}, \left(\bar{Y}_n - z_{\alpha/2} \sqrt{\frac{\bar{Y}_n}{n}} \right)^{-1} \right)$$

and

$$\left(e^{-\bar{Y}_n - z_{\alpha/2} \sqrt{\frac{\bar{Y}_n}{n}}}, e^{-\bar{Y}_n + z_{\alpha/2} \sqrt{\frac{\bar{Y}_n}{n}}} \right)$$

(see Proposition 3.1).

Alternatively, we can use the asymptotic normality of $\hat{\theta}_n = \bar{Y}_n^{-1}$ and $\hat{p}_n = e^{-\bar{Y}_n}$ established in Example 5.7. Corollary 5.1 ensures that $\hat{\theta}_n^{-3}$ and $\hat{p}_n^2 \ln(\hat{p}_n^{-1})$ are consistent estimators of their asymptotic variances and the corresponding asymptotic confidence intervals for θ and p are then

$$\hat{\theta}_n \pm z_{\alpha/2} \sqrt{\frac{\hat{\theta}_n^3}{n}} = \frac{1}{\bar{Y}_n} \pm z_{\alpha/2} \sqrt{\frac{1}{\bar{Y}_n^3 n}}$$

and

$$\hat{p}_n \pm z_{\alpha/2} \hat{p}_n \sqrt{\frac{\ln \hat{p}_n^{-1}}{n}} = e^{-\bar{Y}_n} \left(1 \pm z_{\alpha/2} \sqrt{\frac{\bar{Y}_n}{n}} \right).$$

It should not be surprising that the two approaches led to different results—there is no unique way to construct a confidence interval as we have argued in Chapter 3. Which one is "better"? In Chapter 3, we discussed that one may prefer a confidence interval with a shorter expected length. For asymptotic confidence intervals, another (probably even more important!) issue is the goodness of normal approximation for distributions of the underlying statistics. In particular, for the example at hand, a question is whether the distribution of $\bar{Y}_n$ is closer to normal on the original scale or on the transformed scales $\bar{Y}_n^{-1}$ or $e^{-\bar{Y}_n}$. These issues usually require considering higher order normal approximations, involve a more delicate technical analysis, and are beyond the scope of this book.

Another possible approach to deal with an unknown asymptotic variance $\sigma^2(\theta)$ of a CAN estimator $\hat{\theta}_n$ when constructing an asymptotic confidence interval for θ is by the use of the *variance stabilizing transformation* $g(\theta)$ whose asymptotic variance $g'(\theta)^2 \sigma^2(\theta)$ (see Theorem 5.6) does not depend on θ. Evidently, this transformation $g(\theta)$ should satisfy $g'(\theta) = c/\sigma(\theta)$ for a chosen constant c. A $(1 - \alpha)100\%$ asymptotic confidence interval for the resulting $g(\theta)$ is then

$$g(\hat{\theta}_n) \pm z_{\alpha/2} \frac{c}{\sqrt{n}}, \tag{5.12}$$

which can be used to construct an asymptotic confidence interval for θ itself.

Example 5.14 (continuation of Example 5.13) We have already shown that for Poisson data, $\bar{Y}_n$ is a CAN estimator of $EY = \lambda$ with the asymptotic variance $Var(Y) = \lambda$. Thus, the variance stabilizing transformation $g(\lambda)$ satisfies $g'(\lambda) = \frac{c}{\sqrt{\lambda}}$ or $g(\lambda) = 2c\sqrt{\lambda}$. For convenience, let $c = 1/2$. Then, the $(1 - \alpha)100\%$ asymptotic confidence interval for $\sqrt{\lambda}$ is

$$\sqrt{\bar{Y}_n} \pm z_{\alpha/2} \frac{1}{2\sqrt{n}}$$

and due to the monotonicity, the corresponding $(1 - \alpha)100\%$ asymptotic confidence intervals for λ, $\theta = 1/\lambda$ and $p = e^{-\lambda}$ are respectively

$$\lambda: \qquad \left(\left(\sqrt{\bar{Y}_n} - z_{\alpha/2} \frac{1}{2\sqrt{n}} \right)^2, \left(\sqrt{\bar{Y}_n} + z_{\alpha/2} \frac{1}{2\sqrt{n}} \right)^2 \right),$$

$$\theta = 1/\lambda: \qquad \left(\left(\sqrt{\bar{Y}_n} + z_{\alpha/2} \frac{1}{2\sqrt{n}} \right)^{-2}, \left(\sqrt{\bar{Y}_n} - z_{\alpha/2} \frac{1}{2\sqrt{n}} \right)^{-2} \right),$$

$$p = e^{-\lambda}: \qquad \left(e^{-\left(\sqrt{\bar{Y}_n} + z_{\alpha/2} \frac{1}{2\sqrt{n}} \right)^2}, e^{-\left(\sqrt{\bar{Y}_n} - z_{\alpha/2} \frac{1}{2\sqrt{n}} \right)^2} \right).$$

5.8 Asymptotic properties of MLEs, Wald confidence intervals, and tests

Maximum likelihood estimators were introduced in Section 2.2, but the motivation for them there was mostly heuristic. The real justification for the MLE, however, is not just "philosophical." It is genuinely a good estimation approach and its theoretical justification is based on its strong asymptotic properties.

We first show that under very mild conditions on the parameter space and the distribution function the MLE is a consistent estimator:

Theorem 5.9 Let $Y_1, ..., Y_n \sim f_\theta(y)$, $\theta \in \Theta$ and $\hat{\theta}_n$ is the unique MLE of θ. Let θ_0 be the true (unknown) value of θ. Assume that[2]

1. Θ is compact

2. $f_\theta(y)$ is a continuous function of θ for all y

3. $E_{\theta_0}(\sup_{\theta \in \Theta} |\ln f_\theta(Y)|) < \infty$

Then, $\hat{\theta}_n$ is consistent, i.e. $\hat{\theta}_n \xrightarrow{p} \theta_0$.

Proof. By its definition, $\hat{\theta}_n$ maximizes $\frac{1}{n} l(\theta, Y)$ w.r.t. $\theta \in \Theta$. On the other hand, for any $\theta \in \Theta$, by the WLLN we have

$$\frac{1}{n} l(\theta, Y) = \frac{1}{n} \sum_{i=1}^{n} \ln f_\theta(Y_i) \xrightarrow{p} E_{\theta_0} \ln f_\theta(Y)$$

We now show that $E_{\theta_0} \ln f_\theta(Y)$ is maximized at the true θ_0. Indeed, by Jensen's inequality (see Section A.3.3), for any $\theta \in \Theta$ we have

$$E_{\theta_0} \ln f_\theta(y) - E_{\theta_0} \ln f_{\theta_0}(y) = E_{\theta_0} \ln \frac{f_\theta(Y)}{f_{\theta_0}(Y)} < \ln E_{\theta_0} \frac{f_\theta(Y)}{f_{\theta_0}(Y)} = \ln \int \frac{f_\theta(y)}{f_{\theta_0}(y)} f_{\theta_0}(y) dy = \ln \int f_\theta(y) dy$$

$$= \ln 1 = 0$$

Since $\frac{1}{n} l(\theta, Y) \xrightarrow{p} E_{\theta_0} \ln f_\theta(Y)$, it is tempting to conclude that their points of maximum $\hat{\theta}_n$ converge to the point of maximum θ_0 of $E_{\theta_0} \ln f_\theta(Y)$. However, we need a somewhat stronger condition for

---

[2]These conditions are only sufficient but not necessary and can be replaced by some others.

such a claim to be true. Namely, we need $\frac{1}{n}l(\theta, Y)$ to converge to $E_{\theta_0} \ln f_\theta(Y)$ *uniformly* over $\theta \in \Theta$, i.e. $\sup_{\theta \in \Theta}(\frac{1}{n}l(\theta, Y) - E_{\theta_0} \ln f_\theta(Y)) \overset{p}{\longrightarrow} 0$. The assumptions of the theorem are required to ensure the above uniform convergence. $\qquad \square$

We now verify that under some stronger (but still mild) regularity conditions, MLEs are also CAN. We have already seen this, for example, in Example 5.6 and Example 5.7. The following theorem shows that this is a general fundamental property of MLEs. Moreover, the MLE of θ is asymptotically consistently normal with the minimal possible asymptotic variance among all "regular" estimators:

Theorem 5.10 Let $Y_1, ..., Y_n \sim f_\theta(y)$, $\theta \in \Theta$ and $\hat{\theta}_n$ is the unique MLE of θ. Then, under general "regularity conditions" on f_θ, $\hat{\theta}_n$ is a CAN estimator of θ with the asymptotic variance $I^*(\theta)^{-1}$, where $I^*(\theta) = E((\ln f_\theta(Y))'_\theta)^2$ is the Fisher information number of $f_\theta(y)$ based on a single observation (see Section 2.7.2), that is,

$$\sqrt{n}(\hat{\theta}_n - \theta) \overset{\mathscr{D}}{\longrightarrow} \mathscr{N}(0, I^*(\theta)^{-1})$$

as $n \to \infty$.

Simply put, Theorem 5.9 states that for a sufficiently large sample size n, the MLE $\hat{\theta}_n$ of θ is approximately normally distributed with the mean θ and variance $I(\theta)^{-1} = (nI^*(\theta))^{-1}$ (see Corollary 2.1), which is exactly the Cramer–Rao variance lower bound for unbiased estimators (see Theorem 2.2). [3]

We do not present the precise required regularity conditions but only mention that, in particular, they include those that appeared for the Cramer–Rao lower bound in Theorem 2.2 and the assumption that the true (unknown) value of θ is an *interior* point (not at the boundary) of the parameter space Θ. We omit the tedious proof of Theorem 5.9 and provide only a sketch for it. The interested reader can find regularity conditions and a rigorous proof, for example, in Cox and Hinkley (1974, Section 9.2 (iii)) and Bickel and Doksum (2015, Theorem 5.4.3).

Sketch of the proof. We need to show that $\sqrt{nI^*(\theta_0)}(\hat{\theta}_n - \theta_0) \overset{\mathscr{D}}{\longrightarrow} \mathscr{N}(0,1)$, where θ_0 is the true value of θ. Consider Taylor's expansion of the derivative of the log-likelihood w.r.t. θ:

$$l'(\hat{\theta}_n; \mathbf{Y}) = l'(\theta_0; \mathbf{Y}) + l''(\theta_n^*; \mathbf{Y})(\hat{\theta}_n - \theta_0),$$

where θ_n^* lies between $\hat{\theta}_n$ and θ_0. But $\hat{\theta}_n$ is an MLE and, therefore, $l'(\hat{\theta}_n; y) = 0$. Thus, $\hat{\theta}_n - \theta_0 = -\frac{l'(\theta_0; y)}{l''(\theta_n^*)}$ and

$$\sqrt{nI^*(\theta_0)}(\hat{\theta}_n - \theta_0) = \sqrt{n} \frac{l'(\theta_0; \mathbf{Y})}{n\sqrt{I^*(\theta_0)}} \cdot \frac{l''(\theta_0; \mathbf{Y})}{l''(\theta_n^*; \mathbf{Y})} \cdot \left(-\frac{l''(\theta_0; \mathbf{Y})}{nI^*(\theta_0)}\right)^{-1} \qquad (5.13)$$

We now treat the three terms in the RHS of (5.13) separately. Note that $\frac{1}{n}l'(\theta_0; \mathbf{Y}) = \frac{1}{n}\sum_{i=1}^n (\ln f_{\theta_0}(Y_i))'$ and is, in fact, the sample mean of i.i.d. random variables $\ln f_{\theta_0}(Y_i)$ with zero means and variances $I^*(\theta_0)$ (see (2.11) and (2.13) in Section 2.7.2). The CLT implies then that

$$\sqrt{n} \frac{1}{n}l'(\theta_0; \mathbf{Y}) \overset{\mathscr{D}}{\longrightarrow} \mathscr{N}(0, I^*(\theta_0))$$

and, therefore, the first term in the RHS of (5.13) converges in distribution to the standard normal variable $\mathscr{N}(0,1)$.

---

[3] However, following the arguments at the end of Section 5.6 and Example 5.9, it still does not necessarily mean that an MLE $\hat{\theta}_n$ is asymptotically UMVUE of θ. In fact, the theorem does not say anything about the variance of the MLE, only about its asymptotic variance (see the discussion at the end of Section 5.6), and hence, it is not directly related to the Cramer–Rao bound. However, no "regular" estimator can have a more concentrated asymptotic distribution.

Similarly, by the WLLN,

$$\frac{1}{n}l''(\theta_0;\mathbf{Y}) = \frac{1}{n}\sum_{i=1}^{n}(\ln f_{\theta_0}(Y_i))'' \xrightarrow{\text{p}} E(\ln f_{\theta_0}(Y))'' = -I^*(\theta_0)$$

(see Proposition 2.1) and the last term converges in probability to 1.

Finally, the required regularity conditions ensure that the middle term

$$\frac{l''(\theta_0;\mathbf{Y})}{l''(\theta_n^*;\mathbf{Y})} \xrightarrow{\text{p}} 1$$

Slutsky's Theorem 5.7 completes the proof.

□

Theorem 5.10 implies that under the required regularity conditions, all the results on CAN estimators from previous sections are directly applicable to MLEs. In particular, an MLE $\hat{\theta}_n$ of θ is $\sqrt{n}$-consistent and

$$\hat{\theta}_n \pm z_{\alpha/2}\sqrt{I^{-1}(\hat{\theta}_n)} \tag{5.14}$$

is an asymptotic $100(1-\alpha)\%$ confidence interval for θ aka *Wald confidence interval*.

Example 5.15 (exponential distribution—continuation of Example 3.6) In Example 3.6 of Section 3.3, we derived a 95% (exact) confidence interval for θ from a data sample of exponentially distributed $\exp(\theta)$ waiting times for a bus collected by John during 30 days with an average waiting time of $\bar{t} = 8$ *min*:

$$\left(\frac{\chi^2_{2n,0.975}}{2n\bar{t}_n}, \frac{\chi^2_{2n,0.025}}{2n\bar{t}_n}\right) = (0.084, 0.173)$$

Construct now the corresponding *asymptotic* 95% Wald confidence interval. The MLE $\hat{\theta} = \bar{t}^{-1} = 0.125$ and $l''_\theta(\theta;\mathbf{t}) = -\frac{n}{\theta^2}$ (see Example 2.4). Hence, the Fisher information number $I(\theta) = -El''_\theta(\theta;\mathbf{t}) = \frac{n}{\theta^2}$ and from (5.14) we have

$$\frac{1}{\bar{t}_n} \pm z_{0.025}\frac{1}{\bar{t}_n\sqrt{n}} = \frac{1}{\bar{t}_n}\left(1 \pm z_{0.025}\frac{1}{\sqrt{n}}\right) = (0.080, 0.170)$$

Both confidence intervals are very close.

Asymptotically normal consistency can also be used for deriving asymptotic tests in hypothesis testing. Thus, exploiting the duality between hypothesis testing and confidence intervals (see Proposition 4.3 in Section 4.4), the asymptotic Wald confidence interval (5.14) for θ leads to the following *Wald test* for testing $H_0 : \theta = \theta_0$ vs. $H_1 : \theta \neq \theta_0$ for a fixed θ_0: the null hypothesis is rejected if

$$\sqrt{I(\hat{\theta}_n)}|\hat{\theta}_n - \theta_0| > z_{\alpha/2} \tag{5.15}$$

This test is of asymptotic level α.

Example 5.16 (Wald test for exponential distribution—continuation of Example 5.15) Given a random sample $t_1,\ldots,t_n \sim \exp(\theta)$ construct the asymptotic Wald test for testing $H_0 : \theta = \theta_0$ vs. $H_1 : \theta \neq \theta_0$ at level α. The Wald confidence interval for θ is $\bar{t}_n^{-1}(1 \pm z_{\alpha/2}\frac{1}{\sqrt{n}})$ (see Example 5.15) and the null hypothesis is rejected if $|\theta_0 - \frac{1}{\bar{t}_n}| > \frac{1}{\bar{t}_n\sqrt{n}}z_{\alpha/2}$, that is, if

$$\sqrt{n}|\bar{t}_n\theta_0 - 1| > z_{\alpha/2} \tag{5.16}$$

Although the regularity conditions on the data distribution f_θ in Theorem 5.9 are quite weak and satisfied for a majority of the commonly used distributions, there still exist "non-regular" counterexamples where the asymptotically normal consistency of an MLE does not hold (e.g., a shifted exponential distribution in Exercise 5.9).

5.9 *Multiparameter case

Most of the results for a single parameter from the previous sections can be directly extended to the multiparameter case as well.

Let $Y_1, ..., Y_n \sim f_\theta(y)$, where $\theta = (\theta_1, ..., \theta_p) \in \Theta$ is a p-dimensional parameter. The convergence in MSE or in probability and the corresponding consistency of estimators is understood component-wise, that is, we say that a (p-dimensional) estimator $\hat{\theta}_n$ is a consistent in MSE/probability estimator of θ if $\hat{\theta}_{jn} \xrightarrow{\text{MSE/P}} \theta_j$ for every component $j = 1, ..., p$. The notion of convergence in distribution is generalized in a natural way, but it is important to note that more than a component-wise convergence is needed. The distributions of the vectors of random variables should convergence to the distribution of the limiting random vector at each continuity point of the latter.

The definition of asymptotically normal consistency is extended as follows:

Definition 5.7 (multiparameter asymptotically normal consistency) A p-dimensional estimator $\hat{\theta}_n$ is called a consistently asymptotically normal (CAN) estimator of θ if the distribution of $\sqrt{n}(\hat{\theta}_n - \theta)$ converges to a multinormal distribution with a zero mean vector and some asymptotic (nonsingular) covariance matrix $V(\theta)$, that is,

$$\sqrt{n}(\hat{\theta}_n - \theta) \xrightarrow{\mathscr{D}} \mathscr{N}(\mathbf{0}, V(\theta))$$

and, therefore,

$$n(\hat{\theta}_n - \theta)^\top V^{-1}(\theta)(\hat{\theta}_n - \theta) \xrightarrow{\mathscr{D}} \chi_p^2.$$

Question 5.8 Show that if $\hat{\theta}_n$ is a CAN estimator of θ with an asymptotic covariance matrix $V(\theta)$, then $\hat{\theta}_{jn}$ is a CAN estimator of θ_j for all $j = 1, ..., p$. What are the corresponding asymptotic variances?

There is a multivariate analogue of Theorem 5.6 (multivariate delta method):

Theorem 5.11 Let $\hat{\theta}_n$ be a CAN estimator of $\theta \in \Theta \subseteq \mathbb{R}^p$ with an asymptotic covariance matrix $V(\theta)$ and $\mathbf{g} : \mathbb{R}^p \to \mathbb{R}^d$ be a differentiable function. Let $\nabla \mathbf{g}(\theta)$ be the Jacobian matrix of $\mathbf{g}$, i.e. $\nabla \mathbf{g}_{kj} = \frac{\partial g_k(\theta)}{\partial \theta_j}$, $k = 1, ..., d; j = 1, ..., p$. Then, $\mathbf{g}(\hat{\theta}_n)$ is a CAN estimator of $\mathbf{g}(\theta)$ with the asymptotic covariance matrix $\nabla \mathbf{g}(\theta) V(\theta) \nabla \mathbf{g}(\theta)^\top$. That is, if $\sqrt{n}(\hat{\theta}_n - \theta) \xrightarrow{\mathscr{D}} \mathscr{N}(\mathbf{0}, V(\theta))$, then $\sqrt{n}(\mathbf{g}(\hat{\theta}_n) - \mathbf{g}(\theta)) \xrightarrow{\mathscr{D}} \mathscr{N}(\mathbf{0}, \nabla \mathbf{g}(\theta) V(\theta) \nabla \mathbf{g}(\theta)^\top)$.

The proof repeats the arguments of that of Theorem 5.6 but with multivariate Taylor's expansion.

Like in the single parameter case, asymptotic normality allows one to construct approximate normal *confidence regions* for unknown parameters. More specifically, let $\hat{\theta}_n$ be a CAN estimator of θ with an asymptotic covariance matrix $V(\theta)$ so that $n(\hat{\theta}_n - \theta)^\top V^{-1}(\theta)(\hat{\theta}_n - \theta) \xrightarrow{\mathscr{D}} \chi_p^2$. Then,

$$\{\theta : n(\hat{\theta}_n - \theta)^\top V^{-1}(\theta)(\hat{\theta}_n - \theta) \leq \chi_{p;\alpha}^2\} \tag{5.17}$$

is a $(1 - \alpha)100\%$ approximate confidence region for θ.

The geometry of the confidence region in (5.17) might be quite complicated. However, similar to the single parameter case, one can use an additional approximation replacing $V(\theta)$ in (5.17) by any of its consistent estimators, in particular, by $V(\hat{\theta}_n)$. A corresponding $(1 - \alpha)100\%$ approximate confidence region for θ will then be

$$\{\theta : n(\hat{\theta}_n - \theta)^\top V^{-1}(\hat{\theta}_n)(\hat{\theta}_n - \theta) \leq \chi_{p;\alpha}^2\}. \tag{5.18}$$

The asymptotic confidence region (5.18) may be less accurate than (5.17) but, unlike the latter, has clear geometry: it is a p-dimensional *ellipsoid* with a center at $\hat{\theta}_n$ and axes defined by the eigenvectors and eigenvalues of $V(\hat{\theta}_n)$.

The following theorem extends Theorem 5.9 and shows that an MLE is a CAN estimator for the multiparameter case as well:

Theorem 5.12 Let $Y_1, ..., Y_n \sim f_\theta(y)$, $\theta = (\theta_1, ..., \theta_p) \in \Theta$ and $\hat{\theta}_n$ is the MLE of θ. Then, under general "regularity conditions" on f_θ, $\hat{\theta}_n$ is a CAN estimator of θ with the asymptotic covariance matrix $I^*(\theta)^{-1}$, where $I^*(\theta)$ is the Fisher information matrix based on a single observation: $I^*(\theta) = E\left(\nabla \ln f_\theta(y) \cdot \nabla \ln f_\theta(y)^\top\right) = -E\left(\frac{\partial^2 \ln f_\theta(y)}{\partial\theta\partial\theta^\top}\right)$ (see Section 2.7.3).

We refer to Bickel and Doksum (2015, Theorem 6.2.2) for formulation of regularity conditions and the proof.

Exploiting Theorem 5.12 in (5.17) and (5.18) one can construct approximate confidence regions for θ based on its MLE estimator $\hat{\theta}_n$, namely,

$$\{\theta : (\hat{\theta}_n - \theta)^\top I(\theta)(\hat{\theta}_n - \theta) \leq \chi^2_{p;\alpha}\}$$

or a Wald (ellipsoid-type) confidence region

$$\{\theta : (\hat{\theta}_n - \theta)^\top I(\hat{\theta}_n)(\hat{\theta}_n - \theta) \leq \chi^2_{p;\alpha}\}. \tag{5.19}$$

Example 5.17 (continuation of Example 3.10) Let $Y_1, ..., Y_n \sim \mathcal{N}(\mu, \sigma^2)$. In Example 3.10, we derived the conservative Bonferroni CR_B and the exact CR_E $(1 - \alpha)100\%$ confidence regions for the unknown μ and σ. Now we construct the corresponding approximate confidence regions for large samples.

Recall that the MLE $\hat{\mu}_n = \bar{Y}_n$ and $\hat{\sigma}_n^2 = \frac{1}{n}\sum_{i=1}^n(Y_i - \bar{Y}_n)^2$ (see Example 2.7), and the Fisher information matrix $I(\mu, \sigma) = \begin{pmatrix} \frac{n}{\sigma^2} & 0 \\ 0 & \frac{2n}{\sigma^2} \end{pmatrix}$ (see Example 2.21). The resulting approximate $(1 - \alpha)100\%$ confidence regions are then

$$CR_{A1} = \left\{(\mu, \sigma) : \frac{n}{\sigma^2}(\bar{Y}_n - \mu)^2 + \frac{2n}{\sigma^2}(\hat{\sigma}_n - \sigma)^2 \leq \chi^2_{2,\alpha}\right\} \tag{5.20}$$

and

$$CR_{A2} = \left\{(\mu, \sigma) : \frac{n}{\hat{\sigma}_n^2}(\bar{Y}_n - \mu)^2 + \frac{2n}{\hat{\sigma}^2}(\hat{\sigma}_n - \sigma)^2 \leq \chi^2_{2,\alpha}\right\}. \tag{5.21}$$

See Figure 5.4 for comparison of the approximate CR_{A1} and Wald CR_{A2} confidence regions with CR_B and CR_E confidence regions from Example 3.10.

Example 5.18 (continuation of Example 5.17) We continue the previous Example 5.17 to derive an asymptotic $(1 - \alpha)\%$ Wald confidence interval for the ratio μ/σ. We have

$$\sqrt{n}\begin{pmatrix} \bar{Y}_n - \mu \\ \hat{\sigma}_n - \sigma \end{pmatrix} \xrightarrow{\mathscr{D}} \mathcal{N}\left(\mathbf{0}, I^*(\mu, \sigma)^{-1}\right),$$

where $I^*(\mu, \sigma)^{-1} = \begin{pmatrix} \sigma^2 & 0 \\ 0 & \frac{\sigma^2}{2} \end{pmatrix}$. Applying Theorem 5.11 for $g(\mu, \sigma) = \frac{\mu}{\sigma}$, one has $\nabla g(\mu, \sigma) = \left(\frac{1}{\sigma}, -\frac{\mu}{\sigma^2}\right)^\top$ and

$$\sqrt{n}\left(\frac{\bar{Y}_n}{\hat{\sigma}_n} - \frac{\mu}{\sigma}\right) \xrightarrow{\mathscr{D}} \mathcal{N}\left(0, 1 + \frac{\mu^2}{2\sigma^2}\right)$$

Thus, the $(1 - \alpha)\%$ Wald confidence interval for μ/σ is

$$\frac{\bar{Y}_n}{\hat{\sigma}_n} \pm z_{\alpha/2}\frac{1}{\sqrt{n}}\sqrt{1 + \frac{1}{2}\left(\frac{\bar{Y}_n}{\hat{\sigma}_n}\right)^2}$$

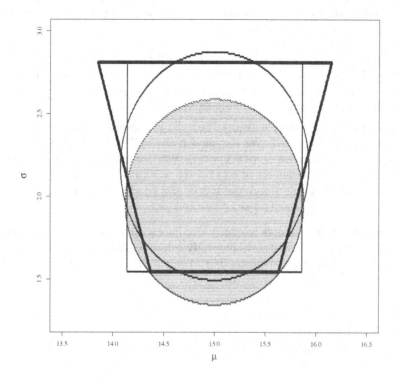

Figure 5.4: Four confidence regions for a $\mathcal{N}(\mu, \sigma^2)$ sample, where $n = 30$, $\bar{Y}_{30} = 15$, $s^2 = 4$, and $\alpha = 0.05$: the Bonferroni's CR_B from (3.5) (rectangle), the exact CR_E from (3.6) (trapezoid), the asymptotic CR_{A1} from (5.20), and the asymptotic Wald CR_{A2} from (5.21) (gray ellipsoid).

To conclude this section we mention that inverting the asymptotic confidence region (5.19) yields the multiparameter Wald test of asymptotic level α for testing

$$H_0 : \theta = \theta_0 \quad vs. \quad H_1 : \theta_j \neq \theta_{0j} \text{ for at least one } j$$

that rejects the null hypothesis if

$$(\hat{\theta}_n - \theta_0)^\top I(\hat{\theta}_n)(\hat{\theta}_n - \theta_0) > \chi^2_{p;\alpha}. \tag{5.22}$$

5.10 *Asymptotic properties of *M*-estimators

In this section we extend the asymptotic results for MLEs to general *M*-estimators. Recall that for a given function $\rho(\theta, y)$, an *M*-estimator $\hat{\theta}_n$ is the minimizer of $\sum_{i=1}^n \rho(\theta, y_i)$. As we have seen in Section 2.5, a family of *M*-estimators includes various known estimators corresponding to different ρ and an MLE, in particular, for $\rho(\theta, y) = -\ln f_\theta(y)$.

Consider $Y_1, \ldots, Y_n \sim f_\theta(y)$ and let θ_0 be the (unknown) true value of θ. For a given $\rho(\theta, y)$ define

$$\theta^* = \arg\min_{\theta \in \Theta} E_{\theta_0} \rho(\theta, Y) \tag{5.23}$$

We now establish that under mild conditions, the *M*-estimator $\hat{\theta}_n$ is a consistent and CAN estimator of θ^*. In particular, for MLE, $\theta^* = \theta_0$ (see the proof of Theorem 5.9) but it is not true in a general case.

Theorem 5.13 Let $Y_1, \ldots, Y_n \sim f_\theta(y)$, $\theta \in \Theta$ and $\hat{\theta}_n$ is the unique M-estimator of θ w.r.t. $\rho(\theta, y)$. Under mild assumptions similar to those required for consistency of MLEs in Theorem 5.9, $\hat{\theta}_n$ is a consistent estimator of θ^* defined in (5.23).

Proof. The proof goes along the lines of Theorem 5.9 for MLEs.

By the WLLN, $\frac{1}{n}\sum_{i=1}^{n} \rho(\theta, Y_i) \xrightarrow{p} E_{\theta_0}\rho(\theta, Y)$ for any $\theta \in \Theta$. The required assumptions ensure a stronger uniform convergence and imply that a minimizer of $\frac{1}{n}\sum_{i=1}^{n} \rho(\theta, Y_i)$ converges in probability to the minimizer of $E_{\theta_0}\rho(\theta, Y)$, which is, by its definition, θ^*. $\qquad\square$

Example 5.19 (consistency of a sample mean) From the WLLN we already know that a sample mean $\bar{Y}_n$ is a consistent estimator of EY. We now show it also by Theorem 5.13.

Recall that $\bar{Y}_n$ is an M-estimator w.r.t. $\rho(\theta, y) = (y - \theta)^2$ (see Example 2.11). To find the corresponding θ^* we need to find a minimizer of $E_{\theta_0}(Y - \theta)^2 = E_{\theta_0}Y^2 - 2\theta E_{\theta_0}Y + \theta^2$ w.r.t. θ. Differentiating and equating the derivative to zero yields $2(\theta - E_{\theta_0}Y) = 0$ and, therefore, $\theta^* = EY$. From Theorem 5.13 it then follows that $\bar{Y}_n \xrightarrow{p} EY$.

Example 5.20 (consistency of a sample median) Consider a sample median $Y_{(0.5)n}$. For simplicity assume that the sample size n is odd (asymptotically it does not really matter), so that the sample median is unique. $Y_{(0.5)n}$ is an M-estimator for $\rho(\theta, y) = |y - \theta|$ (see Example 2.12). We now find θ^* that minimizes $E_{\theta_0}|Y - \theta|$ w.r.t. θ. We have

$$
\begin{aligned}
E_{\theta_0}|Y - \theta| &= \int |y - \theta| f_{\theta_0}(y)dy \\
&= \int_{-\infty}^{\theta} (\theta - y) f_{\theta_0}(y)dy + \int_{\theta}^{\infty} (y - \theta) f_{\theta_0}(y)dy \\
&= \theta F_{\theta_0}(\theta) - \int_{-\infty}^{\theta} y f_{\theta_0}(y)dy + \int_{\theta}^{\infty} y f_{\theta_0}(y)dy - \theta(1 - F_{\theta_0}(\theta)),
\end{aligned}
\tag{5.24}
$$

where $F_{\theta_0}(\cdot)$ is the cdf of Y. Differentiating the RHS of (5.24) w.r.t. θ, a straightforward calculus implies that θ^* satisfies $2F_{\theta_0}(\theta^*) - 1 = 0$ and, therefore, $\theta^* = Med(Y)$. Hence, under the regularity conditions of Theorem 5.13, the sample median is a consistent estimator of the population median.

Assume, in addition, that ρ is differentiable w.r.t. θ with $\psi(\theta, y) = \rho'(\theta, y)$.

Theorem 5.14 Let $Y_1, \ldots, Y_n \sim f_\theta(y)$, $\theta \in \Theta$ and $\hat{\theta}_n$ is the M-estimator of θ w.r.t a twice differentiable ρ. Let θ_0 be the true value of θ and θ^* defined in (5.23).

Then, under general regularity conditions on f_θ similar to those in Theorem 5.10 for MLEs and on ρ, $\hat{\theta}_n$ is a CAN estimator of θ^* with the asymptotic variance

$$
v^2 = \frac{E_{\theta_0}(\psi(\theta^*, Y)^2)}{(E_{\theta_0}\psi'(\theta^*, Y))^2},
$$

that is,

$$
\sqrt{n}(\hat{\theta}_n - \theta^*) \xrightarrow{\mathscr{D}} \mathscr{N}(0, v^2)
$$

as $n \to \infty$.

Sketch of the proof. The idea of the proof is similar to that of Theorem 5.10.

By Taylor's expansion we have

$$
\frac{1}{n}\sum_{i=1}^{n} \psi(\hat{\theta}_n, Y_i) = \frac{1}{n}\sum_{i=1}^{n} \psi(\theta^*, Y_i) + \frac{1}{n}\sum_{i=1}^{n} \psi'(\theta^*, Y_i)(\hat{\theta}_n - \theta^*) + R_n,
$$

where R_n is the remainder term. $\hat{\theta}_n$ is an M-estimator w.r.t. ρ and, thus, $\frac{1}{n}\sum_{i=1}^{n} \psi(\hat{\theta}_n, Y_i) = 0$. Hence,

$$
\sqrt{n}(\hat{\theta}_n - \theta^*) = -\sqrt{n}\frac{\frac{1}{n}\sum_{i=1}^{n} \psi(\theta^*, Y_i)}{\frac{1}{n}\sum_{i=1}^{n} \psi'(\theta^*, Y_i)} + \sqrt{n}R_n
$$

Note that since θ^* is the minimizer of $E_{\theta_0}\rho(\theta,Y)$, $E_{\theta_0}\psi(\theta^*,Y) = 0$ and, therefore, applying the CLT

$$\sqrt{n}\frac{1}{n}\sum_{i=1}^{n}\psi(\theta^*,Y_i) \xrightarrow{\mathscr{D}} \mathscr{N}(0, E_{\theta_0}\psi(\theta^*,Y)^2),$$

while by the WLLN,

$$-\frac{1}{n}\sum_{i=1}^{n}\psi'(\theta^*,Y_i) \xrightarrow{\text{p}} -E_{\theta_0}\psi'(\theta^*,Y)$$

The regularity conditions ensure that $\sqrt{n}R_n = o_P(1)$ and is negligible, and, thus, by Slutsky's Theorem 5.7,

$$\sqrt{n}(\hat{\theta}_n - \theta^*) \xrightarrow{\mathscr{D}} \mathscr{N}\left(0, \frac{E_{\theta_0}(\psi(\theta^*,Y)^2)}{(E_{\theta_0}\psi'(\theta^*,Y))^2}\right)$$

$\square$

Consider again an MLE $\hat{\theta}_n$. In this case $\rho(\theta,y) = -\ln f_{\theta}(y)$, $\theta^* = \theta_0$ (see above) and, therefore, $E_{\theta_0}\psi(\theta^*,Y)^2 = E_{\theta_0}((\ln f_{\theta_0}(Y))')^2 = I^*(\theta_0)$ and $E_{\theta_0}\psi'(\theta^*,Y) = E_{\theta_0}(\ln f_{\theta_0}(Y))'' = I^*(\theta_0)$, where $I^*(\theta_0)$ is the Fisher information number of f_{θ_0} (see Section 2.7.2). Theorem 5.14 yields then $\sqrt{n}(\hat{\theta}_n - \theta_0) \xrightarrow{\mathscr{D}} \mathscr{N}(0, I^*(\theta_0)^{-1})$ and essentially restores Theorem 5.10.

Example 5.21 (asymptotic normality of a sample mean) Asymptotic normality of a sample mean $\bar{Y}_n$ follows directly from the CLT. We now obtain it in a different way applying Theorem 5.14.

As we have seen, a sample mean $\bar{Y}_n$ is an *M*-estimator w.r.t. $\psi(\theta,y) = 2(\theta - y)$ and $\theta^* = EY$. In this case,

$$E_{\theta_0}\psi(\theta^*,Y)^2 = \int (y - \theta_0)^2 f_{\theta_0}(y)dy = Var(Y)$$

and

$$E_{\theta_0}\psi'(\theta^*,Y) = \int 1 f_{\theta_0}(y)dy = 1$$

implying $\sqrt{n}(\bar{Y}_n - EY) \xrightarrow{\mathscr{D}} \mathscr{N}(0, Var(Y))$.

Example 5.22 (asymptotic normality of a sample median) Recall that a sample median corresponds to $\psi(\theta,y) = sign(\theta - y)$ and $\theta^* = Med(Y)$. Then,

$$E_{\theta_0}\psi(\theta^*,Y)^2 = \int (sign(Med(Y) - y))^2 f_{\theta_0}(y)dy = \int f_{\theta_0}(y)dy = 1$$

and by regularity conditions,

$$E_{\theta_0}\psi'(\theta^*,Y) = \left(E_{\theta_0}\psi(\theta,Y)\right)'_{\theta=\theta^*} = \left(\int sign(\theta - y)f_{\theta_0}(y)dy\right)'_{\theta=\theta^*}$$

$$= \left(\int_{-\infty}^{\theta} f_{\theta_0}(y)dy - \int_{\theta}^{\infty} f_{\theta_0}(y)dy\right)'_{\theta=\theta^*} = (2F_{\theta_0}(\theta) - 1)'_{\theta=\theta^*}$$

$$= 2f_{\theta_0}(Med(Y))$$

Thus, a sample median $Y_{(0.5)n}$ is a CAN estimator of $Med(Y)$ with the asymptotic variance $\frac{1}{4f_{\theta_0}(Med(Y))^2}$.

In particular, for a normal sample $Y_1,\ldots,Y_n \sim \mathscr{N}(\mu,\sigma^2)$, $EY = Med(Y) = \mu$ and we can compare a sample mean $\bar{Y}_n$ and a sample median $Y_{(0.5)n}$ as estimators of μ by their variances. Note that $f_{\mu}(\mu) = \frac{1}{\sqrt{2\pi}\sigma}$ and, therefore, $Var(Y_{(0.5)n}) \sim \frac{\pi}{2}\frac{\sigma^2}{n}$, which is $\frac{\pi}{2} \approx 1.57$ times larger than $Var(\bar{Y}_n) = \frac{\sigma^2}{n}$. Hence, estimating μ by a sample mean is about 1.57 times more accurate than by a sample median.

Similar to MLE, consistency and asymptotic normality of M-estimators can be straightforwardly extended to the multiparameter case for $\theta \in \mathbb{R}^p$. Similar to the one-parameter case, let $\Psi(\theta, y) = \nabla \rho(\theta, y)$ be the gradient of $\rho(\theta, Y)$ and $\theta^* = \arg\min_{\theta \in \Theta} E_{\theta_0} \rho(\theta, Y)$ as in (5.23). Then, under the required regularity conditions,

$$\sqrt{n}(\hat{\theta}_n - \theta^*) \xrightarrow{\mathscr{D}} \mathscr{N}(\mathbf{0}, V)$$

with the asymptotic covariance matrix

$$V = \left(E_{\theta_0} \nabla \Psi(\theta^*, Y)\right)^{-1} E_{\theta_0} \left(\Psi(\theta^*, Y)\Psi^T(\theta^*, Y)\right) \left(\left(E_{\theta_0} \nabla \Psi(\theta^*, Y)\right)^{-1}\right)^T,$$

where $\nabla \Psi(\theta, y)$ is the Jacobian matrix of $\Psi(\theta, y)$ (or, equivalently, the Hessian matrix of $\rho(\theta, y)$). See Bickel and Doksum, 2015, Section 6.2.1 for more details.

5.11   Score (Rao) asymptotic tests and confidence regions

In Section 5.8 we considered Wald asymptotic test for testing $H_0 : \theta = \theta_0$ vs. $H_1 : \theta \neq \theta_1$, where the resulting test statistic exploited the asymptotic normality of the MLE. In this section we consider another approach based on asymptotic distribution of a score function $l'(\theta, \mathbf{y}) = \sum_{i=1}^{n}(\ln f_\theta(y_i))'$ defined in Section 2.7.2.

In the proof of the Cramer-Rao lower bound (see Section 2.7.2) we showed that for the true θ_0, $El'(\theta_0, \mathbf{Y}) = 0$ and $Var(l'(\theta_0, \mathbf{Y})) = I(\theta_0) = nI^*(\theta_0)$. Thus, by the CLT, under $H_0 : \theta = \theta_0$,

$$\sqrt{n} \frac{1}{n} l'(\theta_0, \mathbf{Y}) \xrightarrow{\mathscr{D}} \mathscr{N}(0, I^*(\theta_0))$$

In other words, under the null hypothesis, $l'(\theta_0, \mathbf{Y})$ is approximately normal with zero mean and the variance $I(\theta_0)$, and H_0 is rejected if

$$\frac{|l'(\theta_0)|}{\sqrt{I(\theta_0)}} > z_{\alpha/2} \tag{5.25}$$

The test (5.25) is called the *score* (aka *Rao*) test and has an asymptotic level α.

Example 5.23 (score test for the binomial parameter) Let $Y_1, \ldots, Y_n$ be a series of Bernoulli trials $\mathscr{B}(1, p)$ and we want to test $H_0 : p = p_0$ vs. $H_1 : p \neq p_0$. Recall that the score function $l'(p) = \frac{\sum_{i=1}^{n} y_i - np}{p(1-p)}$ and the Fisher information number $I(p) = \frac{n}{p(1-p)}$. Applying the score test we reject the null hypothesis if

$$\frac{|\sum_{i=1}^{n} Y_i - np_0|}{p_0(1-p_0)} \cdot \sqrt{\frac{p_0(1-p_0)}{n}} = \frac{|\sum_{i=1}^{n} Y_i - np_0|}{\sqrt{np_0(1-p_0)}} = \frac{|\hat{p}_n - p_0|}{\sqrt{\frac{p_0(1-p_0)}{n}}} > z_\alpha/2, \tag{5.26}$$

where $\hat{p}_n = \frac{\sum_{i=1}^{n} Y_i}{n}$ is the MLE of p.

Compare (5.26) with the corresponding Wald test that rejects the null if

$$\sqrt{I(\hat{p}_n)}|\hat{p}_n - p_0| = \frac{|\hat{p}_n - p_0|}{\sqrt{\frac{\hat{p}_n(1-\hat{p}_n)}{n}}} > z_\alpha/2$$

If H_0 is true, the MLE $\hat{p}_n$ is expected to be close to p_0 and both tests should behave similarly.

In fact, similarity between Wald and score tests demonstrated in Example 5.23 is a common phenomenon. Indeed, since $l'(\hat{\theta}_n) = 0$,

$$l'(\theta_0) = l'(\theta_0) - l'(\hat{\theta}_n) \approx l''(\hat{\theta}_n)(\theta_0 - \hat{\theta}_n) \approx I(\hat{\theta}_n)(\hat{\theta}_n - \theta_0)$$

and $I(\hat{\theta}_n) \approx I(\theta_0)$ under the null. Hence,

$$\frac{|l'(\theta_0)|}{\sqrt{I(\theta_0)}} \approx \sqrt{I(\hat{\theta}_n)}|\hat{\theta}_n - \theta_0|$$

implying that the score and Wald test statistic are asymptotically close (first-order equivalent).

Sometimes both tests may even coincide:

Example 5.24 (score test for exponential distribution) Let $t_1,\ldots,t_n \sim \exp(\theta)$ and derive the score test for testing $H_0 : \theta = \theta_0$ vs. $H_1 : \theta \neq \theta_0$ at level α. We already know that $l'(\theta) = n(\frac{1}{\theta} - \bar{t}_n)$ and $I(\theta) = \frac{n}{\theta^2}$ (see Example 5.15). Thus, the score test rejects the null hypothesis if

$$\frac{n|\frac{1}{\theta_0} - \bar{t}_n|}{\sqrt{n/\theta_0^2}} = \sqrt{n}|\bar{t}_n\theta_0 - 1| > z_{\alpha/2}$$

and coincides with the corresponding Wald test (5.16).

Similar to Wald tests, the duality between hypothesis testing and confidence regions can be used to derive the score (Rao) confidence (generally) region from (5.25) as

$$\left\{\theta : \frac{|l'(\theta)|}{\sqrt{I(\theta)}} \leq z_{\alpha/2}\right\}.$$

Score tests allow a direct extension to the multiparameter case, where $\theta \in \mathbb{R}^p$: by the same arguments, $E\nabla l(\theta_0, \mathbf{Y}) = \mathbf{0}$ and $Var(\nabla l(\theta_0, \mathbf{Y})) = I(\theta_0)$, where $I(\theta)$ is the Fisher information matrix (see Section 2.7.3). Thus, by the CLT, under $H_0 : \theta = \theta_0$, $\sqrt{n}\frac{1}{n}\nabla l(\theta_0, \mathbf{Y}) \xrightarrow{\mathscr{D}} \mathscr{N}(0, I^*(\theta_0))$, or, equivalently, $n\nabla l(\theta_0, \mathbf{Y})^\top I^*(\theta_0)^{-1}\nabla l(\theta_0, \mathbf{Y}) \xrightarrow{\mathscr{D}} \chi_p^2$, where $I(\theta_0) = nI^*(\theta_0)$, and the score test rejects the null hypothesis if

$$n\nabla l(\theta_0, \mathbf{Y})^\top I^*(\theta_0)^{-1}\nabla l(\theta_0, \mathbf{Y}) > \chi_{p,\alpha}^2.$$

5.12 Asymptotic distribution of the GLRT, Wilks' theorem

In Section 4.3.3 we presented the GLRT for testing two general composite hypotheses

$$H_0 : \theta \in \Theta_0 \quad vs. \quad H_1 : \theta \in \Theta_1,$$

where $\Theta_0 \cap \Theta_1 = \emptyset$ and $\Theta_0 \cup \Theta_1 = \Theta$. Recall that the null hypotheses is rejected if the GLR

$$\lambda(\mathbf{y}) = \frac{\sup_{\theta \in \Theta} L(\theta, \mathbf{y})}{\sup_{\theta \in \Theta_0} L(\theta, \mathbf{y})} \geq C,$$

where the critical value C satisfies $\sup_{\theta \in \Theta_0} P_\theta(\lambda(\mathbf{Y}) \geq C) = \alpha$ (with the usual modifications for a discrete distribution).

The GLR $\lambda(\mathbf{Y})$ might be a quite complicated statistic and the main difficulty in practical implementation of the GLRT becomes the derivation of its distribution under the null hypothesis to calculate a critical value C. As we have seen in Section 4.3.3 sometimes it is possible to find an equivalent test statistic $T(\mathbf{Y})$ whose distribution is "sufficiently simple." Unfortunately, in a variety of situations it is not the case. However, it happens that for quite a general setup, the *asymptotic* distribution of $\lambda(\mathbf{Y})$ under the null hypothesis is remarkably simple:

Theorem 5.15 (Wilks' theorem) Let $Y_1, ..., Y_n$ be a random sample from $f_\theta(\mathbf{y})$, where the p-dimensional parameter $\theta = (\theta_1, ..., \theta_p) \in \Theta$, and test a null hypothesis on its r components (assumed to be the first without loss of generality) of the form

$$H_0 : \theta_1 = \theta_{10}, ..., \theta_r = \theta_{r0} \quad (1 \le r \le p)$$

against a general alternative H_1 with no restrictions in $\theta_1, ..., \theta_p$.

Consider the corresponding GLR

$$\lambda_n(\mathbf{y}) = \frac{\sup_{\theta \in \Theta} L(\theta; \mathbf{y})}{\sup_{\theta \in \Theta: \theta_1 = \theta_{10}, ..., \theta_r = \theta_{r0}} L(\theta; \mathbf{y})}.$$

Then, if H_0 is true, under general "regularity conditions" on $f_\theta(\mathbf{Y})$ similar to those required for the normal consistency of the MLE in Theorem 5.9 of Section 5.8, [4]

$$2 \ln \lambda_n(\mathbf{Y}) \xrightarrow{\mathscr{D}} \chi_r^2$$

as $n \to \infty$.

Sketch of the proof. For simplicity of exposition we consider in detail the single parameter case $p = r = 1$. Let $\hat{\theta}_n$ be the MLE of θ. Using Taylor's expansion for the log-likelihood $l(\theta; \mathbf{y})$ we have

$$2 \ln \lambda_n(\mathbf{y}) = 2 \left(l(\hat{\theta}_n; \mathbf{y}) - l(\theta_0; \mathbf{y}) \right) = 2l'(\hat{\theta}_n; \mathbf{y})(\theta_0 - \hat{\theta}_n) - l''(\theta_n^*; \mathbf{y})(\theta_0 - \hat{\theta}_n)^2,$$

where θ_n^* lies between $\hat{\theta}_n$ and θ_0. Since $\hat{\theta}_n$ is an MLE, $l'(\hat{\theta}_n; \mathbf{y}) = 0$ and, therefore,

$$2 \ln \lambda(\mathbf{y}) = -l''(\theta_n^*; \mathbf{y})(\theta_0 - \hat{\theta}_n)^2 = -\frac{l''(\theta_0; \mathbf{y})}{n} \cdot \frac{l''(\theta_n^*; \mathbf{y})}{l''(\theta_0; \mathbf{y})} \cdot \left(\sqrt{n}(\hat{\theta}_n - \theta_0) \right)^2.$$

As we have shown in the sketch of the proof of Theorem 5.9, $-\frac{l''(\theta_0; \mathbf{y})}{n} \xrightarrow{\text{p}} I^*(\theta_0)$, while under the required regularity conditions $\frac{l''(\theta_n^*; \mathbf{y})}{l''(\theta_0; \mathbf{y})} \xrightarrow{\text{p}} 1$. Finally, under the null hypothesis $\theta = \theta_0$, Theorem 5.9 implies that $\sqrt{nI^*(\theta_0)}(\hat{\theta}_n - \theta_0)$ is asymptotically standard normal $\mathcal{N}(0, 1)$ and, therefore, $nI^*(\theta_0)(\hat{\theta}_n - \theta_0)^2 \xrightarrow{\mathscr{D}} \chi_1^2$. The proof is completed by applying Slutsky's theorem (Theorem 5.7).

The extension to the multiparameter case $p > 1$ is straightforward by applying the corresponding multivariate Taylor's expansion and asymptotic normality of MLE (see Theorem 5.12). $\square$

Wilks' Theorem 5.15 immediately yields the asymptotic GLRT for testing $H_0 : \theta_1 = \theta_{10}, ..., \theta_r = \theta_{r0}$ against a general alternative H_1: the null is rejected if

$$2 \ln \lambda_n(\mathbf{Y}) \ge \chi_{r;\alpha}^2.$$

Example 5.25 (test for comparing two Poisson means) Let $X_1, ..., X_n \sim Pois(\mu_1)$ and $Y_1, ..., Y_m \sim Pois(\mu_2)$ be two independent Poisson samples. We want to test that equality of their means: $H_0 : \mu_1 = \mu_2$ against $H_1 : \mu_1 \neq \mu_2$.

Note that the null hypothesis can be equivalently re-written in the form of Theorem 5.15 by one-to-one reparametrization: define $\theta_1 = \mu_1 / \mu_2$ and $\theta_2 = \mu_1$. Then, $H_0 : \theta_1 = 1$ vs. $H_1 : \theta_1 \neq 1$. The joint likelihood $L(\mu_1, \mu_2; \mathbf{x}, \mathbf{y})$ of two the independent Poisson samples is

$$L(\mu_1, \mu_2; \mathbf{x}, \mathbf{y}) = \frac{e^{-n\mu_1} \mu_1^{\sum_{i=1}^n x_i}}{\prod_{i=1}^n x_i!} \cdot \frac{e^{-m\mu_2} \mu_2^{\sum_{j=1}^m y_j}}{\prod_{j=1}^m y_j!}.$$

---

[4] In particular, the theorem can fail if the null value θ_0 is on the boundary of Θ.

Simple calculus yields the obvious MLE $\hat{\mu}_1 = \bar{x}$, $\hat{\mu}_2 = \bar{y}$ and

$$L(\hat{\mu}_1, \hat{\mu}_2; \mathbf{x}, \mathbf{y}) = \frac{e^{-(n\bar{x}+m\bar{y})} \bar{x}^{n\bar{x}} \bar{y}^{m\bar{y}}}{\prod_{i=1}^{n} x_i! \prod_{j=1}^{m} y_j!}.$$

Under the null hypothesis, the data can essentially be viewed as a single Poisson sample of size $n+m$ with a common mean μ_0. Thus,

$$\hat{\mu}_0 = \frac{\sum_{i=1}^{n} x_i + \sum_{j=1}^{m} y_j}{n+m} = \frac{n\bar{x}+m\bar{y}}{n+m}$$

and

$$L(\hat{\mu}_0; \mathbf{x}, \mathbf{y}) = \frac{e^{-(n\bar{x}+m\bar{y})}}{\prod_{i=1}^{n} x_i! \prod_{j=1}^{m} y_j!} \left(\frac{n\bar{x}+m\bar{y}}{n+m} \right)^{n\bar{x}+m\bar{y}}.$$

Finally,

$$2\ln\lambda(\mathbf{x}, \mathbf{y}) = 2 \left(n\bar{x}\ln\bar{x} + m\bar{y}\ln\bar{y} - (n\bar{x}+m\bar{y})\ln\left(\frac{n\bar{x}+m\bar{y}}{n+m} \right) \right) \tag{5.27}$$

and the null hypothesis is rejected if the RHS of (5.27) is larger than $\chi^2_{1;\alpha}$. Wilks' Theorem 5.15 ensures that for large sample sizes n and m the significance level of such a test is approximately α.

Example 5.26 (goodness-of-fit test) Let $Y_1, ..., Y_n$ be a random sample from a discrete distribution with a finite set of k values $s_1, ..., s_k$ and the corresponding unknown probabilities $p_j = P(Y = s_j)$. We want to test whether a particular (discrete) distribution with $P(Y = s_j) = \pi_j$, $j = 1, ..., k$ fits the data (goodness-of-fit test), that is,

$$H_0 : p_j = \pi_j, \ j = 1, ..., k,$$

against a general alternative of any other distribution.

Similar to the previous Example 5.25, H_0 can be equivalently re-written in the form of Wilks' Theorem 5.15 by re-parametrization $\theta_j = p_j - \pi_j$: $H_0 : \theta_1 = ... = \theta_k = 0$.

Define the *observed frequencies* $O_j = \#\{Y_i : Y_i = s_j\}$, $j = 1, ..., k$-the number of observations Y_i's that took the value s_j, where obviously $\sum_{j=1}^{k} O_j = n$. In fact, since $\sum_{j=1}^{k} p_j = 1$, then $p_k = 1 - \sum_{j=1}^{k-1} p_j$ and the likelihood

$$L(\mathbf{p}; \mathbf{y}) = \prod_{j=1}^{k} p_j^{O_j} = \prod_{j=1}^{k-1} \pi_j^{O_j} \cdot (1 - \sum_{j=1}^{k-1} p_j)^{O_k}$$

has only $k-1$ parameters $p_1, ..., p_{k-1}$. To find the MLE for p_j's we need to maximize the log-likelihood $l(\mathbf{p}; \mathbf{y}) = \sum_{j=1}^{k-1} O_j \ln p_j + O_k \ln(1 - \sum_{j=1}^{k-1} p_j)$ w.r.t. $p_1, ..., p_{k-1}$. We have

$$\frac{\partial l(\mathbf{p}; \mathbf{y})}{\partial p_j} = \frac{O_j}{p_j} - \frac{O_k}{1 - \sum_{j=1}^{k-1} p_j} = \frac{O_j}{p_j} - \frac{O_k}{p_k} = 0, \ j = 1, ..., k-1$$

Solving the above equations yields $\hat{p}_j = \frac{O_j}{n}$, $j = 1, ..., k$ with a clear intuitive meaning: the unknown probability is estimated by the corresponding sample proportion. Hence,

$$L(\hat{\mathbf{p}}; \mathbf{y}) = \prod_{j=1}^{k} \left(\frac{O_j}{n} \right)^{O_j}.$$

Under the null hypothesis, $p_j = \pi_j$ and

$$L(\pi; \mathbf{y}) = \prod_{j=1}^{k} \pi_j^{O_j}.$$

The resulting GLR

$$\lambda(\mathbf{y}) = \frac{L(\hat{\mathbf{p}};\mathbf{y})}{L(\pi;\mathbf{y})} = \prod_{j=1}^{k} \left(\frac{O_j}{n\pi_j}\right)^{O_j} = \prod_{j=1}^{k} \left(\frac{O_j}{E_j}\right)^{O_j},$$

where $E_j = n\pi_j$ is the *expected frequency* of s_j under H_0, and therefore

$$2\ln\lambda(\mathbf{y}) = 2\sum_{j=1}^{k} O_j \ln\left(\frac{O_j}{E_j}\right). \tag{5.28}$$

Note that since $\sum_{j=1}^{k}\pi_j = 1$, there are only $k-1$ restrictions under H_0 and the (asymptotic) GLRT for the goodness-of-fit rejects then the null if

$$2\sum_{j=1}^{k} O_j \ln\left(\frac{O_j}{E_j}\right) \geq \chi^2_{k-1;\alpha}. \tag{5.29}$$

The RHS in (5.28) can be approximated by Taylor's expansion of $\ln O_j$ around E_j:

$$2\sum_{j=1}^{k} O_j(\ln O_j - \ln E_j)$$

$$\approx 2\sum_{j=1}^{k} \left((O_j - E_j) + E_j\right)\left(\frac{O_j - E_j}{E_j} - \frac{(O_j - E_j)^2}{2E_j^2}\right) \tag{5.30}$$

$$= 2\sum_{j=1}^{k} O_j - 2\sum_{j=1}^{k} E_j + \sum_{j=1}^{k} \frac{(O_j - E_j)^2}{E_j} - \sum_{j=1}^{k} \frac{(O_j - E_j)^3}{E_j^2},$$

where $\sum_{j=1}^{k} O_j = \sum_{j=1}^{k} E_j = n$. Ignoring the last term in the RHS of (5.30), asymptotically "negligible" under H_0, implies

$$2\ln\lambda(\mathbf{y}) \approx \sum_{j=1}^{k} \frac{(O_j - E_j)^2}{E_j}$$

which is the well-known *Pearson's statistic* X^2. The corresponding Pearson's χ^2-test for goodness-of-fit rejects H_0 if

$$X^2 = \sum_{j=1}^{k} \frac{(O_j - E_j)^2}{E_j} \geq \chi^2_{k-1;\alpha}. \tag{5.31}$$

Thus, the Pearson's χ^2-test (5.31) is just an approximation of the asymptotic GLRT (5.29).

Moreover, after some algebra it can be shown that (5.31) equals also to a score (Rao) goodness-of-fit test in this case (see Bickel and Doksum, 2015, Section 6.4.1 for details). In practice, both GLR and Peasron's X^2 test statistics are usually similar.

Consider an example, where we want to check a computer's random digit generator, which is supposed to provide uniformly distributed digits from 0 to 9. We generated a random sample of 200 digits on the computer and the obtained observed frequencies are summarized in Table 5.2. The null hypothesis is that the distribution of generated digits is uniform, that is, the probability of each digit is 0.1 and all the expected frequencies are therefore equal to 20.

Table 5.2: Observed and expected frequencies for the random digits generator.

Digit	0	1	2	3	4	5	6	7	8	9
Observed frequency	25	17	19	23	21	17	14	17	21	26
Expected frequency	20	20	20	20	20	20	20	20	20	20

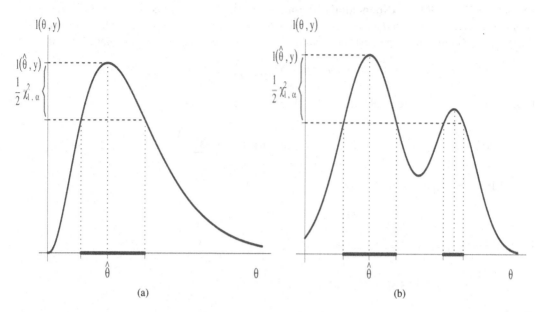

Figure 5.5: Asymptotic confidence interval (a) and region in a more general case (b)", i.e., (a) corresponds to an interval, while (b) to a more general region.

The GLRT statistic $2\ln\lambda(\mathbf{y}) = 2\sum_{j=0}^{9} O_j \ln\left(\frac{O_j}{E_j}\right) = 6.81$ and Pearson's statistic $X^2 = \sum_{j=0}^{9} \frac{(O_j - E_j)^2}{E_j} = 6.80$ are very close. The resulting p-value $= P(\chi_9^2 \geq 6.80) = 0.66$ and the null hypothesis is not rejected at any reasonable level of α.

We can now use again the duality between confidence intervals and two-sided hypothesis testing together with Wilks' Theorem 5.15 to construct an asymptotic confidence interval for θ based on the GLR $\lambda(\mathbf{Y})$. Indeed, the asymptotic GLRT test does not reject $H_0 : \theta = \theta_0$ against $H_1 : \theta \neq \theta_0$ at level α if $2\ln\lambda(\mathbf{y}) = 2(l(\hat{\theta};\mathbf{y}) - l(\theta_0;\mathbf{y})) \leq \chi_{1;\alpha}^2$ and, therefore, the corresponding $(1-\alpha)100\%$ asymptotic confidence interval (region in a general case) is then

$$S(\mathbf{y}) = \left\{ \theta : l(\hat{\theta};\mathbf{y}) - l(\theta;\mathbf{y}) \leq \frac{1}{2}\chi_{1;\alpha}^2 \right\} \qquad (5.32)$$

(see Figure 5.5).

Which of the Wald, score (Rao) and GLR approaches for deriving asymptotic confidence intervals and tests is more accurate? There is no general answer. They are based on different asymptotical approximations. The first one exploits the asymptotic normality of the MLEs, the second is based on asymptotic normality of the score function, while the third uses the asymptotic approximation for a distribution of the GLR. Typically, however, all the three approximations are asymptotically first-order equivalent. In Section 5.11 we showed the similarity between Wald and score tests. Following arguments similar to those in the proof of Wilks' Theorem 5.15, under the regularity conditions one also has

$$2\ln\lambda_n(\mathbf{y}) = 2(l(\hat{\theta}_n;\mathbf{y}) - l(\theta;\mathbf{y})) \approx -l''(\hat{\theta}_n)(\hat{\theta}_n - \theta)^2 \approx I(\hat{\theta}_n)(\hat{\theta}_n - \theta)^2.$$

Hence, the asymptotic GLR confidence interval (5.32) is close to

$$\left\{ \theta : I(\hat{\theta}_n)(\hat{\theta}_n - \theta)^2 \leq \chi_{1;\alpha}^2 \right\} = \left\{ \theta : \sqrt{I(\hat{\theta}_n)}|\hat{\theta}_n - \theta| \leq z_{\alpha/2} \right\},$$

which is the Wald confidence interval (5.14). To compare the goodness of the three approximations one should investigate the corresponding higher-order asymptotics.

Example 5.27 (GLRT for exponential distribution) Test $H_0 : \theta = \theta_0$ vs. $H_1 : \theta \neq \theta_0$ from a random sample $t_1, \ldots, t_n \sim \exp(\theta)$ at level α. As we have already seen in Example 5.24, both Wald and score tests reject the null hypothesis if $\sqrt{n}|\bar{t}_n \theta_0 - 1| > z_\alpha$ or, re-writing it in equivalent form, if

$$n(\bar{t}_n \theta_0 - 1)^2 > \chi^2_{1,\alpha} \tag{5.33}$$

Derive now the asymptotic GLRT. The MLE $\hat{\theta}_n = \bar{t}_n^{-1}$ and, therefore,

$$2\ln \lambda_n(\mathbf{t}) = 2(l(\bar{t}_n^{-1}) - l(\theta_0)) = 2\left(-n\ln \bar{t}_n - n - n\ln \theta_0 + \theta_0 \sum_{i=1}^{n} t_i \right) = 2n\left(\bar{t}_n \theta_0 - 1 - \ln(\bar{t}_n \theta_0) \right)$$

The GLRT rejects then the null hypothesis if

$$2n\left(\bar{t}_n \theta_0 - 1 - \ln(\bar{t}_n \theta_0) \right) > \chi^2_{1,\alpha}$$

Using Taylor's approximation we have

$$\ln(\bar{t}_n \theta_0) = \ln((\bar{t}_n \theta_0 - 1) + 1) \approx \bar{t}_n \theta_0 - 1 - \frac{(\bar{t}_n \theta_0 - 1)^2}{2}$$

Thus,

$$2\ln \lambda_n(\mathbf{t}) \approx 2n\frac{(\bar{t}_n \theta_0 - 1)^2}{2} = n(\bar{t}_n \theta_0 - 1)^2$$

which is the Wald and score tests (5.33).

5.13   Exercises

Exercise 5.1 Let $Y_1, \ldots, Y_n$ be a random sample.

1. Show that the j-th sample moment $m_j = \frac{1}{n}\sum_{i=1}^{n} Y_i^j$ is a consistent estimator of the j-th moment $\mu_j = EY^j$.

2. Assume that $\mu_{2j} < \infty$. Show that m_j is a CAN estimator of μ_j with asymptotic variance $\mu_{2j} - \mu_j^2$.

Exercise 5.2 Continue Exercise 2.12. Check whether the MLE $\hat{\theta}_{MLE} = Y_{max}$ and the MME $\hat{\theta}_{MME} = 2\bar{Y}_n$ of θ in the uniform distribution $\mathcal{U}(0, \theta)$ are consistent in MSE. Which estimator converges faster?

Exercise 5.3 Let $Y_1, \ldots, Y_n \sim Pois(\lambda)$. In Example 2.23, we found the UMVUE $\hat{p} = 1 - (1 - \frac{1}{n})^{\sum_{i=1}^{n} Y_i}$ for $p = P(Y \geq 1)$. Is it consistent?

Exercise 5.4 Let $Y_1, \ldots, Y_n$ be a random sample with finite mean μ and variance σ^2. Find the asymptotic distributions of $\bar{Y}_n^2$ when $\mu \neq 0$ and $\mu = 0$, and $e^{-\bar{Y}_n}$.

Exercise 5.5 Construct an asymptotic $100(1 - \alpha)\%$ confidence interval for the odds $\theta = \frac{p}{1-p}$ from the results of a series of n Bernoulli trials.

Exercise 5.6 Continue Exercises 1.4 and 2.15. Find the asymptotic distribution of the MLE of p and construct the asymptotic $100(1 - \alpha)\%$ Wald confidence interval for p.

Exercise 5.7 The number of power failures in an electrical network per day is Poisson distributed with an unknown mean λ. During the last month (30 days), 5 power failures have been registered. Let p be the probability that there are no power failures during a single day.

1. Find the MLE for p.

2. Derive a 95% confidence interval for p (if you use any kind of approximations—explain).

Exercise 5.8 Continue Exercise 2.21. Find an asymptotic $100(1 - \alpha)\%$ confidence interval for p.

Exercise 5.9 Let $Y_1, ..., Y_n$ be a random sample from a shifted exponential distribution with a density $f_\tau(y) = e^{-(y-\tau)}$, $y \geq \tau$.

1. Find the MLE $\hat{\tau}_n$ for τ (see also Exercise 2.1).

2. Is $\hat{\tau}_n$ consistent in MSE?

3. Show that $n(\hat{\tau}_n - \tau) \sim \exp(1)$.

4. Find the asymptotic distribution of $\sqrt{n}(\hat{\tau}_n - \tau)$.

5. Why does the asymptotic normality of MLE not hold in this case?

Exercise 5.10 Let $Y_1, ..., Y_n \sim \mathcal{N}(\mu, a^2\mu^2)$ with a known a.

1. Construct an asymptotic $100(1 - \alpha)\%$ confidence interval for μ based on the distribution of $\bar{Y}_n$.

2. Repeat the previous paragraph but this time using a variance stabilization transformation $g(\mu)$.

3. Compare the resulting asymptotic confidence intervals with the exact one from Exercise 3.4.

Exercise 5.11 Let $Y_1, \cdots, Y_n \sim Pois(\lambda)$. Derive asymptotic Wald, score (Rao) and GLRT tests for testing $H_0 : \lambda = \lambda_0$ vs. $H_1 : \lambda \neq \lambda_0$ at level α. Show that all the three tests are asymptotically similar.

Exercise 5.12 Consider a series of Bernoulli trials $Y_1, ..., Y_n \sim \mathscr{B}(1, p)$.

1. Show that the variance stabilizing transformation is $g(p) = \arcsin(\sqrt{p})$. What is the resulting asymptotic variance?

2. Use the results of the previous paragraph to construct an asymptotic $100(1 - \alpha)\%$ confidence interval for p.

3. Show that the above asymptotic confidence interval is first-order equivalent to (5.11) from Example 5.12. Compare them for the newborn babies data from Example 1.1.

Exercise 5.13 Extend the results of Exercise 4.13 for testing the equality of variances for two independent normal samples to k samples.

1. Derive the corresponding GLR.

2. The exact distribution of the GLR under the null is difficult to compute for $k > 2$. Apply Wilks' theorem to find an asymptotic GLRT at level α.

Exercise 5.14 * In Exercise 2.22 we established that the sample $\alpha \cdot 100\%$-quantile $Y_{(\alpha)}$ is the M-estimator corresponding to $\rho(\theta, y) = \alpha(y - \theta)_+ + (1 - \alpha)(\theta - y)_+$. Show that under the required regularity conditions, $Y_{(\alpha)}$ is a consistent estimator of a $\alpha \cdot 100\%$-quantile $F_{\theta_0}^{-1}(\alpha)$ of the distribution of Y_i's with a true θ_0 and find its asymptotic distribution.

Chapter 6

Bayesian Inference

6.1 Introduction

Consider a general statistical model, where one observes a random sample $Y_1, ..., Y_n \sim f_\theta(y)$ and the parameter of interest is $\theta \in \Theta$. So far we have treated θ as a fixed (unknown) constant. Its estimation and inference was based solely on the observed data, which was considered as a random sample from the set of all possible samples of a given size from a population (sample space). The inference about θ was essentially based on *averaging* over all those (hypothetical!) samples that might raise conceptual questions about its appropriateness for a *particular* single observed sample. Recall, for example, the meaning of the MSE, unbiasedness, the confidence interval or the significance level of tests discussed in previous chapters.

In this chapter, we consider the alternative *Bayesian* approach. Within the Bayesian framework, the initial uncertainty about θ is modeled by considering θ as a *random* variable with a certain *prior* distribution that expresses one's prior beliefs about the plausibility of every possible value of $\theta \in \Theta$. At first glance, it might seem somewhat confusing to treat θ as a random variable—one does not really assume that its value randomly changes from experiment to experiment. The problem is with the interpretation of randomness. One typically associates probability with a relative frequency meaning in a long series of trials. Saying that the probability that a fair coin turns up heads is 0.5 actually means that in a long series of tosses one will get heads about half the trial. Such a frequentist interpretation of probability is indeed reasonable in such contexts, but might be inadequate in other situations. Listening to the weather forecast on the radio that there is a probability of 80% for rain tomorrow, one would unlikely think about a series of daily trials under the same weather conditions, where in about 80% of them it will be rain on the next day, rather than an expert's evaluation of chances for having rain based on his knowledge and previous experience, analysis of various meteorological processes, etc. A similar situation occurs in gambling when one tries, for example, to predict the result of a sport event based on his (subjective) judgment of chances of a team to win the game. In such cases, it is more appropriate to think of a probability as a measure of belief. Philosophical disputes on the nature of probability have a long history but, luckily, the corresponding formal probability calculus does not depend on its interpretation.

Thus, in addition to the observed data $\mathbf{y} = (y_1, ..., y_n)$, a Bayesian specifies a prior distribution $\pi(\theta)$, $\theta \in \Theta$ for the unknown parameter θ. The model distribution for the data $f_\theta(\mathbf{y})$ is treated as a *conditional* distribution $f(\mathbf{y}|\theta)$ of $\mathbf{Y}$ given θ. Thus, the Bayesian specifies before observing the data a joint distribution for the parameter and the data. The inference on θ is based then on a *posterior* distribution $\pi(\theta|\mathbf{Y} = \mathbf{y})$ ($\pi(\theta|\mathbf{y})$ for brevity) calculated by Bayes' theorem:

$$\pi(\theta|\mathbf{y}) = \frac{f(\mathbf{y}|\theta)\pi(\theta)}{f(\mathbf{y})},$$

where the marginal distribution $f(\mathbf{y}) = \int_\Theta f(\mathbf{y}|\theta')\pi(\theta')d\theta'$ is a normalizing constant that does not depend on θ. Note also that $f(\mathbf{y}|\theta) = f_\theta(\mathbf{y})$ is a likelihood function $L(\theta; \mathbf{y})$ and, therefore, the posterior distribution of $\theta|\mathbf{y}$ is proportional to the multiplication of the likelihood $L(\theta, \mathbf{y})$ and the prior distribution $\pi(\theta)$:

$$\pi(\theta|\mathbf{y}) \propto L(\theta; \mathbf{y})\pi(\theta) \tag{6.1}$$

DOI: 10.1201/9781003175407-6

Figure 6.1: Various Beta-densities: $Beta(1,1)$ (solid), $Beta(2,2)$ (dashed), $Beta(.5,.5)$ (dotted), $Beta(3,2)$ (dotdash), $Beta(.8,2)$ (longdash).

In fact, since $\pi(\theta|\mathbf{y})$ is a density, there is no need to calculate the normalizing constant $f(\mathbf{y})$ by integration (or summation for the discrete case) once one identifies the type of a posterior distribution from (6.1) and can use then its known formula.

Example 6.1 (Bernoulli trials) Suppose we observe a series of coin tosses $Y_1,...,Y_n$ with an unknown probability p for heads. Assume a prior Beta-distribution $Beta(\alpha,\beta)$ for p with the density

$$\pi_{\alpha,\beta}(p) = \frac{p^{\alpha-1}(1-p)^{\beta-1}}{B(\alpha,\beta)}, \quad 0 \le p \le 1, \ \alpha,\beta > 0 \tag{6.2}$$

where the normalizing constant $B(\alpha,\beta) = \frac{\Gamma(\alpha)\Gamma(\beta)}{\Gamma(\alpha+\beta)}$ (see Appendix A.4.9). For different α and β, (6.2) defines a family of distributions. In particular, the uniform distribution $U[0,1]$ corresponds to $\alpha = \beta = 1$. Generally, if $\alpha = \beta$, the distributions are symmetric around 0.5. Figure 6.1 shows some typical distributions from this family. Following Bayes' theorem and ignoring meanwhile all multiplicative terms that do not depend on p, (6.1) and (6.2) yield

$$\pi(p|\mathbf{y}) \propto p^{\sum_{i=1}^{n} y_i}(1-p)^{n-\sum_{i=1}^{n} y_i}p^{\alpha-1}(1-p)^{\beta-1} = p^{\sum_{i=1}^{n} y_i+\alpha-1}(1-p)^{n-\sum_{i=1}^{n} y_i+\beta-1}$$

which is, up to a normalizing constant, the density of the Beta-distribution $Beta(\sum_{i=1}^{n} y_i + \alpha, n - \sum_{i=1}^{n} y_i + \beta)$. Once the posterior distribution is identified, the normalizing constant must be $B(\sum_{i=1}^{n} y_i + \alpha, n - \sum_{i=1}^{n} y_i + \beta)$, to have a function integrating to 1. The posterior distribution $p|\mathbf{y} \sim Beta(\sum_{i=1}^{n} y_i + \alpha, n - \sum_{i=1}^{n} y_i + \beta)$ depends both on the data and the parameters of the prior distribution.

Example 6.2 (normal distribution with a known variance) Consider a sample $Y_1,...,Y_n$ from a normal distribution $\mathcal{N}(\mu,\sigma^2)$ with a known σ and assume a normal prior $\mathcal{N}(\mu_0,\tau^2)$ for an unknown mean μ. One has

$$\pi(\mu|\mathbf{y}) \propto \exp\left\{ -\frac{\sum_{i=1}^{n}(y_i-\mu)^2}{2\sigma^2} \right\} \cdot \exp\left\{ -\frac{(\mu-\mu_0)^2}{2\tau^2} \right\}$$

Using the decomposition $\sum_{i=1}^{n}(y_i - \mu)^2 = \sum_{i=1}^{n}(y_i - \bar{y})^2 + n(\bar{y} - \mu)^2$ (see (1.6)) and ignoring all multiplicative terms that do not involve μ, straightforward calculus implies

$$\pi(\mu|\mathbf{y}) \propto \exp\left\{ -\frac{1}{2}\mu^2\left(\frac{n}{\sigma^2} + \frac{1}{\tau^2}\right) + \mu\left(\frac{n}{\sigma^2}\bar{y} + \frac{\mu_0}{\tau^2}\right)\right\}$$

$$\propto \exp\left\{ -\frac{1}{2}\left(\frac{n}{\sigma^2} + \frac{1}{\tau^2}\right)\left(\mu - \frac{(n/\sigma^2)y + \mu_0/\tau^2}{n/\sigma^2 + 1/\tau^2}\right)^2\right\}$$

that corresponds to a normal density with the mean

$$E(\mu|\mathbf{y}) = \frac{(n/\sigma^2)\bar{y} + \mu_0/\tau^2}{n/\sigma^2 + 1/\tau^2} = \frac{\tau^2}{\frac{\sigma^2}{n} + \tau^2}\bar{y} + \frac{\frac{\sigma^2}{n}}{\frac{\sigma^2}{n} + \tau^2}\mu_0 \tag{6.3}$$

and the variance

$$Var(\mu|\mathbf{y}) = \left(\frac{n}{\sigma^2} + \frac{1}{\tau^2}\right)^{-1} \tag{6.4}$$

The posterior mean in (6.3) has a clear meaning: it combines information from the two sources by taking a weighted average of a sample mean $\bar{y}$ (data contribution) and a prior mean μ_0 (prior contribution). The weights are defined by the ratio between the prior variance τ^2 and the sample mean's variance $\frac{\sigma^2}{n}$. The more accurate is a source (the less is its variance), the larger weight it is given.

In contrast to the frequentist approach (sometimes called classical, Neyman–Pearson, or non-Bayesian), Bayesian inference is inherently conditional on a specific observed sample rather than the result of averaging over the entire sample space of the experiment. In this sense some consider it as having a more "natural" interpretation and as more "conceptually appealing." The main criticism with respect of Bayesians, however, is due to the use of a prior which is the core of the approach. A choice for $\pi(\theta)$ is subjective and represents one's beliefs or information about θ before the experiment. Different priors can change the results of Bayesian analysis while in many practical cases, subjective quantification of uncertainty in terms of a prior distribution might be difficult. Moreover, sufficient prior information about a parameter is not always available. Some Bayesian counterarguments are presented below in this chapter.

Although the two approaches are conceptually different, they are nevertheless not "complete strangers" to each other and may benefit from cooperation. Frequentist analysis of Bayesian procedures can be very useful for better understanding and validating their results. Similarly, Bayesian interpretation of a frequentist procedure is often helpful in providing the intuition behind it.

A detailed presentation of Bayesian analysis or a comprehensive survey is beyond the scope of this book and we refer interested readers to, e.g. Berger (1985), Box and Tiao (1973), Gelman et al. (1995) and Ghosh et al. (2006). We first briefly present several approaches for choosing a prior and then, similar to the classical setup, consider the Bayesian perspective on the three main statistical problems, namely, point estimation, interval estimation and hypothesis testing.

6.2   Choice of priors

6.2.1   Conjugate priors

As we have argued, although a prior distribution should ideally be defined solely on one's prior beliefs and knowledge, such situations are rare in the real world. On the other hand, one can often specify a (usually parametric) *class* or *family* of possible priors. This class should be sufficiently "rich" to include priors that reflect various possible scenarios for the problem at hand and at the same time to allow simple computation of the corresponding posterior distributions. Conjugate priors are probably the most known and used of these classes:

Definition 6.1 (conjugate prior) A (parametric) class of priors $\mathscr{P}$ is called conjugate for the data if for any prior $\pi(\theta) \in \mathscr{P}$, the corresponding posterior distribution $\pi(\theta|\mathbf{y})$ also belongs to $\mathscr{P}$.

The main motivation for conjugate classes of priors is that the posteriors can be calculated in closed form. Still, they leave enough flexibility to express a variety of possibilities.

Example 6.3 From the results of Example 6.1 one verifies that the family of distributions $Beta(\alpha, \beta)$ is conjugate for the binomial data. As we have demonstrated, this family includes different types of functions to represent various prior beliefs on the unknown binomial parameter p.

Example 6.4 The Gaussian family is conjugate for the normal data (see Example 6.2).

When choosing a parametric (e.g., conjugate) class of priors, one should still specify the corresponding parameters of a prior (called *hyperparameters*). Often it can be done subjectively—to define parameters of a distribution is after all a much easier task than to elicit the entire prior distribution itself!

Otherwise, within the Bayesian paradigm, one can place a prior distribution (*hyperprior*) on the unknown hyperparameters that leads to a *hierarchical* prior model. Let $\pi_\alpha(\theta)$ be a prior distribution with hyperparameters α and $\psi(\alpha)$ be the hyperprior for α. Thus,

$$\theta|\alpha \sim \pi_\alpha(\theta)$$

$$\alpha \sim \psi(\alpha)$$

and, therefore, the marginal prior for θ is

$$\pi(\theta) = \int \pi(\theta|\alpha)\psi(\alpha)d\alpha = \int \pi_\alpha(\theta)\psi(\alpha)d\alpha$$

The hyperprior $\psi(\alpha)$ is assumed to be known and not dependent on any other unknown parameters. If it is not the case, we can place a prior on them as well adding another level to the hierarchical prior model and so on. Evidently, at some bottom level we should stop with a hyperprior assumed to be completely known.

Example 6.5 Let $Y_1, \ldots, Y_n \sim Pois(\lambda)$ with the exponential prior on λ, i.e., $\lambda \sim \exp(\theta)$ but with the unknown hyperparameter θ. Assume the hyperprior $Gamma(\alpha, \beta)$ (with known α and β) on θ (see Section A.4.8). The marginal prior of λ is then

$$\begin{aligned}
\pi(\lambda) &= \int_0^\infty \theta e^{-\theta\lambda} \frac{\beta^\alpha}{\Gamma(\alpha)} \theta^{\alpha-1} e^{-\beta\theta} d\theta \\
&= \frac{\beta^\alpha}{\Gamma(\alpha)} \int_0^\infty \theta^{\alpha+1-1} e^{-(\lambda+\beta)\theta} d\theta \\
&= \frac{\beta^\alpha}{\Gamma(\alpha)} \frac{\Gamma(\alpha+1)}{(\lambda+\beta)^{\alpha+1}} \\
&= \alpha\beta^\alpha \frac{1}{(\lambda+\beta)^{\alpha+1}}
\end{aligned}$$

Unfortunately, in most cases the resulting marginal prior $\pi(\theta)$ and the posterior $\pi(\theta|\mathbf{y})$ for hierarchical prior models cannot be obtained in closed form. However, the existing Bayes computational techniques allow one to derive them numerically in efficient way.

Alternatively, one can estimate the unknown hyperparameters of the prior from the data. Such an approach is known as *empirical Bayes* and utilizes the marginal distribution of the data

$$f_\alpha(\mathbf{y}) = \int_\Theta f(\mathbf{y}|\theta)\pi_\alpha(\theta)d\theta$$

The density $f_\alpha(\mathbf{y})$ does not depend on θ but only on the unknown hyperparameters α of the prior that can be estimated by any classical method, e.g, maximum likelihood or method of moments.

Example 6.6 Consider again the Poisson sample $Y_1, \ldots, Y_n \sim Pois(\lambda)$, where $\lambda \sim \exp(\theta)$ with the unknown hyperparameter θ (see Example 6.5), and estimate now θ from the data following the empirical Bayes approach. The marginal distribution of the data

$$
\begin{aligned}
f_\theta(\mathbf{y}) &= \int_0^\infty e^{-n\lambda} \frac{\lambda^{\sum_{i=1}^n y_i}}{\Pi y_i!} \theta e^{-\theta\lambda} d\lambda \\
&= \frac{\theta}{\Pi y_i!} \int_0^\infty \lambda^{\sum_{i=1}^n y_i + 1 - 1} e^{-(n+\theta)\lambda} d\lambda \\
&= \frac{\theta}{(n+\theta)^{\sum_{i=1}^n y_i + 1}} \frac{(\sum_{i=1}^n y_i)!}{\Pi y_i!}
\end{aligned}
$$

(see Section A.4.8). Estimate θ by the maximum likelihood by maximizing $\ln f_\theta(\mathbf{y})$ w.r.t. θ, i.e., find

$$
\hat{\theta} = \arg\max_\theta \left\{ \ln\theta - (\sum_{i=1}^n y_i + 1)\ln(n+\theta) \right\}.
$$

Differentiating w.r.t. θ, a straightforward calculus yields $\hat{\theta} = \frac{1}{\bar{y}}$.

Although empirical Bayes essentially contradicts the core Bayesian principle that a prior should not depend on the data, it is nevertheless a useful approach in many practical situations.

6.2.2 Noninformative (objective) priors

Another approach for choosing a suitable prior is by using the so-called noninformative or objective priors. It is especially useful when there is very little prior information on the unknown parameter θ available.

Suppose we know nothing a priori about $\theta \in \Theta$. Somewhat paradoxically, even complete ignorance is also knowledge! It essentially means that we cannot prefer any specific value of θ over any other one and all values of θ in Θ are equally likely. A natural way to model ignorance would then be to place a *uniform* prior distribution for θ over the parameter space Θ. Such a prior is called *noninformative* or *objective*.

The idea looks appealing but unfortunately it cannot be applied in common situations, where the parameter space Θ has an *infinite* measure, e.g. for the normal mean, where $\theta = \mu$ and $\Theta = (-\infty, \infty)$. Anyone who ever took a basic course in probability learned that there is no uniform distribution over such a set since one cannot define a uniform distribution $\mathscr{U}(-\infty, \infty)$. Nevertheless, let's "ignore" this basic probabilistic law (at least for a while) and consider a prior density $\pi(\theta) = c$ over Θ for some fixed $c > 0$. It is obviously an *improper* density since $\int_\Theta \pi(\theta)d\theta = \infty$. However, the resulting posterior density

$$
\pi(\theta|\mathbf{y}) = \frac{f(\mathbf{y}|\theta)c}{\int_\Theta f(\mathbf{y}|\theta)c d\theta} = \frac{L(\theta; \mathbf{y})}{\int_\Theta L(\theta; \mathbf{y})d\theta}
$$

can still be proper (provided that the integral in the denominator is finite for all $\mathbf{y}$), and, moreover, it does not depend on a specific c! Thus, although the proposed prior is improper, it can be used in Bayesian inference, as long as the corresponding posterior is proper.

There is, however, another, probably a more serious problem with using uniform distributions (proper or improper) as a general remedy for noninformative priors. Let θ be uniformly distributed with $\pi_\theta(\theta) = c$, but then θ^2, $1/\theta$, $\ln\theta$ or, in fact, almost any function $\xi = \xi(\theta)$ is no longer uniformly distributed! Indeed, assume for simplicity that $\xi(\cdot)$ is a one-to-one differentiable function and denote $\theta'_\xi = \frac{d\theta}{d\xi}$. Then, the corresponding prior $\pi_\xi(\cdot)$ for ξ is

$$
\pi_\xi(\xi) = c \cdot |\theta'_\xi|
$$

which is not constant in general. Thus, it seems as if we know nothing about θ but do know something about $\xi(\theta)$. The lack of invariance of a uniform distribution under transformations leads to the logical paradox and raises then the question whether one should impose a uniform distribution on the original scale of θ or the transformed one (and if using a transformation, which scale to employ?).

Sometimes one can elicit a noninformative prior exploiting special properties of the model at hand. For example, suppose that the parameter of interest θ is a *location* parameter, that is, $f(y|\theta) = f_0(y - \theta)$ for some density $f_0(\cdot)$. In such a case it is reasonable to require a noninformative prior $\pi(\theta)$ to be invariant under translations, so that $\pi(\theta) = \pi(\theta - \theta_0)$ for all θ_0. The only possible choice then is $\pi(\theta) = c$ or $\pi(\theta) \propto 1$, which is a (improper) uniform distribution over Θ.

Example 6.7 Consider again a sample $Y_1, ..., Y_n$ from a normal distribution $\mathcal{N}(\mu, \sigma^2)$ with a known variance σ^2. The mean μ is a location parameter and the corresponding noninformative prior is then $\pi(\mu) \propto 1$. Straightforward calculus similar to that in Example 6.2 yields

$$\pi(\mu|\mathbf{y}) \propto L(\mu; \mathbf{y}) \propto e^{-\frac{n(\bar{y}-\mu)^2}{2\sigma^2}}$$

and, thus, $\mu|\mathbf{y} \sim \mathcal{N}(\bar{y}, \sigma^2/n)$. One can also look at this result as a limiting case of a (proper) Gaussian prior $\mu \sim \mathcal{N}(\mu_0, \tau^2)$ when $\tau \to \infty$ considered in Example 6.2.

Consider now the situation where $\theta > 0$ is a *scale* parameter, that is, $f(y|\theta) = \theta^{-1} f_0(y/\theta)$. Examples include the standard deviation σ in the normal distribution $\mathcal{N}(\mu, \sigma^2)$, the expectation $\mu = 1/\theta$ in the exponential distribution $\exp(\theta)$ and the interval's length θ in the uniform distribution $\mathcal{U}(0, \theta)$. Consider a new transformed random variable $X = \ln Y$-we assume, for simplicity, that $Y > 0$. Using standard probabilistic calculus, the density function (or, respectively, the probability function for the discrete case) of X is

$$f_X(x|\theta) = e^x f_Y(e^x|\theta) = e^x \frac{1}{\theta} f_0\left(\frac{e^x}{\theta}\right) = e^{x - \ln\theta} f_0(e^{x - \ln\theta}),$$

and, therefore, $\xi = \ln\theta \in (-\infty, \infty)$ is a *location* parameter of the distribution of the transformed variable $X = \ln Y$. Following our previous arguments, the noninformative prior for ξ should then be $\pi_\xi(\xi) \propto 1$ which implies that the corresponding (also improper) noninformative prior for the original parameter θ is

$$\pi_\theta(\theta) \propto \frac{1}{\theta}, \ \theta > 0 \tag{6.5}$$

The choice for the prior $\pi_\theta(\theta)$ in (6.5) can be also justified by the requirement of scale invariance $\pi(\theta) = c^{-1} \pi(\theta/c)$ for all $c > 0$.

Example 6.8 (exponential distribution) Consider a sample $Y_1, .., Y_n$ from an exponential distribution $\exp(\theta)$, where $\mu = E(Y) = 1/\theta$ is a scale parameter. The noninformative prior for μ is therefore $\pi_\mu(\mu) \propto 1/\mu$ and the corresponding prior for $\theta = 1/\mu$ is $\pi_\theta(\theta) \propto \theta \cdot 1/\theta^2 = 1/\theta$. Thus, the posterior distribution

$$\pi(\theta|\mathbf{y}) \propto L(\theta; \mathbf{y})\pi(\theta) \propto \theta^n e^{-\theta \sum_{i=1}^n y_i} \cdot \theta^{-1} = \theta^{n-1} e^{-\theta \sum_{i=1}^n y_i}$$

which is a (proper) Gamma-density $Gamma(n, \sum_{i=1}^n y_i)$.

Example 6.9 Continue Example 6.7 with a normal sample but in a more realistic scenario, where the variance σ^2 is also unknown. The parameter of interest is still the mean μ and σ can be considered a nuisance parameter. We should now specify a *joint* noninformative prior distribution $\pi_{\mu,\sigma}(\mu, \sigma)$. According to our previous results, the *marginal* noninformative priors for μ and σ are $\pi_\mu(\mu) \propto 1$ and $\pi_\sigma(\sigma) \propto 1/\sigma$ respectively. Assuming in addition that μ and σ are a priori independent, we have

$$\pi_{\mu,\sigma}(\mu, \sigma) = \pi_\mu(\mu) \cdot \pi_\sigma(\sigma) \propto 1/\sigma \tag{6.6}$$

We now want to derive the posterior (marginal) distribution of $\mu|\mathbf{y}$. We start from the joint posterior distribution $(\mu, \sigma)|\mathbf{y}$:

$$\pi_{\mu,\sigma}((\mu,\sigma)|\mathbf{y}) \propto L(\mu,\sigma;\mathbf{y}) \cdot \pi_{\mu,\sigma}(\mu,\sigma) \propto \frac{1}{\sigma^n} e^{-\frac{\Sigma_{i=1}^n (y_i-\mu)^2}{2\sigma^2}} \frac{1}{\sigma} = \frac{1}{\sigma^{n+1}} e^{-\frac{W+n(\bar{y}-\mu)^2}{2\sigma^2}}, \qquad (6.7)$$

where $W = \sum_{i=1}^n (y_i - \bar{y})^2$ does not depend on the unknown μ and σ. Integrating out the nuisance parameter σ in (6.7) yields the marginal posterior distribution of μ:

$$\pi_\mu(\mu|\mathbf{y}) = \int_0^\infty \pi_{\mu,\sigma}((\mu,\sigma)|\mathbf{y})d\sigma \propto \int_0^\infty \frac{1}{\sigma^{n+1}} e^{-\frac{A}{2\sigma^2}} d\sigma,$$

where $A = W + n(\bar{y} - \mu)^2$ does not depend on σ. Changing the variables to $\psi = 1/\sigma^2$ within the integral one has

$$\pi_\mu(\mu|\mathbf{y}) \propto \int_0^\infty \psi^{\frac{n}{2}-1} e^{-\frac{1}{2}A\psi} d\psi \qquad (6.8)$$

which is, up to a normalizing constant $(\frac{A}{2})^{n/2}/\Gamma(\frac{n}{2})$, is the integral of the Gamma-density $Gamma(\frac{n}{2}, \frac{A}{2})$. Multiplying and dividing (6.8) by this normalizing constant yields then

$$\pi_\mu(\mu|\mathbf{y}) \propto \frac{\Gamma(\frac{n}{2})}{(\frac{A}{2})^{n/2}} \propto \left(W + n(\bar{y}-\mu)^2\right)^{-\frac{n}{2}} \propto \left(1 + \frac{n(\bar{y}-\mu)^2}{s^2(n-1)}\right)^{-\frac{n}{2}},$$

where recall that $s^2 = \frac{\Sigma_{i=1}^n (y_i-\bar{y})^2}{n-1}$. Consider a standardized variable $T = \frac{\bar{y}-\mu}{s/\sqrt{n}}$. Then,

$$\pi_T(t|\mathbf{y}) \propto \left(1 + \frac{t^2}{n-1}\right)^{-\frac{(n-1)+1}{2}}$$

which is the Student's distribution t_{n-1} with $n-1$ degrees of freedom. Although this result seems identical to that for the classical approach, where also $\frac{\bar{Y}-\mu}{s/\sqrt{n}} \sim t_{n-1}$ (see (3.2)), the difference lies in its interpretation. Within classical framework, the random variable is $\bar{Y}$, while μ is an unknown (constant) parameter. For Bayesians, it is the conditional distribution of the random μ given $\bar{Y} = \bar{y}$.

Consider now the general case. Let $Y_1, ..., Y_n \sim f(y|\theta)$. Recall that the problem with the uniform noninformative prior was the lack of invariance under transformations of the original parameter θ. We would like to construct a noninformative prior that will be invariant under one-to-one differentiable transformations $\xi = \xi(\theta)$. By standard probabilistic arguments, a prior for ξ is

$$\pi_\xi(\xi) = \pi_\theta(\theta(\xi)) \cdot |\theta'_\xi| \qquad (6.9)$$

The *Jeffreys prior* for θ is

$$\pi(\theta) \propto \sqrt{I_\theta^*(\theta)}, \qquad (6.10)$$

where $I_\theta^*(\theta) = E((\ln f(y|\theta))'_\theta)^2) = -E((\ln f(y|\theta))''_\theta)$ is the Fisher information number of $f(y|\theta)$ (see Section 2.7.2). Note that in terms of ξ,

$$I_\xi^*(\xi) = E\left(\left(\ln f(y|\xi))'_\xi\right)^2\right) = E\left(\left((\ln f(y|\theta))'_\theta \cdot |\theta'_\xi|\right)^2\right) = I_\theta^*(\theta) \cdot (\theta'_\xi)^2$$

and, therefore,

$$\sqrt{I_\xi^*(\xi)} = \sqrt{I_\theta^*(\theta)} \cdot |\theta'_\xi|$$

which is, by (6.9), also a Jeffreys prior for ξ. Hence, such class of priors is transformation invariant.

An immediate extension of Jeffreys noninformative prior (6.10) for the multiparameter case is

$$\pi(\theta) \propto \sqrt{\det(I^*(\theta))}, \qquad (6.11)$$

where $I^*(\theta)$ is the Fisher information matrix based on a single observation (see Section 2.7.3).

Example 6.10 (Bernoulli trials) We leave as an exercise to show that for a series of Bernoulli trials, $I^*(p) = \frac{1}{p(1-p)}$. The corresponding Jeffreys prior is then $\pi(p) \propto p^{-1/2}(1-p)^{-1/2}$, which is a (proper!) $Beta(\frac{1}{2}, \frac{1}{2})$ distribution (see Figure 6.1).

Example 6.11 (Normal data) Consider several possible scenarios for a normal sample $Y_1, ..., Y_n \sim \mathcal{N}(\mu, \sigma^2)$.

1. The parameter of interest is μ, σ is known. In Example 2.18, we derived $I^*(\mu) = 1/\sigma^2$ which is constant for a known σ. Thus, the Jeffreys prior $\pi(\mu) \propto \sqrt{I^*(\mu)} \propto 1$ is a uniform noninformative prior for the location parameter (see Example 6.7).

2. The parameter of interest is σ, μ is known. From Example 2.20 it follows that $I^*(\sigma) = 2/\sigma^2$. Hence, $\pi(\sigma) \propto 1/\sigma$ which is a noninformative prior for the scale parameter.

3. Both μ and σ are unknown. In this case, the Fisher information matrix is

$$I^*(\mu, \sigma) = \begin{pmatrix} \frac{1}{\sigma^2} & 0 \\ 0 & \frac{2}{\sigma^2} \end{pmatrix}$$

(see Example 2.21). The resulting joint Jeffreys prior for (μ, σ) is, therefore,

$$\pi(\mu, \sigma) \propto \sqrt{\det(I^*(\mu, \sigma))} \propto 1/\sigma^2$$

(compare with $\pi(\mu, \sigma) \propto 1/\sigma$ in (6.6) of Example 6.9 under the additional assumption of independence).

6.3 Point estimation

Given the posterior distribution $\pi(\theta|\mathbf{y})$, a Bayes point estimator $\hat{\theta}$ for θ depends on the measure of goodness-of-estimation. Similar to non-Bayesian estimation, the most conventional measure is the squared error $(\hat{\theta} - \theta)^2$. The difference between the two approaches is that Bayesians consider the expectation of a squared error w.r.t. the posterior distribution $\pi(\theta|\mathbf{y})$ rather than the data distribution $f(\mathbf{y}|\theta)$ as in the MSE (see Section 2.6).

We define the posterior expected squared error $E\left((\hat{\theta} - \theta)^2|\mathbf{y}\right)$ and find an estimator $\hat{\theta}$ that minimizes it. We have

$$E\left((\hat{\theta} - \theta)^2|\mathbf{y}\right) = \hat{\theta}^2 - 2\hat{\theta}E(\theta|\mathbf{y}) + E(\theta^2|\mathbf{y})$$

($\hat{\theta}$ is a function of the data, and hence it is a "constant" given the data), which is evidently minimized for $\hat{\theta} = E(\theta|\mathbf{y})$ that leads us to the following proposition:

Proposition 6.1 The Bayes point estimator for θ with respect to the squared error is the posterior mean $\hat{\theta} = E(\theta|\mathbf{y})$.

Question 6.1 Show that the corresponding minimal posterior expected squared error is the posterior variance $Var(\theta|\mathbf{y})$.

Example 6.12 A farmer is interested in the average weight μ of adult cows in his large herd. He weighed 10 randomly chosen cows from the herd and obtained an average weight $\bar{y} = 410\,kg$ (about $925\,lbs$). For simplicity assume that the standard deviation σ is known and $\sigma = 20\,kg$. Regardless of the distribution of the weights, a (non-Bayesian) method of moments estimate for μ is $\bar{Y} = 410\,kg$. Typically, it is quite reasonable to assume the normality of the distribution of the weights. In this case $\bar{Y}$ is also the MLE and the UMVUE (see Chapter 2). However, based on his previous experience with herds of the same sort, the farmer also believes that μ typically should be around $420\,kg$ and is very unlikely to differ from this value by more than $30\,kg$. Since he is not a complete *amateur* in statistics, he exploited this knowledge using a Bayesian approach. The farmer modeled the prior information about μ by assuming a conjugate normal prior $\mathcal{N}(420, 10^2)$ on μ, where he utilized the empirical *three-sigma* rule that for the normal distribution nearly all values lie within three standard deviations of the mean (99.7% by the exact calculus). The corresponding posterior distribution of μ is also normal (see Example 6.2) and from (6.3), the Bayes estimate is the weighted average of the sample and prior means:

$$\hat{\mu} = E(\mu|y) = \frac{100}{400/10 + 100} 410 + \frac{400/10}{400/10 + 100} 420 = 412.9 \ kg$$

Example 6.13 (exponential distribution—continuation of Example 2.4) We continue Example 2.4 where John recorded waiting times $t_1, ..., t_{30} \sim \exp(\theta)$ for a bus during 30 days in order to estimate the unknown θ but now within the Bayesian framework. John does not have any prior knowledge about θ and assumes a noninformative prior $\pi(\theta) \propto 1/\theta$. The posterior distribution of $\theta|\mathbf{t}$ is then $Gamma(n, T)$, where $T = \sum_{i=1}^{n} t_i$ (see Example 6.8), and the Bayes estimator $\hat{\theta} = E(\theta|\mathbf{t}) = 1/\bar{t}$. However, as we have mentioned in Example 2.4, John's primary interest is in the expected waiting time $\mu = Et = 1/\theta$ rather than in θ itself. We can use the posterior distribution of $\theta|\mathbf{t}$ to find $\hat{\mu} = E(\mu|\mathbf{t}) = E(\theta^{-1}|\mathbf{t})$:

$$\hat{\mu} = \int_0^\infty \theta^{-1} \frac{T^n}{\Gamma(n)} \theta^{n-1} e^{-\theta T} d\theta = \frac{T^n}{\Gamma(n)} \int_0^\infty \theta^{n-2} e^{-\theta T} d\theta \tag{6.12}$$

The integral in the RHS of (6.12), up to the normalizing constant $\frac{T^{n-1}}{\Gamma(n-1)}$, is the integral of the Gamma-density $Gamma(n-1, T)$ and, therefore,

$$\hat{\mu} = \frac{T^n}{\Gamma(n)} \cdot \frac{\Gamma(n-1)}{T^{n-1}} = \frac{T}{n-1} = \frac{n}{n-1} \bar{t}$$

(note that $\hat{\mu} \neq \hat{\theta}^{-1}$!). Substituting $\bar{t} = 8 \ min$ and $n = 30$ implies $\hat{\mu} = 8.28 \ min \sim 8 \ min. 17 \ sec$. Recall that the MLE estimate $\hat{\mu}_{MLE} = \bar{t} = 8 \ min$ (see Example 2.4).

Similarly, the Bayes estimate for the probability $p = e^{-5\theta}$ that John will wait for a bus for more than 5 minutes and will be late for an exam (see Example 2.5) is

$$\begin{aligned} \hat{p} &= E(p|\mathbf{t}) = E(e^{-5\theta}|\mathbf{t}) = \int_0^\infty e^{-5\theta} \frac{T^n}{\Gamma(n)} \theta^{n-1} e^{-\theta T} d\theta = \frac{T^n}{\Gamma(n)} \int_0^\infty \theta^{n-1} e^{-\theta(T+5)} d\theta = \left(\frac{T}{T+5}\right)^n \\ &= 0.539 \end{aligned}$$

For large n, $\hat{p} = \left(1 - \frac{5}{T+5}\right)^n \sim e^{-5/\bar{t}} = 0.535$, which is the MLE estimate of $\hat{p}$ (see Example 2.5). In both cases the Bayes and MLE estimates are close since the prior for θ was noninformative.

Another possible measure of goodness-of-estimation is the absolute error $|\hat{\theta} - \theta|$:

Proposition 6.2 The Bayes point estimator for θ with respect to the absolute error is the posterior median $\hat{\theta} = Med(\theta|\mathbf{y})$.

Proof. We have

$$\begin{aligned} E_{\theta|\mathbf{y}}|\hat{\theta} - \theta| &= \int_\Theta |\hat{\theta} - \theta| \pi(\theta|\mathbf{y}) d\theta = \int_{\theta \leq \hat{\theta}} (\hat{\theta} - \theta) \pi(\theta|\mathbf{y}) d\theta + \int_{\theta > \hat{\theta}} (\theta - \hat{\theta}) \pi(\theta|\mathbf{y}) d\theta \\ &= \hat{\theta} \Pi(\hat{\theta}|\mathbf{y}) - \int_{\theta \leq \hat{\theta}} \theta \pi(\theta|\mathbf{y}) d\theta - \hat{\theta}(1 - \Pi(\hat{\theta}|\mathbf{y})) + \int_{\theta > \hat{\theta}} \theta \pi(\theta|\mathbf{y}) d\theta, \tag{6.13} \end{aligned}$$

where $\Pi(\cdot|\mathbf{y})$ is the posterior cdf of $\theta|\mathbf{y}$. Differentiating the RHS of (6.13) with respect to $\hat{\theta}$ after a simple calculus yields

$$\frac{\partial E_{\theta|\mathbf{y}}|\hat{\theta} - \theta|}{\partial \hat{\theta}} = 2\Pi(\hat{\theta}|\mathbf{y}) - 1$$

Thus, the minimizer of $E_{\theta|\mathbf{y}}|\hat{\theta} - \theta|$ should satisfy $\Pi(\hat{\theta}|\mathbf{y}) = 1/2$, which is exactly the definition of the median. $\qquad \square$

As we saw in Example 6.13, Bayes estimators generally are not invariant w.r.t. transformations: it depends on the measure of goodness-of-estimation and a specific transformation at hand.

Question 6.2 Show that for the squared error, Bayes estimators are invariant w.r.t. linear transformations, while for the absolute error, they are invariant w.r.t. any monotonic transformation.

6.4   Interval estimation: Credible sets

Bayesian approach to interval estimation is conceptually straightforward and simple. The posterior distribution of the unknown parameter of interest essentially contains all the information available on it from the data and prior knowledge. A Bayesian analog of confidence intervals (or sets in general) are credible sets:

Definition 6.2 (credible set) A $100(1-\alpha)\%$ credible set for θ is a subset (usually interval) $S_{\mathbf{y}} \subset \Theta$ such that

$$P(\theta \in S_{\mathbf{y}}|\mathbf{y}) = 1 - \alpha \qquad (6.14)$$

(for a discrete θ, the equality in (6.14) is relaxed by the inequality "$\geq$" being as close to an equality as possible).

Credible sets can usually be easily calculated from a posterior distribution in contrast to their frequentist counterparts. It is important to emphasize that the probability in (6.14) is on θ rather than on a set that makes probabilistic interpretation of Bayesian credible sets more straightforward.

Example 6.14 (continuation of Example 6.12) In addition to a point estimate for the average weight of cows μ obtained in Example 6.12, the farmer wants to construct a 95% credible set for it. The posterior mean $E(\mu|\mathbf{y}) = 412.9 \, kg$ (see Example 6.12), while from (6.4) the posterior variance $Var(\mu|\mathbf{y}) = (10/400 + 1/100)^{-1} = 28.6 \, kg^2$. Thus, the posterior distribution $\mu|\mathbf{y} \sim \mathcal{N}(412.9, 28.6)$ and the 95% credible interval for μ is

$$412.9 \pm z_{0.025}\sqrt{28.6} = (402.4, 423.4) \qquad (6.15)$$

Question 6.3 Do you think that assuming his prior and based on the data, the farmer can now claim that with probability 95% the average weight of cows is within the interval $(402.4, 423.4)$?

Example 6.15 Consider again a normal sample $Y_1, ..., Y_n \sim \mathcal{N}(\mu, \sigma^2)$ but with the unknown variance σ^2. Assume the independent noninformative prior $\pi(\mu, \sigma) \propto 1/\sigma$ from (6.6). Then, as we saw in Example 6.9, the posterior distribution of the standardized $\frac{\mu - \bar{y}}{s/\sqrt{n}}$ is t_{n-1}, and the credible $100(1-\alpha)\%$ interval for μ is

$$\bar{y} \pm t_{n-1,\alpha/2}\frac{s}{\sqrt{n}}, \qquad (6.16)$$

where $s^2 = \frac{1}{n-1}\sum_{i=1}^{n}(y_i - \bar{y})^2$. Although it coincides with the corresponding $100(1-\alpha)\%$ confidence interval for μ in Example 3.3, their probabilistic interpretations are different!

Table 6.1: Distribution of alcohol consumption in the UK and France and the corresponding likelihood ratio λ.

y	Beer	Brandy	Whisky	Wine
$P_{France}(y)$	0.1	0.2	0.1	0.6
$P_{UK}(y)$	0.5	0.1	0.2	0.2
$\lambda(y)$	5	1/2	2	1/3

Similar to confidence intervals, a credible set can usually be chosen in a nonunique way. For example, any interval of the form $(\bar{y} - t_{n-1,\alpha_1}\frac{s}{\sqrt{n}}, \bar{y} + t_{n-1,\alpha_2}\frac{s}{\sqrt{n}})$, where $\alpha_1 \geq 0.5$ and $\alpha_1 + \alpha_2 = \alpha$, is a $100(1-\alpha)\%$ credible set for μ in Example 6.15. It may be desirable then to find a credible set of a minimal size. Such a set would evidently include the "most likely" values of θ. This leads one to the definition of *highest posterior density* sets:

Definition 6.3 (highest posterior density set) A $100(1 - \alpha)\%$ highest posterior density (HPD) set for θ is a subset $S_\mathbf{y} \subset \Theta$ of the form

$$S_\mathbf{y} = \{\theta \in \Theta : \pi(\theta|\mathbf{y}) \geq c_\alpha\},$$

where c_α is such that $P(\theta \in S_\mathbf{y}|\mathbf{y}) = 1 - \alpha$ (with obvious modifications for a discrete θ).

Question 6.4 Show that if the posterior density $\pi(\theta|\mathbf{y})$ is unimodal and continuous, the HPD set for θ is an *interval* (a, b) and $\pi(a|\mathbf{y}) = \pi(b|\mathbf{y})$.

Evidently, for a *symmetric* posterior density, the corresponding HPD set is also symmetric around the posterior mean. Hence, the HPD interval for μ in Example 6.15 is indeed given by (6.16), as (6.15) is the 95% HPD interval for μ in Example 6.14.

Finally note that the definitions of credible and HPD sets are immediately extended for the multiparameter case.

6.5   Hypothesis testing

We will discuss Bayesian hypothesis testing within a general compound decision theory framework in Chapter 7 and present now its basic ideas. We start from two simple hypotheses and then consider the general case.

6.5.1   *Simple hypotheses*

Suppose that given the data $\mathbf{Y} \sim f_\theta(\mathbf{y})$, we want to test two simple hypotheses

$$H_0 : \theta = \theta_0 \quad \text{vs.} \quad H_1 : \theta = \theta_1$$

In such a setup, the parameter space Θ is essentially reduced to a set of two points: $\Theta = \{\theta_0, \theta_1\}$. A Bayesian defines a prior distribution $\pi(\cdot)$ on Θ, where $\pi(\theta_0) = \pi_0$ and $\pi(\theta_1) = 1 - \pi_0$, and calculates the corresponding posterior probabilities $\pi(\theta_0|\mathbf{y})$ and $\pi(\theta_1|\mathbf{y}) = 1 - \pi(\theta_0|\mathbf{y})$. In fact, it is often more convenient to work with the posterior *odds ratio* (or simply posterior odds) of H_1 to H_0, that is, $\pi(\theta_1|\mathbf{y})/\pi(\theta_0|\mathbf{y})$. In particular, one does not need then to calculate the normalizing constant in the posterior distribution. According to Bayes' theorem

$$\frac{P(\theta = \theta_1|\mathbf{y})}{P(\theta = \theta_0|\mathbf{y})} = \frac{f(\mathbf{y}|\theta_1)}{f(\mathbf{y}|\theta_0)} \cdot \frac{1 - \pi_0}{\pi_0} = \frac{L(\theta_1; \mathbf{y})}{L(\theta_0; \mathbf{y})} \cdot \frac{1 - \pi_0}{\pi_0}, \tag{6.17}$$

where $f(\mathbf{y}|\theta_0)$ and $f(\mathbf{y}|\theta_1)$ are distributions of the data under H_0 and H_1 respectively. Thus, for two simple hypotheses, the posterior odds is the multiplication of the likelihood ratio and the prior odds. In particular, assuming both hypotheses to be equally likely a priori ($\pi_0 = 0.5$), the posterior odds is identical to the likelihood ratio.

Clearly, the larger the posterior odds, the larger $P(\theta_1|\mathbf{y})$ and the more evidence there is against H_0 in favor of H_1. A Bayesian then sets a threshold C and rejects the null, if the posterior odds is larger than C. For example, if $C = 1$, the null is rejected if $P(\theta_1|\mathbf{y}) > 0.5$ which mimics a simple rule of choosing the hypothesis with higher posterior probability. Setting $C = 9$ or $C = 19$ corresponds to rejecting the null if $P(\theta_1|\mathbf{y}) > 0.9$ or $P(\theta_1|\mathbf{y}) > 0.95$ respectively.

In terms of the likelihood ratio $\lambda(\mathbf{y}) = \frac{L(\theta_1; \mathbf{y})}{L(\theta_0; \mathbf{y})}$, the null hypothesis is rejected if $\lambda(\mathbf{y}) > C^*$, where $C^* = \frac{\pi_0}{1 - \pi_0} C$. Hence, for simple hypotheses, from a non-Bayesian perspective, the Bayes test is a likelihood ratio test with a critical value C^*, which is the MP test of size $\alpha = P_{\theta_0}(\lambda(\mathbf{y}) > C^*)$ (see Section 4.2.4).

To be on a more rigorous ground for choosing C, suppose that one pays penalties of K_0 and K_1 on erroneous rejection of H_0 and H_1 respectively. If the null is rejected, the posterior expectation of payment for a false rejection is then $K_0 P(\theta_0|\mathbf{y})$, while if it is accepted, the corresponding expectation of payment for a wrong decision is $K_1 P(\theta_1|\mathbf{y})$. Choosing between these two options in order to minimize the expected posterior loss, one will then reject H_0 if $K_0 P(\theta_0|\mathbf{y}) < K_1 P(\theta_1|\mathbf{y})$ or, equivalently, the posterior odds $P(\theta_1|\mathbf{y})/P(\theta_0|\mathbf{y}) > K_0/K_1$, and accept otherwise. Thus, the threshold C is, in fact, a ratio between penalties for the two types of erroneous decisions. In particular, if the penalties are equal, $C = 1$.

Example 6.16 Consider testing two simple hypotheses for a normal mean with known variance from the Bayesian perspective. Let $Y_1, \dots, Y_n \sim \mathcal{N}(\mu, \sigma^2)$ and we want to test $H_0 : \mu = \mu_0$ against $H_1 : \mu = \mu_1$, where $\mu_1 > \mu_0$. Assume prior probabilities $\pi_0 = P(\mu = \mu_0)$ and $\pi_1 = 1 - \pi_0 = P(\mu = \mu_1)$. The posterior odds ratio is then

$$\frac{P(\mu = \mu_1|\mathbf{y})}{P(\mu = \mu_0|\mathbf{y})} = \lambda(\mathbf{y})\frac{\pi_1}{\pi_0} = \exp\left\{\frac{1}{\sigma^2}(\mu_1 - \mu_0)n\left(\bar{y} - \frac{\mu_1 + \mu_0}{2}\right)\right\} \cdot \frac{\pi_1}{\pi_0}$$

(see Example 4.8). Following the simplest Bayes rule and choosing a more likely a posterior hypothesis, the null is rejected if $\frac{P(\mu=\mu_1|\mathbf{y})}{P(\mu=\mu_0|\mathbf{y})} > 1$, that is, if

$$\bar{y} > \frac{\mu_0 + \mu_1}{2} + \frac{\sigma^2}{(\mu_1 - \mu_0)n} \cdot \ln\frac{\pi_0}{\pi_1}$$

In particular, if $\pi_0 = \pi_1 = 0.5$, we reject it if

$$\bar{y} > \frac{\mu_0 + \mu_1}{2}$$

Example 6.17 (continuation of Example 4.7) Recall that in Example 4.7 we tried to identify the origin of a man ($H_0 : France$ or $H_1 : UK$) according to his choice of a favorite drink. We continue now this example using the Bayesian approach. In addition to the distributions of alcohol consumption in the UK and France (see Table 4.4), we assume some prior probabilities $\pi_0 = P(UK)$ and $\pi_1 = P(France)$. To specify π_0 and π_1 we may try to utilize some other available (even subjective) information, observing, say, the manner in which the man is dressed or his behavior. Otherwise, we can use the noninformative prior setting $\pi_0 = \pi_1 = 0.5$. We apply the simplest Bayes rule for the latter case and reject then the null if the likelihood ratio $\lambda(y) = P_{UK}(y)/P_{France}(y) > 1$. From Example 4.7 we have:

We decide then that the man is from the UK if his favorite drink is *beer* or *whisky* and from France if he prefers *wine* or *brandy* (compare with the corresponding MP test from Example 4.7). Note that unlike the MP test, the above Bayesian test is symmetric: if we exchange the two hypotheses and test $H_0 = UK$ against $H_1 = France$, the result remains the same.

6.5.2 Composite hypotheses

The ideas for testing simple hypotheses can be extended to composite hypotheses. Let

$$H_0 : \theta \in \Theta_0 \quad \text{vs.} \quad H_1 : \theta \in \Theta_1,$$

where $\Theta_0 \cap \Theta_1 = \emptyset$, $\Theta_0 \cup \Theta_1 = \Theta$. The prior $\pi(\theta)$ can be defined either directly over the entire parameter space Θ or by defining first the prior probabilities $P(\theta \in \Theta_0) = \pi_0$ and $P(\theta \in \Theta_1) = 1 - \pi_0$ for both hypotheses and then the (conditional) distributions $p_0(\theta)$ and $p_1(\theta)$ on Θ_0 and Θ_1 respectively. In this case

$$\pi(\theta) = \pi_0 p_0(\theta)I(\theta \in \Theta_0) + (1 - \pi_0)p_1(\theta)I(\theta \in \Theta_1) \tag{6.18}$$

As we shall see, such a form is more convenient, for example, for testing the point null hypothesis. In addition, it is more flexible and, in particular, allows Θ_0 and Θ_1 to be of different dimensions.

The corresponding posterior odds can be written then as

$$\frac{P(\theta \in \Theta_1|\mathbf{y})}{P(\theta \in \Theta_0|\mathbf{y})} = \frac{\int_{\Theta_1} \pi(\theta|\mathbf{y})d\theta}{\int_{\Theta_0} \pi(\theta|\mathbf{y})d\theta} = \frac{\int_{\Theta_1} f(\mathbf{y}|\theta)p_1(\theta)d\theta}{\int_{\Theta_0} f(\mathbf{y}|\theta)p_0(\theta)d\theta} \cdot \frac{1-\pi_0}{\pi_0} = \frac{\int_{\Theta_1} L(\theta;\mathbf{y})p_1(\theta)d\theta}{\int_{\Theta_0} L(\theta;\mathbf{y})p_0(\theta)d\theta} \cdot \frac{1-\pi_0}{\pi_0}$$

(6.19)

and, as for the simple hypotheses case, the null hypothesis is rejected if it is greater than a fixed threshold C. The choice of C is discussed in the previous Section 6.5.1; in particular, in the simplest case, $C = 1$.

Example 6.18 (one-sided test for normal data with unknown variance) Continue Example 6.15, where $Y_1, ..., Y_n \sim \mathcal{N}(\mu, \sigma^2)$ and $\pi(\mu, \sigma) \propto 1/\sigma$. Consider testing one-sided hypotheses on μ: $H_0 : \mu \le \mu_0$ versus $H_1 : \mu > \mu_0$. The posterior distribution of a standardized variable $T = \frac{\mu - \bar{y}}{s/\sqrt{n}}$ is t_{n-1} (see Example 6.9). The posterior probability of the null is then

$$P(\mu \le \mu_0|\mathbf{y}) = P\left(T \le \frac{\mu_0 - \bar{y}}{s/\sqrt{n}}\right) = T_{n-1}\left(\frac{\mu_0 - \bar{y}}{s/\sqrt{n}}\right),$$

where T_{n-1} is the cdf of Student's distribution with $n-1$ degrees of freedom. For a fixed threshold C, we reject the null hypothesis if the posterior odds ratio is larger than C, that is, if $T_{n-1}\left(\frac{\mu_0 - \bar{y}}{s/\sqrt{n}}\right) < \frac{C}{C+1}$.

Compare the above Bayesian test with its classical GLRT counterpart, which is a one-sided t-test (see Example 4.18 in Section 4.3.3). The corresponding p-value is

$$p - \text{value} = P_{\mu_0}\left(\frac{\bar{Y} - \mu_0}{s/\sqrt{n}} \ge \frac{\bar{y} - \mu_0}{s/\sqrt{n}}\right) = 1 - T_{n-1}\left(\frac{\bar{y} - \mu_0}{s/\sqrt{n}}\right) = T_{n-1}\left(\frac{\mu_0 - \bar{y}}{s/\sqrt{n}}\right)$$

due to the symmetry of the t-distribution. Thus, the posterior probability of H_0 equals to the p-value. Such connections between both approaches tend to happen when testing one-sided hypotheses for a location parameter of a distribution imposing noninformative priors.

6.5.3   Testing a point null hypothesis

Consider testing a point null hypothesis about a *continuous* parameter θ:

$$H_0 : \theta = \theta_0 \quad \text{vs.} \quad H_1 : \theta \ne \theta_0$$

Defining a continuous prior $\pi(\theta)$ on the entire parameter space Θ is meaningless in this case since it assigns a zero prior (and, therefore, zero posterior!) probability for the *point* null hypothesis $\theta = \theta_0$. We consider several Bayesian approaches to tackle this problem.

In many practical situations one may argue that, in fact, the point null hypothesis is just an idealization of an *interval* hypothesis on a small interval. For example, testing the null hypothesis that a new drug has no effect, one actually does not really mean that the true effect is *exactly* zero rather that it is so small that it cannot be distinguishable from zero for a given measurement accuracy or from a clinical point of view. In this case one can essentially consider the *interval* hypothesis $H_0 : \theta \in (\theta - \varepsilon, \theta + \varepsilon)$ versus $H_1 : \theta \notin (\theta - \varepsilon, \theta + \varepsilon)$ and apply the standard Bayesian testing technique. Such an approach, however, requires one to know ε, which is often unspecified and the inference might strongly depend on its choice.

A more reasonable alternative is to consider a point null hypothesis with prior probability π_0 and take $p_0(\cdot)$ to be a *point mass* at θ_0 in (6.18). It means that with probability π_0, θ is exactly θ_0, while otherwise it is spread over all $\theta \ne \theta_0$ according to the (proper) density $p_1(\theta)$. The resulting prior $\pi(\theta)$ is

$$\pi(\theta) = \pi_0 I(\theta = \theta_0) + (1 - \pi_0)p_1(\theta)I(\theta \ne \theta_0)$$

and has both discrete and continuous parts. It follows then that the posterior odds in (6.19) is

$$\frac{P(\theta \ne \theta_0|\mathbf{y})}{P(\theta = \theta_0|\mathbf{y})} = \frac{\int f(\mathbf{y}|\theta)p_1(\theta)d\theta}{f(\mathbf{y}|\theta_0)} \cdot \frac{1 - \pi_0}{\pi_0},$$

(6.20)

where $\int f(\mathbf{y}|\theta)p_1(\theta)d\theta$ is the marginal density of the data with respect to the prior $p_1(\cdot)$. Note that $p_1(\theta)$, $\theta \in \Theta_1$ should necessarily be a *proper* density since otherwise the posterior odds in (6.20) would depend on a specific constant.

Example 6.19 (two-sided test for normal data with known variance) Let $Y_1,...,Y_n \sim \mathcal{N}(\mu,\sigma^2)$ with known variance σ^2. We want to test $H_0 : \mu = \mu_0$ versus $H_1 : \mu \neq \mu_0$. Set a prior probability π_0 for the null hypothesis (in particular, $\pi_0 = 0.5$ if both hypotheses are assumed to be a priori equally likely). Unless we possess some prior information that μ tends to be larger or smaller than μ_0 under the alternative, it is reasonable to choose $p_1(\mu)$ to be symmetric around μ_0. A convenient choice is a conjugate normal prior $\mathcal{N}(\mu_0,\tau^2)$. Then, after straightforward calculus we have

$$
\frac{P(\mu \neq \mu_0|\mathbf{y})}{P(\mu = \mu_0|\mathbf{y})} = \frac{\int_{-\infty}^{\infty} \left(\frac{1}{\sqrt{2\pi\sigma^2}}\right)^n \exp\left(-\frac{\sum_{i=1}^{n}(y_i-\mu_0)^2}{2\sigma^2}\right) \frac{1}{\sqrt{2\pi\tau^2}} \exp\left(-\frac{(\mu-\mu_0)^2}{2\tau^2}\right) d\mu}{\left(\frac{1}{\sqrt{2\pi\sigma^2}}\right)^n \exp\left(-\frac{\sum_{i=1}^{n}(y_i-\mu_0)^2}{2\sigma^2}\right)} \cdot \frac{1-\pi_0}{\pi_0}
$$

$$
= \frac{1}{\sqrt{1+\gamma}} \exp\left(\frac{n(\bar{y}-\mu_0)^2}{2\sigma^2(1+1/\gamma)}\right) \cdot \frac{1-\pi_0}{\pi_0}, \tag{6.21}
$$

where the variance ratio $\gamma = n\tau^2/\sigma^2$ and the point $\mu = \mu_0$ does not matter in the integration since the density of $\mathcal{N}(\mu_0,\tau^2)$ is continuous. Note that the posterior odds (6.21) is a monotonically increasing function of $|z| = \frac{|\bar{y}-\mu_0|}{\sigma/\sqrt{n}}$, which is also a test statistic for the analogous non-Bayesian test (see Example 4.14 in Section 4.3). Thus, both Bayesian and classical tests will reject the null when $|z|$ is "too large" but the corresponding critical values are determined by different arguments.

Question 6.5 Is there duality between Bayesian hypothesis testing a point null hypothesis and credible sets similar to that between frequentist hypothesis testing and confidence intervals established in Section 4.4?

6.6 *Asymptotic properties of the posterior distribution

Similar to frequentist asymptotics considered in Chapter 5, it is interesting to consider asymptotic properties of Bayesian approach when a sample size n tends to infinity. Intuitively, the more data we have, the more information we can elicit from it, the less should be the impact of a prior belief on inference. In this sense Bayesian and frequentist inferences should somehow merge as $n \to \infty$.

Thus, a natural question is what is an asymptotic approximation of a posterior distribution? Whether it is concentrated closer and closer around the true (unknown) θ_0 as a sample size n increases?

The following theorem (aka Bernstein–von Mises theorem) shows that under general regularity conditions, the posterior distribution $\pi(\theta|\mathbf{y})$ approaches a normal distribution centered at the MLE $\hat{\theta}_n$ and the variance $I^{-1}(\theta_0)$, where $I(\theta_0)$ is the Fisher information number at the true θ_0:

Theorem 6.1 (Bernstein–von Mises) Let $Y_1,...,Y_n \sim f(y|\theta)$ and $\theta \sim \pi(\theta)$. Let θ_0 be the true (unknown) value and $\pi(\theta_0) > 0$ (otherwise, θ_0 is impossible a priori). Let $\hat{\theta}_n$ be a MLE of θ.

Then under "general regularity" conditions on $f(y|\theta)$ (similar to those required for asymptotic normality of MLE) and a prior $\pi(\theta)$, as $n \to \infty$,

$$
\sqrt{n}(\theta - \hat{\theta}_n)|\mathbf{y} \xrightarrow{\mathscr{D}} \mathcal{N}(0,I^*(\theta_0)^{-1})
$$

a.s. under $f(y|\theta_0)$, that is,

$$
\sup_t | P\left(\sqrt{n}(\theta - \hat{\theta}_n) \leq t|\mathbf{y}\right) - \Phi(\sqrt{I^*(\theta_0)}\,t) | \xrightarrow{\text{a.s.}} 0 \tag{6.22}
$$

Exact regularity conditions and the proof can be found in Bickel and Doksum (2015, Theorem 5.5.2).

The (6.22) implies the following corollary:

Corollary 6.1 Under the conditions of Theorem 6.1,

$$\sup_t | P\left(\sqrt{n}(\theta - \hat{\theta}_n) \leq t | \mathbf{y}\right) - \Phi(\sqrt{I^*(\hat{\theta}_n)}\, t) | \xrightarrow{a.s.} 0$$

Simply stated, Corollary 6.1 says that for large samples, the posterior distribution of θ is approximately normal with the mean $\hat{\theta}_n$ and variance $I(\hat{\theta}_n)^{-1} = (nI^*(\hat{\theta}_n))^{-1}$.

In particular, Corollary 6.1 allows one to construct an asymptotic normal $100(1-\alpha)\%$ credible interval for θ of the form

$$\hat{\theta}_n \pm z_{\alpha/2}\sqrt{I^{-1}(\hat{\theta}_n)} \tag{6.23}$$

It is important to emphasize again that although the asymptotic credible interval (6.23) coincides with the asymptotic Wald confidence interval (5.14), they have conceptually different probabilistic interpretations.

Example 6.20 (continuation of Example 6.2) Consider a normal sample $Y_1,...,Y_n \sim \mathcal{N}(\mu, \sigma^2)$ with a known σ and a normal prior $\mathcal{N}(\mu_0, \tau^2)$ for an unknown mean μ.

The posterior distribution of $\mu|\mathbf{y}$ is normal (even not asymptotically)

$$\pi(\mu|\mathbf{y}) \sim \mathcal{N}\left(\frac{\tau^2}{\frac{\sigma^2}{n} + \tau^2}\bar{y}_n + \frac{\frac{\sigma^2}{n}}{\frac{\sigma^2}{n} + \tau^2}\mu_0, \left(\frac{n}{\sigma^2} + \frac{1}{\tau^2}\right)^{-1}\right)$$

(see Example 6.2) and as $n \to \infty$, it approaches to $\mathcal{N}(\bar{y}_n, \frac{\sigma^2}{n})$. Recall that the MLE $\hat{\mu}_n = \bar{y}_n$ (see Example 2.7) and $I(\mu) = \frac{n}{\sigma^2}$ (see Example 2.18).

Example 6.21 Let $Y_1,...,Y_n \sim Pois(\lambda)$ and assume an exponential prior $\exp(\theta)$ on λ. The MLE $\hat{\lambda}_n = \bar{Y}_n$ and $I(\lambda) = \frac{n}{\lambda}$ (see Exercise 2.16). The posterior distribution of $\lambda|\mathbf{y}$ is

$$\pi(\lambda|\mathbf{y}) \propto e^{-n\lambda}\lambda^{\sum_{i=1}^n y_i}e^{-\theta\lambda} = \lambda^{\sum_{i=1}^n y_i}e^{-(n+\theta)\lambda}$$

Hence, $\lambda|\mathbf{y} \sim Gamma(\sum_{i=1}^n y_i + 1, n + \theta)$ and is approximated by $Gamma(\sum_{i=1}^n y_i, n)$ for large n. Using the properties of Gamma distribution, $Gamma(\sum_{i=1}^n y_i, n) = \frac{1}{n}\sum_{i=1}^n Gamma(y_i, 1)$ (see Section A.4.8) which by the CLT, is asymptotically normal $\mathcal{N}(\bar{y}_n, \frac{\bar{y}_n}{n}) = \mathcal{N}(\hat{\lambda}_n, I(\hat{\lambda}_n)^{-1})$. Thus, the asymptotic $100(1-\alpha)\%$ credible set for λ is $\bar{y}_n \pm z_{\alpha/2}\sqrt{\bar{y}_n/n}$ that coincides with the asymptotic $100(1-\alpha)\%$ confidence interval for λ from Example 5.13.

A more detailed study of Bayesian asymptotics is beyond the scope of this book and we refer the interested reader to, e.g., Bickel and Doksum (2015, Chapter 5.5).

6.7 Exercises

Exercise 6.1 Assume that $Y \sim f_\theta(y)$ and a priori $\theta \sim \pi(\theta)$. After obtaining the first observation y_1 from $f_\theta(y)$ we update the distribution of θ by calculating its posterior distribution $\pi(\theta|y_1)$ and use it as a prior distribution for θ before we observe y_2 independent of y_1. Having y_2, we update the distribution of θ again by deriving the corresponding posterior distribution. Show that one will get the same posterior distribution of θ if instead of calculating it sequentially, he would use the entire data (i.e. y_1 and y_2) and the original prior $\pi(\theta)$.

Exercise 6.2 Three friends, Optimist, Realist and Pessimist, go to a casino. They decide to play a gambling game for which they do not know the probability p of winning. Motivated by an exciting lecture on Bayesian statistics, they decided to apply Bayesian analysis to the problem. Optimist chose a prior on p from the conjugate family of distributions. In addition, he believes that the chances are 50%-50% (i.e. $E(p) = 1/2$) with $Var(p) = 1/36$. Realist chose a uniform prior $\mathcal{U}(0,1)$.

1. Show that both priors belong to the family of Beta distributions and find the parameters for the priors of Optimist and Realist.

2. Pessimist does not have any prior beliefs and decided to choose the noninformative Jeffreys prior. What is his prior? Does it also belong to the Beta family?

3. Being poor students of Statistics, Optimist, Realist and Pessimist do not have enough money to gamble separately, so they decided to play together. They played the game 25 times and won 12 times. What is the posterior distribution for each one of them?

4. Find the corresponding posterior means and calculate the 95% Bayesian credible intervals for p for each student.

   (*Hint*: use the fact that if $p \sim Beta(\alpha, \beta)$ and $\rho = \frac{p}{1-p}$ is the odds ratio, then $\frac{\beta}{\alpha}\rho \sim F_{2\alpha, 2\beta}$)

5. The three friends told their classmate Skeptic about the exciting Bayesian analysis each one of them has performed. However, Skeptic is, naturally, skeptical about the Bayesian approach. He does not believe in any priors and decided to perform a "classical" (non-Bayesian) analysis of the same data. What is his estimate and the 95% confidence interval for p? Compare the results and comment on them.

Exercise 6.3 Let $Y_1, \ldots, Y_n \sim f_\theta(y)$, where $f_\theta(y)$ belongs to the (one-parameter) exponential family, i.e. the support $\text{supp}(f_\theta)$ does not depend on θ and $f_\theta(y) = \exp\{c(\theta)T(y) + d(\theta) + S(y)\}$, $y \in \text{supp}(f_\theta)$.

1. Show that the conjugate prior family for θ is $\pi_{\alpha, \beta}(\theta) \propto \exp\{\alpha c(\theta) + \beta d(\theta)\}$.

2. Find a conjugate prior for the parameter λ of the Poisson distribution $Pois(\lambda)$.

Exercise 6.4 The waiting time for a bus at a given bus stop at a certain time of day is known to have a uniform distribution $\mathcal{U}(0, \theta)$. From other similar routes it is known that θ has the Pareto distribution $Pa(7, 4)$, where the density of the Pareto distribution $Pa(\alpha, \beta)$ is $f_{\alpha, \beta}(\theta) \propto \theta^{-(\alpha+1)}$, $\theta \geq \beta$ (see Exercise 1.2). Waiting times of 10, 3, 2, 5 and 14 minutes have been observed at the given bus stop during the last 5 days.

1. Show that the Pareto distribution provides a conjugate prior for uniform data and find the posterior distribution of θ.

2. Estimate θ w.r.t. the squared error (recall that $E(Pa(\alpha, \beta)) = \frac{\alpha\beta}{\alpha-1}$, $\alpha > 1$).

3. Find a 95% HPD for θ.

4. Test the hypothesis $H_0 : 0 \leq \theta \leq 15$ vs. $H_1 : \theta > 15$ by choosing the more likely (a posteriori) hypothesis.

Exercise 6.5 Let $Y_1, \ldots, Y_n \sim Geom(p)$.

1. Show that a family of Beta priors $Beta(\alpha, \beta)$ for p is a conjugate family of priors for geometric data, derive the corresponding posterior distribution and find the Bayes estimator of p w.r.t. the squared error.

2. Find Jeffreys noninformative prior for p and repeat the previous question for it.

Exercise 6.6 Continue Exercise 2.3. Assume a Gamma prior $Gamma(\alpha, \beta)$ for θ.

1. Is it a conjugate prior for the data?

2. Find the Bayes estimators w.r.t. the squared error of

   (a) θ

   (b) expected lifetimes of a single element and an entire device containing k elements

   (c) probability that a device will operate for at least a hours

3. Are the above Bayes estimators consistent from a frequentist point of view?

Exercise 6.7 An accurate calculation of the IQ (intelligence level) μ of a child requires a long series of various tests. Instead, psychologists suggest an express method using a less accurate intelligence test whose result X is related to the true IQ level μ as $X \sim \mathcal{N}(\mu, 100)$. It is known that in the population as a whole, $\mu \sim \mathcal{N}(100, 225)$. A young Genius got 115 in the test.

1. Find the Bayes estimate for the Genius' IQ w.r.t. the squared error and 95% HPD set for it.

2. Genius claims that he belongs to the group of children with an especially high IQ (above 110). Test his claim.

Chapter 7

*Elements of Statistical Decision Theory

7.1 Introduction and notations

What is statistical inference? In fact, it can generally be viewed as making a conclusion about an underlying random process from the data generated by it. Statistical inference can, therefore, be considered as a decision making under uncertainty based on analysis of data. Each decision has consequences as there is a price for an error, and the goal is then to find a proper rule for making a "good" decision. In particular, statistical decisions, by being statistical, are always subject to random errors that might be larger or smaller. A "good" statistical rule is thus the one whose price for error is stochastically "small." The choice of such a rule should take into account the following information:

a) the data;

b) other available information, whether it is just a model specification for the data or some prior knowledge about the problem at hand;

c) consequences of all possible inference errors.

More formally, and similar to the previous chapters, we consider a parametric setup, where the decision should be made on an unknown parameter of the model. The main components of a statistical decision problem are then

1. A *parameter space* Θ. Usually it would be a subset of $\mathbb{R}^p$.

2. The *data* $\mathbf{y} \in \Omega$ which lies in the *sample space* Ω, $\mathbf{y} \in \mathbb{R}^n$.

3. A family of distributions $\mathscr{F} = \{f_\theta(\mathbf{y}), \theta \in \Theta\}$ on Ω.

4. An *action set* $\mathscr{A}$ containing a collection of all possible actions (decisions).

5. A *loss function* $L(\theta, a) : \Theta \times \mathscr{A} \to \mathbb{R}$ specifying the loss (the price) of choosing an action $a \in \mathscr{A}$ when the true value of the unknown parameter is $\theta \in \Theta$.

6. A *decision rule* chooses an action $a \in \mathscr{A}$ based on the observed data $\mathbf{y} \in \Omega$ and can thus be defined as a function $\delta = \delta(\mathbf{y}) : \Omega \to \mathscr{A}$, where $\delta \in \Delta$, a class of functions $\Omega \to \mathscr{A}$.

All three main problems of statistical inference, namely, point estimation, confidence intervals and hypothesis testing can be put within the general statistical decision theory framework.

Example 7.1 (point estimation) Estimation of the unknown parameter $\theta \in \Theta$ from a random sample $Y_1, ..., Y_n \sim f_\theta(y)$ can be viewed as a decision making procedure, where any estimator $\hat{\theta}$ is an action and, therefore, $\mathscr{A} = \Theta$. A loss function is a measure of goodness-of-estimation. The most conventional one is the quadratic loss corresponds to $L(\theta, \hat{\theta}) = (\hat{\theta} - \theta)^2$, although one can consider other losses as well, e.g. weighted quadratic losses $L(\theta, \hat{\theta}) = w(\theta)(\hat{\theta} - \theta)^2$ for various weights functions $w(\theta)$, or the absolute error loss $L(\theta, \hat{\theta}) = |\hat{\theta} - \theta|$ (see Section 2.6).

Example 7.2 (confidence intervals) In this setup, an action is any interval (a_1, a_2) and the action set $\mathscr{A} = \{(a_1, a_2) : -\infty < a_1 < a_2 < \infty\}$. An interval (a_1, a_2) either covers the true parameter $\theta \in \Theta$ (correct decision) or not (wrong decision) but the length of an interval should also count. There is no commonly used loss function for confidence intervals. A suitable choice would be, for example,

$$L(\theta, (a_1, a_2)) = c_1 \mathbb{I}\{\theta \notin (a_1, a_2)\} + c_2(a_2 - a_1)$$

DOI: 10.1201/9781003175407-7

for some $c_1, c_2 > 0$. That is, the loss has c_1 when θ is not in the interval, and there is a penalty on the length of the interval. Alternatively, one can consider

$$L(\theta, (a_1, a_2)) = \begin{cases} \frac{a_2 - a_1}{1 + a_2 - a_1} & a_1 \leq \theta \leq a_2 \\ \\ 1 & \text{otherwise.} \end{cases} \tag{7.1}$$

Thus, the loss is large if the interval does not cover the true parameter and small if it does and the interval is short. Therefore, a decision maker faces the dilemma of having a short interval that might not include the parameter, or a longer one that even if it does cover the parameter, leaves the decision maker with a substantial loss.

Example 7.3 (hypothesis testing) Consider testing two general hypotheses $H_0 : \theta \in \Theta_0$ vs. $H_1 : \theta \in \Theta_1$, where $\Theta_0 \cap \Theta_1 = \emptyset$ and $\Theta_0 \cup \Theta_1 = \Theta$. An action set consists of two possible actions $\mathscr{A} = \{0, 1\}$, where $a = 0$ and $a = 1$ corresponds respectively to acceptance or rejection of the null. The standard loss function in this case is the so-called 0-1 loss, which is zero if the decision is correct and one otherwise. It can be written as

$$L(\theta, a) = \begin{cases} a, & \theta \in \Theta_0 \\ 1 - a, & \theta \in \Theta_1. \end{cases}$$

One can consider a more general loss function, where the prices K_0 and K_1 for erroneous rejection and acceptance of the null are not necessarily the same. In this case,

$$L(\theta, a) = \begin{cases} K_0\, a, & \theta \in \Theta_0 \\ K_1\, (1 - a), & \theta \in \Theta_1. \end{cases}$$

The set of possible rules Δ may be restricted to a certain class. For example, for point estimation under quadratic loss in Example 7.1 one can consider only the class of unbiased estimators or the class of affine functions $\delta(y) = a + b\bar{y}$, where a and b are real.

7.2   Risk function and admissibility

The main goal of statistical decision theory is to find the "best" decision rule $\delta = \delta(\mathbf{Y})$. How to measure the "goodness" of a rule? A first naive way would be to consider its loss $L(\theta, \delta)$ but it is a random variable since $\delta(\mathbf{Y})$ depends on a random sample. We define then the *expected loss* of a rule δ called its *risk function*:

Definition 7.1 (risk) The risk function of a decision rule $\delta = \delta(\mathbf{Y})$ w.r.t. the loss function $L(\theta, \delta)$ is $R(\theta, \delta) = EL(\theta, \delta(\mathbf{Y}))$, $\theta \in \Theta$, where the expectation is taken w.r.t. the distribution $f_\theta(\mathbf{y})$ of $\mathbf{Y}$.

Example 7.4 (point estimation under the quadratic loss) Consider again the estimation of the unknown parameter $\theta \in \Theta$ from a random sample (see Example 7.1) under the quadratic loss $L(\theta, \hat{\theta}) = (\hat{\theta} - \theta)^2$. In this case the risk function $R(\theta, \hat{\theta}) = E(\hat{\theta} - \theta)^2$, which is exactly the mean squared error $MSE(\theta, \hat{\theta})$ (see Section 2.6).

Example 7.5 (testing two simple hypotheses) Consider testing two simple hypotheses $H_0 : \theta = \theta_0$ vs. $H_1 : \theta = \theta_1$. A rule (test statistic) $\delta(\mathbf{Y})$ is either zero (accept H_0) or one (reject H_0). For the 0-1 loss,

$$L(\theta, \delta(\mathbf{Y})) = \begin{cases} \delta(\mathbf{Y}), & \theta = \theta_0 \\ 1 - \delta(\mathbf{Y}), & \theta = \theta_1 \end{cases}$$

(see Example 7.3) and the corresponding risk function is, therefore,

$$R(\theta, \delta) = \begin{cases} P_{\theta_0}(\delta(\mathbf{Y}) = 1), & \theta = \theta_0 \\ P_{\theta_1}(\delta(\mathbf{Y}) = 0), & \theta = \theta_1, \end{cases}$$

where $P_{\theta_0}(\delta(\mathbf{Y}) = 1)$ and $P_{\theta_1}(\delta(\mathbf{Y}) = 0)$ are respectively the probabilities of Type I and Type II errors. Similarly, for non-symmetric losses K_0 and K_1 (see Example 7.3), the risk is

$$R(\theta, \delta) = \begin{cases} K_0 P_{\theta_0}(\delta(\mathbf{Y}) = 1), & \theta = \theta_0 \\ K_1 P_{\theta_1}(\delta(\mathbf{Y}) = 0), & \theta = \theta_1 \end{cases}$$

We can now compare two rules δ_1 and δ_2 and choose the one with the smaller risk. The problem is, however, that similar to MSE in the estimation setup (see Section 2.6), the risk function depends also on the specific value of the unknown θ. Hence, we can claim that δ_1 outperforms δ_2 only if its risk is *uniformly* smaller for all $\theta \in \Theta$. In this case, the rule δ_2 is clearly inferior to δ_1, which is uniformly better. This argument leads us to the following definition:

Definition 7.2 (**admissibility**) A decision rule δ is *inadmissible* (w.r.t. a given loss and class of rules) if there exists another decision rule $\tilde{\delta}$ from the class such that $R(\theta, \tilde{\delta}) \leq R(\theta, \delta)$ for all $\theta \in \Theta$ and there exists $\theta_0 \in \Theta$ for which the inequality is strict: $R(\theta_0, \tilde{\delta}) < R(\theta_0, \delta)$. A decision rule is *admissible*, if it is not inadmissible, that is, it cannot be uniformly improved.

Example 7.6 Suppose we want to estimate an unknown normal mean μ from a single observation $Y \sim N(\mu, 1)$ w.r.t. the quadratic loss within a class of affine estimators (rules) of the form $\delta_{a,b}(Y) = a + bY$, $a, b \in \mathbb{R}$. We will use this example throughout the entire Chapter 7.

Estimating a mean μ from a single observation might seem a trivial, somewhat toy example. Suppose, however, that there is a random sample $X_1, \ldots, X_n \sim \mathcal{N}(\theta, \sigma^2)$ of a size n, where σ^2 is known. Then, $Y = \sqrt{n} \frac{1}{n\sigma} \sum_{i=1}^n X_i$ is a sufficient statistic (see Section 1.3) and $Y \sim N(\sqrt{n} \frac{\theta}{\sigma}, 1)$. Thus, we can ignore the rest of the information in the sample, and consider a single observation $Y \sim N(\mu, 1)$, where $\mu = \sqrt{n}\theta/\sigma$.

We have

$$R(\mu, \delta_{a,b}) = \mathrm{E}\left(\mu - a - bY\right)^2 = \mathrm{Var}(a + bY) + \left(\mu - \mathrm{E}(a + bY)\right)^2$$
$$= b^2 + \left(\mu(1 - b) - a\right)^2, \tag{7.2}$$

since $\mathrm{E}(a + bY) = a + b\mu$ and $\mathrm{Var}(a + bY) = b^2$. In particular, $\delta_{0,1}(Y) = Y$ and $R(\mu, \delta_{0,1}) = 1$. Thus, for any $b > 1$, $R(\mu, \delta_{a,b}) > b^2 > 1 = R(\mu, \delta_{0,1})$ for every μ. On the other hand, for any $b < 0$ and any a we have

$$R(\mu, \delta_{a,b}) > \left(\mu(1 - b) - a\right)^2 = (1 - b)^2 \left(\mu - \frac{a}{1 - b}\right)^2 > \left(\mu - \frac{a}{1 - b}\right)^2 = R(\mu, \delta_{\frac{a}{1-b}, 0}),$$

where $\delta_{\frac{a}{1-b}, 0}(Y)$ is the constant estimator $\frac{a}{1-b}$. Hence, the affine decision rules $\delta_{a,b}$ are inadmissible for all $b < 0$ and $b > 1$. Furthermore, $\delta_{a,1}(Y)$ for any $a \neq 0$ is also inadmissible since $R(\mu, \delta_{a,1}) = 1 + a^2 > 1 = R(\mu, \delta_{0,1})$ for every μ. We demonstrate later in Example 7.16 admissibility of $\delta_{a,b}(Y)$ for all $0 \leq b < 1$ and in Example 7.9 for $\delta_{0,1}(Y) = Y$.

Example 7.7 (**testing two simple hypotheses—continuation of Example 7.5**) Consider testing two simple hypotheses $H_0 : \theta = \theta_0$ vs. $H_1 : \theta = \theta_1$ w.r.t. 0-1 loss. For simplicity assume that the distribution of Y is continuous. We will show that all tests which are not based on the likelihood ratio $\lambda(\mathbf{y}) = \frac{f_{\theta_1}(\mathbf{Y})}{f_{\theta_0}(\mathbf{Y})}$ are inadmissible. Indeed, consider any other test with a rejection region Ω_1. Let $\alpha = P_{\theta_0}(\mathbf{Y} \in \Omega_1)$ be the probability of its Type I error. Consider the LRT that rejects the null if $\lambda(\mathbf{Y}) \geq C$, where C is chosen to satisfy $\alpha = P_{\theta_0}(\lambda(\mathbf{Y}) \geq C)$. By construction, both tests have the same probability α of Type I error (risk at $\theta = \theta_0$), while by the Neyman–Pearson lemma, the LRT is an MP test and, therefore, has a smaller probability of Type II error (risk at $\theta = \theta_1$) than any other test at the same level. Thus, the original test with the rejection region Ω_1 is uniformly improved by the LRT with the correspondingly chosen C and, therefore, is inadmissible. We will establish admissibility of the LRT later in Example 7.19.

Generally, however, it might be hard (if possible) to check the admissibility (or inadmissibility) of a given rule directly by Definition 7.2. Remarkably, admissibility of a wide variety of rules can be easily established through their Bayesian interpretation as we show later in Section 7.6.

It seems obvious that only admissible rule should be of interest though there exist particular examples, where admissible rules are somewhat ad-hoc, while one can find an intuitively clear and simple inadmissible procedure. Usually, however, it is quite rare that one "reasonable" rule is uniformly better than another "reasonable" one. In this sense, admissibility is quite a weak requirement—it essentially declares the absence of a negative property, rather than possession of a positive one. Admissibility actually only means that there exist some value(s) of θ, where the rule cannot be improved, but it still does not tell anything about the goodness of the rule for other values of θ. As an extreme example of a useless admissible rule consider a trivial estimator $\delta_{a,0}(Y) = a$ for a fixed a in Example 7.6, which is just a constant. In this case, the risk function w.r.t. the quadratic loss is $R(\mu, \delta_{a,0}) = (\mu - a)^2$, which is zero if the true μ is equal to a, where no other rule (estimator) can have a zero risk and, therefore, no rule can outperform $\delta_{a,0}(Y)$ at $\mu = a$. Hence, such a trivial estimator is admissible while it is clear that it behaves poorly for μ different from a. Similarly, a degenerate test that never rejects the null hypothesis in Example 7.3 has evidently an unimprovable zero risk for all $\theta \in \Theta_0$ and, hence, is admissible, but at the same time has zero power for any $\theta \in \Theta_1$.

7.3   Minimax risk and minimax rules

Recall that our goal is to find the "best" rule (w.r.t. a given loss and class of rules). We would obviously like to find the rule that outperforms all other counterparts within a given class uniformly over all $\theta \in \Theta$. However, as we have discussed in Section 7.2, in most realistic situations it is impossible unless the action set is very restricted. What can be done then? Instead of looking at a risk function pointwise at each θ we need some *global* criteria over all $\theta \in \Theta$.

One possible approach is to consider the worst case scenario, that is, $\sup_{\theta \in \Theta} R(\theta, \delta)$ and seek the rule that will minimize the maximal possible risk:

Definition 7.3 (minimax rule) A rule $\delta^* = \delta^*(\mathbf{Y})$ is called *minimax* (w.r.t. a given loss and class of rules) if $\sup_{\theta \in \Theta} R(\theta, \delta^*) \leq \sup_{\theta \in \Theta} R(\theta, \delta)$ for any other rule δ within the class, that is, $\sup_{\theta \in \Theta} R(\theta, \delta^*) = \inf_{\delta \in \Delta} \sup_{\theta \in \Theta} R(\theta, \delta)$. The risk $\sup_{\theta \in \Theta} R(\theta, \delta^*)$ is called the *minimax risk*.

Example 7.8 (continuation of Example 7.6) Consider again a class of affine estimators $\delta_{a,b}(Y) = a + bY$ for a normal mean from a single observation Y with the unit variance. From (7.2) we have

$$\sup_{\mu} R(\mu, \delta_{a,b}) = \begin{cases} \infty, & b \neq 1 \\ 1 + a^2, & b = 1 \end{cases} \tag{7.3}$$

The maximal risk (7.3) is evidently minimized for $b = 1$ and $a = 0$. Thus, an estimator (rule) $\delta_{0,1}(Y) = Y$ is minimax within the family of affine estimators.

By its definition, the minimax criterion is conservative, where one wants to ensure a maximal protection in the worst possible case essentially ignoring all other less pessimistic scenarios. Minimax rule may be not unique. Moreover, they are not even necessarily admissible (although usually they are).

Theorem 7.1 If a minimax rule is unique, it is admissible.

Proof. Let δ^* be the unique minimax rule w.r.t. a loss function $L(\theta, \delta)$ and consider any other rule δ. We have

$$\sup_{\theta \in \Theta} R(\theta, \delta^*) < \sup_{\theta \in \Theta} R(\theta, \delta)$$

Hence, for any $\delta \in \Delta$, there exists $\theta_\delta \in \Theta$ such that $R(\theta_\delta, \delta^*) < R(\theta_\delta, \delta)$. Thus, δ^* cannot be improved on all Θ by any other rule and, therefore, is admissible. $\square$

Example 7.9 (continuation of Examples 7.6 and 7.8) In Example 7.8, we showed that the estimator $\delta_{0,1}(Y) = Y$ is a (unique) minimax estimator within the class of affine estimators and, hence, it is admissible within that class.

To find a minimax rule might be a hard problem in a general case. We will return to it later in Section 7.7 using the Bayesian approach.

7.4 Bayes risk and Bayes rules

An alternative global risk criterion to the conservative minimax risk approach is a *weighted average risk* $\int_\Theta R(\theta, \delta) \pi(\theta) d\theta$ (for continuous $\pi(\theta)$) or $\sum_i R(\theta_i, \delta) \pi(\theta_i)$ (for discrete $\pi(\theta)$) of a rule δ, where the weight function $\pi(\theta) \geq 0$ gives more emphasis to more "relevant" values of θ and reduces the influence of others. Assume that the integral (or sum) of $\pi(\theta)$ is finite—without loss of generality, we can assume that it is equal to one (why?). In this sense, $\pi(\theta)$ can be treated as a *prior distribution* of θ over Θ and the resulting weighted average risk is thus called a *Bayes risk* of a rule δ:

Definition 7.4 (Bayes risk) The *Bayes risk* of a rule $\delta = \delta(Y)$ w.r.t. a prior $\pi(\theta)$ and a loss function $L(\theta, \delta)$ is the weighed (or expected) risk $\rho(\pi, \delta) = \int_\Theta R(\theta, \delta) \pi(\theta) d\theta$ if π is continuous, and $\sum_j R(\theta_j, \delta) \pi(\theta_j)$ if π is discrete.

Definition 7.5 (Bayes rule) The *Bayes rule* $\delta_\pi^* = \delta_\pi^*(Y)$ (w.r.t. to a given loss, class of rules and a prior) is a decision rule that minimizes the Bayes risk $\rho(\pi, \delta)$, that is, $\rho(\pi, \delta^*) = \inf_{\delta \in \Delta} \rho(\pi, \delta)$.

Question 7.1 Show that for any prior, the Bayes risk is not larger than the minimax risk w.r.t. the same loss.

Example 7.10 (continuation of Example 7.6) Consider again $Y \sim \mathcal{N}(\mu, 1)$, where we want to estimate μ by an affine estimator $\delta_{a,b}(Y) = a + bY$ w.r.t. the quadratic loss. In addition, assume now a (conjugate) normal prior $\pi(\mu) = \mathcal{N}(\mu_0, \tau^2)$ on μ. From (7.2), the Bayes risk of a rule $\delta_{a,b}(\mathbf{Y})$ is

$$\rho(\pi, \delta_{a,b}) = E_\pi \left(b^2 + (\mu(1-b) - a)^2 \right) = b^2 + (E_\pi (\mu(1-b) - a))^2 + \text{Var}_\pi ((\mu(1-b)) - a))$$
$$= b^2 + (\mu_0(1-b) - a)^2 + \tau^2 (1-b)^2 \tag{7.4}$$

Minimizing $\rho(\pi_0, \delta_{a,b})$ in (7.4) w.r.t. a and b after a simple calculus yields $\hat{a} = \frac{1}{1+\tau^2} \mu_0$, $\hat{b} = \frac{\tau^2}{1+\tau^2}$ and the resulting Bayes rule $\delta^*(Y)$ is then

$$\delta_\pi^*(Y) = \delta_{\hat{a},\hat{b}}(Y) = \frac{1}{1+\tau^2} \mu_0 + \frac{\tau^2}{1+\tau^2} Y = \mu_0 + \frac{\tau^2}{1+\tau^2}(Y - \mu_0), \tag{7.5}$$

which is a linear shrinkage towards the prior mean μ_0. From (7.4) it follows that the minimal Bayes risk is

$$\rho(\pi, \delta_\pi^*) = \left(\frac{\tau^2}{1+\tau^2} \right)^2 + \frac{\tau^2}{(1+\tau^2)^2} = \frac{\tau^2}{1+\tau^2} \tag{7.6}$$

The name "Bayes risk" is, in fact, somewhat misleading since the risk function $R(\theta, \delta)$ being the *average* loss over all possible $\mathbf{y} \in \Omega$ is not considered as a relevant object by the "Bayesians," who restrict attention to quantities that are conditioned on the *observed* sample $\mathbf{y}$ (see Chapter 6). However, this is a common terminology. Besides, one can argue that a weight function $\pi(\theta)$ can be interpreted not necessarily as a prior but as measure of importance of different values of θ. Moreover, as we shall see in the next section, a "pure" Bayesian approach that utilizes a posterior expected loss $E(L(\theta, \delta)|\mathbf{y})$ leads to the same rules. This equivalency is usually used for deriving Bayes rules.

7.5   Posterior expected loss and Bayes actions

We consider now a Bayesian view on the decision making problem. Bayesian approach to statistical inference was discussed in Chapter 6. Recall that within Bayesian framework, the uncertainty about an unknown θ is modeled by considering θ as a random variable with some prior distribution $\pi(\theta)$. The distribution of observed data $f_\theta(\mathbf{y})$ is treated as the conditional distribution $f(\mathbf{y}|\theta)$ and the inference is based on the posterior distribution $\pi(\theta|\mathbf{Y} = \mathbf{y})$. Thus, a natural Bayesian way to find "the best" rule w.r.t. the loss $L(\theta, a)$ and the prior $\pi(\theta)$ is to consider a *Bayesian (posterior) expected loss*

$$r(a, \mathbf{y}) = E\left(L(a, \theta)|\mathbf{y}\right)$$

and to find a rule (also called a *Bayes action*) [1] $\delta_\pi^* = \delta_\pi^*(\mathbf{y})$ that minimizes $r(a, \mathbf{y})$ over $a \in \mathscr{A}$ for a given $\mathbf{y}$.

In particular, for the quadratic loss $L(\theta, a) = (\theta - a)^2$ and the absolute error loss $L(\theta, a) = |\theta - a|$ the corresponding Bayesian actions are respectively the posterior mean $E(\theta|\mathbf{y})$ and the posterior median $Med(\theta|\mathbf{y})$ (see Propositions 6.1 and 6.2 of Chapter 6).

Question 7.2 Show that for the quadratic loss, the posterior expected loss of the corresponding Bayes action is the posterior variance $Var(\theta|\mathbf{y})$.

Example 7.11 Consider estimating an unknown normal mean μ from $Y \sim \mathcal{N}(\mu, 1)$ w.r.t. the quadratic loss with a normal prior $\pi(\mu) = \mathcal{N}(\mu_0, \tau^2)$ for μ. The posterior distribution of $\mu|y$ is then

$$\pi(\mu|y) \sim \mathcal{N}\left(\frac{\tau^2}{1 + \tau^2}y + \frac{1}{1 + \tau^2}\mu_0, (1 + \tau^{-2})^{-1}\right)$$

(see Example 6.2) and the Bayes action (estimator)

$$\delta_\pi^*(y) = E(\mu|y) = \frac{\tau^2}{1 + \tau^2}y + \frac{1}{1 + \tau^2}\mu_0 = \mu_0 + \frac{\tau^2}{1 + \tau^2}(y - \mu_0) \tag{7.7}$$

Question 7.3 Show that the same action $\delta_\pi^*(y)$ in (7.7) is the Bayes action w.r.t. the absolute error loss as well.

Example 7.12 (continuation of Examples 7.6 and 7.11) Continue the previous Example 7.11 of estimating an unknown normal mean w.r.t. quadratic loss and normal prior but this time among the restricted class of affine estimators $\delta_{a,b}(y) = a + by$ considered in Example 7.6. Note that the unrestricted Bayes estimator $\delta_\pi^*(y)$ in (7.7) belongs to this class with $a = \frac{1}{1+\tau^2}\mu_0$ and $b = \frac{\tau^2}{1+\tau^2}$ and, therefore, it is the Bayesian action among affine estimators as well. Interesting, that it also coincides with the Bayes rule from Example 7.10 that minimizes the Bayes risk within this class. As we shall see below in Corollary 7.1 it is not just a coincidence.

Example 7.13 (Bernoulli trials) Suppose we observe a series of Bernoulli trials $Y_1, ..., Y_n$ with an unknown probability p. Assume a conjugate Beta-prior $Beta(\alpha, \beta)$ for p. The posterior distribution of $p|\mathbf{y}$ is then $Beta(\sum_{i=1}^n y_i + \alpha, n - \sum_{i=1}^n y_i + \beta)$ (see Example 6.1) and for the quadratic loss, the Bayes action is

$$\delta_{Beta}^*(\mathbf{y}) = E(p|\mathbf{y}) = \frac{\sum_{i=1}^n y_i + \alpha}{n + \alpha + \beta}$$

Consider now a *weighted* quadratic loss $L(p, a) = \frac{(p-a)^2}{p(1-p)}$, where we pay more attention to accurately estimate p near the boundaries, and find the corresponding Bayes action $\delta_{Beta}^*(\mathbf{y})$. The posterior expected loss of a rule δ decision a is

$$r(a, \mathbf{y}) = E\left(\frac{(p-a)^2}{p(1-p)}\bigg|\mathbf{y}\right)$$

---

[1] Meanwhile, we deliberately do not call it a Bayes rule to distinguish from a definition of a Bayes rule in Section 7.4 . However, we will see later that for the finite Bayes risk they indeed coincide.

We have

$$r(a, \mathbf{y}) = E\left(\frac{(p-a)^2}{p(1-p)}\Big|\mathbf{y}\right) = E\left(\frac{p^2}{p(1-p)}\Big|\mathbf{y}\right) - 2aE\left(\frac{p}{p(1-p)}\Big|\mathbf{y}\right) + a^2 E\left(\frac{1}{p(1-p)}\Big|\mathbf{y}\right)$$
$$= E\left(\frac{p}{1-p}\Big|\mathbf{y}\right) - 2aE\left(\frac{1}{1-p}\Big|\mathbf{y}\right) + a^2 E\left(\frac{1}{p(1-p)}\Big|\mathbf{y}\right) \tag{7.8}$$

and minimizing (7.8) w.r.t. a yields

$$\delta_{Beta}^*(\mathbf{y}) = \frac{E\left(\frac{1}{1-p}\Big|\mathbf{y}\right)}{E\left(\frac{1}{p(1-p)}\Big|\mathbf{y}\right)} = \frac{\int_0^1 (1-p)^{-1} p^{\alpha + \sum_{i=1}^n y_i - 1} (1-p)^{n+\beta - \sum_{i=1}^n y_i - 1} dp}{\int_0^1 p^{-1}(1-p)^{-1} p^{\alpha + \sum_{i=1}^n y_i - 1} (1-p)^{n+\beta - \sum_{i=1}^n y_i - 1} dp}$$
$$= \frac{B(\alpha + \sum_{i=1}^n y_i, n + \beta - \sum_{i=1}^n y_i - 1)}{B(\alpha + \sum_{i=1}^n y_i - 1, n + \beta - \sum_{i=1}^n y_i - 1)} = \frac{\sum_{i=1}^n y_i + \alpha - 1}{\alpha + \beta + n - 2} \tag{7.9}$$

In particular, for $\alpha = \beta = 1$ corresponding to the uniform prior $\mathcal{U}(0,1)$, $\delta_{\mathcal{U}}^*(\mathbf{y}) = \bar{y}$.

Example 7.14 (interval estimation) Interval estimation is an arena in which the Bayesian approach differs considerably from the frequentist one. From a Bayesian view, interval estimation is just a standard decision problem with the loss function given, for example, by (7.1). The Bayes action is the interval (a_1^*, a_2^*) that minimizes the posterior expected loss $E(L(\theta, (a_1, a_2))|\mathbf{y})$. In particular, for the loss function $L(\theta, (a_1, a_2))$ in (7.1), we have

$$E(L(\theta, (a_1, a_2))|\mathbf{y}) = \frac{a_2 - a_1}{1 + a_2 - a_1} P(a_1 < \theta < a_2|\mathbf{y}) + 1 - P(a_1 < \theta < a_2|\mathbf{y}) = 1 - \frac{P(a_1 < \theta < a_2|\mathbf{y})}{1 + a_2 - a_1} \tag{7.10}$$

In the following Lemma 7.1 we argue that if the posterior distribution is unimodal, then the resulting interval (a_1^*, a_2^*) is a highest posterior density (HPD) interval for θ (see Section 6.4):

Lemma 7.1 Assume that $\pi(\theta|\mathbf{y})$ is a unimodal continuous function of θ. Then, the interval (a_1^*, a_2^*) that minimizes the posterior expected loss (7.10) is the HPD interval for θ.

Proof. Let $\hat{d} = a_2^* - a_1^*$ be the length of (a_1^*, a_2^*) and $(\tilde{a}_1, \tilde{a}_2)$ be the HPD interval of length d (its existence for a unimodal and continuous posterior density $\pi(\theta|\mathbf{y})$ was verified in Question 6.4). We will show that the intervals (a_1^*, a_2^*) and $(\tilde{a}_1, \tilde{a}_2)$ coincide. Since $(\tilde{a}_1, \tilde{a}_2)$ is the HPD interval, $\pi(\theta|\mathbf{y}) \geq c(d)$ for all $\theta \in (\tilde{a}_1, \tilde{a}_2)$, where $c(d) = \pi(\tilde{a}_1|\mathbf{y}) = \pi(\tilde{a}_2|\mathbf{y})$ (see also Question 6.4). Thus, if $a_1^* > \tilde{a}_2$ or $a_2^* < \tilde{a}_1$, then $\pi(\theta|\mathbf{y}) \leq c(d)$ for all $\theta \in (a_1^*, a_2^*)$. Hence,

$$P(\tilde{a}_1 < \theta < \tilde{a}_2|\mathbf{y}) \geq c(d)\, d \geq P(a_1^* < \theta < a_2^*|\mathbf{y})$$

contradicting the claim that (a_1^*, a_2^*) minimizes (7.10). If $\tilde{a}_1 < a_1^* < \tilde{a}_2 < a_2^*$, the same arguments imply $P(\tilde{a}_1 < \theta < a_1^*|\mathbf{y}) < P(\tilde{a}_2 < \theta < a_2^*|\mathbf{y})$, contradicting again (7.10). Similarly, if $a_1^* < \tilde{a}_1 < a_2^* < \tilde{a}_2$. $\qquad\square$

Example 7.15 (hypothesis testing) Consider testing two general hypotheses $H_0 : \theta \in \Theta_0$ vs. $H_1 : \theta \in \Theta_1$, where $\Theta_0 \cap \Theta_1 = \emptyset$ and $\Theta_0 \cup \Theta_1 = \Theta$, w.r.t. the 0-1 loss

$$L(\theta, a) = \begin{cases} a, & \theta \in \Theta_0 \\ 1 - a, & \theta \in \Theta_1, \end{cases}$$

where $a = 0$ if H_0 is accepted and $a = 1$ otherwise (see Example 7.3). Assume the prior probabilities $\pi_0 = P(\theta \in \Theta_0)$ and $\pi_1 = 1 - \pi_0 = P(\theta \in \Theta_1)$. The posterior expected loss of a rule δ is then

$$E(L(\theta, \delta)|\mathbf{y}) = \begin{cases} P(\theta \in \Theta_1|\mathbf{y}), & \delta(\mathbf{y}) = 0 \\ P(\theta \in \Theta_0|\mathbf{y}), & \delta(\mathbf{y}) = 1 \end{cases}$$

and the resulting Bayes action will simply select the hypothesis with higher posterior probability:

$$\delta^*_{\pi_0}(\mathbf{y}) = \mathbb{I}\{P(\theta \in \Theta_1|\mathbf{y}) \geq P(\theta \in \Theta_0|\mathbf{y})\}$$

(see also Section 6.5).

We now show that the Bayes rule defined in Section 7.4 that minimizes the Bayes risk w.r.t. a certain loss function and a prior coincides with the Bayes action that minimizes the corresponding expected posterior loss. We start from the following lemma:

Lemma 7.2 (relation between the Bayes risk and the posterior expected loss) Let δ be any rule with a finite Bayes risk w.r.t. the loss $L(\theta, \delta)$ and prior $\pi(\theta)$. Assume for simplicity that all distributions have densities. Then,

$$\rho(\pi, \delta) = \int_{supp(f(\mathbf{y}))} r(\delta, \mathbf{y}) f(\mathbf{y}) d\mathbf{y}, \tag{7.11}$$

where $\rho(\pi, \delta)$ is its Bayes risk, $r(\delta, \mathbf{y})$ is its posterior expected loss and $f(\mathbf{y}) = \int_\Theta f(\mathbf{y}|\theta)\pi(\theta)d\theta$ is the marginal distribution of the data $\mathbf{Y}$.

Proof. By definition,

$$\begin{aligned}
\rho(\pi, \delta) &= \int_\Theta R(\theta, \delta)\pi(\theta)d\theta \\
&= \int_\Theta \left(\int_{supp(f(\mathbf{y}))} L(\theta, \delta(\mathbf{y})) f(\mathbf{y}|\theta)d\mathbf{y} \right) \pi(\theta)d\theta \\
&= \int_\Theta \int_{supp(f(\mathbf{y}))} L(\theta, \delta(\mathbf{y})) \frac{f(\mathbf{y}|\theta)\pi(\theta)}{f(\mathbf{y})} f(\mathbf{y})d\mathbf{y}d\theta \\
&= \int_{supp(f(\mathbf{y}))} \left(\int_\Theta L(\theta, \delta(\mathbf{y})) f(\theta|\mathbf{y})d\theta \right) f(\mathbf{y})d\mathbf{y} \\
&= \int_{supp(f(\mathbf{y}))} r(\delta(\mathbf{y}), \mathbf{y}) f(\mathbf{y})d\mathbf{y},
\end{aligned}$$

and since all the involved integrals are finite, we could interchange the order of integration. $\square$

Note that since $f(\mathbf{y})$ is positive a.s. on its support, the Bayes rule $\delta^*_\pi(\mathbf{y})$ minimizing the Bayes risk $\rho(\pi, \delta)$ in the left hand side of (7.11) also minimizes $r(\delta(\mathbf{y}), \mathbf{y})$ in the right hand side for each $\mathbf{y}$ and, hence, is a Bayes action. Thus, we have the immediate corollary:

Corollary 7.1 (equivalence between Bayes rules and Bayes actions) Let δ^*_π be a Bayes rule w.r.t. a loss $L(\theta, \delta)$ and a prior $\pi(\theta)$ with a finite Bayes risk. Assume that all distributions have densities. Then, δ^*_π is also a Bayes action w.r.t. to $L(\theta, \delta)$ and $\pi(\theta)$.

From now on we will not distinguish between Bayesian rules and actions (though they are still conceptually different!). Based on this equivalence, Bayes rules are typically derived through minimizing the posterior expected loss.

7.6 Admissibility of Bayes rules

The remarkable property of Bayes rules is that they are typically admissible as demonstrated in the following two theorems:

Theorem 7.2 If a Bayes rule δ^*_π is *unique*[2] and the Bayes risk $\rho(\pi, \delta^*_\pi) < \infty$, then δ^*_π is admissible.

---

[2]To be precise, the uniqueness is understood up to a set of probability zero for any $\theta \in \Theta$.

Proof. Suppose that δ_π^* is inadmissible. Then there exists another rule δ such that $R(\theta, \delta_\pi^*) \geq R(\theta, \delta)$ for all $\theta \in \Theta$. Then,

$$\rho(\pi, \delta_\pi^*) = \int_\Theta R(\theta, \delta_\pi^*) \pi(\theta) d\theta \geq \int_\Theta R(\theta, \delta) \pi(\theta) d\theta = \rho(\pi, \delta)$$

which contradicts the uniqueness of a Bayes rule. $\qquad\square$

Theorem 7.3 Let δ_π^* be a Bayes rule w.r.t. a prior $\pi(\theta)$. Suppose that one of the two following conditions holds:

a) Θ is finite and $\pi(\theta) > 0$ for all $\theta \in \Theta$.

b) Θ is an interval (possibly infinite), $\pi(\theta) > 0$ for all $\theta \in \Theta$, the risk function $R(\theta, \delta)$ is a continuous function of θ for all δ, and the Bayes risk $\rho(\pi, \delta_\pi^*) < \infty$.

Then, δ_π^* is admissible.

Proof. We prove b) and leave a) for the reader as an exercise.

The argument is similar to that used in the proof of Theorem 7.2. Suppose δ_π^* is inadmissible. Then there is another decision rule δ such that $R(\theta, \delta_\pi^*) \geq R(\theta, \delta)$ for all $\theta \in \Theta$ and there exists $\theta_0 \in \Theta$ for which the inequality is strict, that is, $R(\theta_0, \delta_\pi^*) > R(\theta_0, \delta)$. Due to continuity of the risk function there should exist an entire interval $(\theta_0 - \gamma, \theta_0 + \gamma)$ and $\varepsilon > 0$ such that, $R(\theta, \delta_\pi^*) \geq R(\theta, \delta) + \varepsilon$ for all $\theta \in (\theta_0 - \gamma, \theta_0 + \gamma)$. Hence,

$$\rho(\pi, \delta_\pi^*) - \rho(\pi, \delta) = \int_\Theta \left(R(\theta, \delta_\pi^*) - R(\theta, \delta)\right) \pi(\theta) d\theta \geq \int_{\theta_0 - \gamma}^{\theta_0 + \gamma} \left(R(\theta, \delta_\pi^*) - R(\theta, \delta)\right) \pi(\theta) d\theta$$

$$\geq \varepsilon \int_{\theta_0 - \gamma}^{\theta_0 + \gamma} \pi(\theta) d\theta > 0$$

that contradicts that δ_π^* is the Bayes rule w.r.t. π. $\qquad\square$

These results show that it is easy to verify admissibility of a rule δ once one can find a prior such that δ is the corresponding Bayes rule. Note that the prior does not have to be proper (i.e., summed or integrated to 1) as long as the corresponding Bayes risk is finite.

Example 7.16 (continuation of Example 7.6) Consider again the class of affine decision rules $\delta_{a,b}(Y) = a + bY$ for estimating μ, where $Y \sim \mathcal{N}(\mu, 1)$. Recall that in Example 7.6 we showed inadmissibility of these rules for all $b < 0$, $b > 1$ and for $b = 1$ when $a \neq 0$. We now consider $0 \leq b < 1$. From the results of Example 7.10 it follows directly that for all $0 < b < 1$, a rule $\delta_{a,b}(Y)$ is the unique Bayes rule for a normal prior $\pi(\mu) = \mathcal{N}(\mu_0, \tau^2)$, where $\mu_0 = \frac{a}{1-b}$ and $\tau^2 = \frac{b}{1-b}$. One can easily verify from (7.4) that in this case $\rho(\pi, \delta_{a,b}) = b < 1 < \infty$ and we conclude that $\delta_{a,b}(Y)$ is admissible for any a and $0 < b < 1$. The case $\delta_{0,1}(Y) = Y$ implies an infinite prior variance and corresponds to an *improper* prior $\pi(\mu) \propto 1$ (see Section 6.2.2) with *infinite* Bayes risk. Nevertheless, the admissibility of $\delta_{0,1}(Y)$ within the class of affine estimators has already been established by different arguments in Example 7.9. Finally, note that $b = 0$ yields a trivial constant estimator $\delta_{a,0}(Y) = a$ with a (unimprovable) zero risk at $\mu = a$ and, therefore, is admissible for any a (see also Section 7.2).

Example 7.17 (extension of Example 7.6) We now extend the results on admissibility of $\delta_{a,b}(Y)$ within the restricted class of affine estimators from Example 7.6 to *all* (not necessarily affine) estimators of μ. Obviously, $\delta_{a,b}(Y)$ remains inadmissible for $b < 0$, $b > 1$ and for $b = 1, a \neq 0$. For $0 < b < 1$, recall that $\delta_{a,b}(Y)$ is also the unique Bayes estimator (posterior mean) w.r.t. normal prior $\mathcal{N}(\mu_0, \tau^2)$ with $\mu_0 = \frac{a}{1-b}$ and $\tau^2 = \frac{b}{1-b}$ within the unrestricted class of estimators (see Examples 7.11 and 7.12) and, therefore, is still admissible. $\delta_{a,0}(Y) = a$ is again admissible since it has a zero risk at $\mu = a$. Although the arguments from Example 7.16 for admissibility of $\delta_{0,1}(Y) = Y$ do not hold for the unrestricted class of estimators, it can still be established by more involved techniques.

Example 7.18 (Bernoulli trials—continuation of Example 7.13) Let $Y_1,...,Y_n \sim \mathscr{B}(1,p)$ and consider the estimation of p under the quadratic loss. Then, any estimator (rule) of the form $\hat{p}_{\alpha,\beta} = \frac{\sum_{i=1}^n Y_i + \alpha}{n+\alpha+\beta}$, $\alpha,\beta > 0$ is a unique Bayes estimator w.r.t. the Beta-prior $Beta(\alpha,\beta)$ (see Example 7.13). We leave to the reader as a technical exercise to verify the finiteness of its Bayes risk and, hence, $\hat{p}_{\alpha,\beta}$ is admissible. Note that the natural estimator $\hat{p} = \bar{Y}$ (which is also the MLE and the UMVUE) corresponds to an *improper* prior. However, one can check that its Bayes risk is finite and the above arguments are still valid to verify its admissibility.

Example 7.19 (admissibility of LRT) In Example 7.7, we showed that testing two simple hypotheses, any test which is not LRT is inadmissible and can be improved by the corresponding LRT. We now demonstrate that any LRT with a rejection region $\Omega_{1C}(\mathbf{y}) = \{\mathbf{y} : \lambda(\mathbf{y}) = \frac{\ln f_{\theta_1}(\mathbf{y})}{\ln f_{\theta_0}(\mathbf{y})} \geq C\}$, $C \geq 0$ is admissible w.r.t. 0-1 loss. Assume any prior probabilities $\pi_0 = P(\theta = \theta_0)$ and $\pi_1 = 1 - \pi_0 = P(\theta = \theta_1)$. The corresponding Bayes test rejects the null hypothesis if the posterior odds $\frac{P(\theta=\theta_1|\mathbf{y})}{P(\theta=\theta_0|\mathbf{y})} \geq 1$ (see Example 7.15). On the other hand, in Section 6.5 we saw that $\frac{P(\theta=\theta_1|\mathbf{y})}{P(\theta=\theta_0|\mathbf{y})} = \lambda(\mathbf{y}) \cdot \frac{1-\pi_0}{\pi_0}$. Thus, LRT with any given $C \geq 0$ is a Bayes test corresponding to the prior probability $\pi_0 = \frac{C}{1+C}$ and is admissible by Theorem 7.3 for finite $\Theta = \{\theta_0, \theta_1\}$. By similar arguments it can be shown that LRTs remain admissible w.r.t. a more general non-symmetric loss (see Exercise 7.5).

7.7   Minimaxity and Bayes rules

As we have mentioned, derivation of a minimax rule is generally a nontrivial problem. It turns out that Bayesian approach can also be useful for finding minimax risks and rules.

Recall that for any prior $\pi(\theta)$, the Bayes risk of the corresponding Bayes rule δ_π^* is not larger than the minimax risk (see Question 7.1): $\rho(\pi, \delta_\pi^*) = \int_\Theta R(\theta_\pi^*, \delta)\pi(\theta)d\theta \leq \inf_\delta \sup_{\theta \in \Theta} R(\theta, \delta)$ and, therefore, $\rho(\pi, \delta_\pi^*)$ is a *lower bound* for the minimax risk.

On the other hand, for any rule, and for δ_π^*, in particular, its maximum risk $\sup_{\theta \in \Theta} R(\theta, \delta_\pi^*)$ is evidently an *upper bound* for the minimax risk: $\inf_\delta \sup_{\theta \in \Theta} R(\theta, \delta) \leq \sup_{\theta \in \Theta} R(\theta, \delta_\pi^*)$.

Hence, for any Bayes rule δ_π^*,

$$\rho(\pi, \delta_\pi^*) \leq \inf_\delta \sup_{\theta \in \Theta} R(\theta, \delta) \leq \sup_{\theta \in \Theta} R(\theta, \delta_\pi^*) \tag{7.12}$$

and we would naturally seek a least favorable prior which implies the largest Bayes risk to make the bounds for the minimax risk in (7.12) as tight as possible:

Definition 7.6 (least favorable prior) A prior $\pi(\theta)$ is called a *least favorable* if $\rho(\pi, \delta_\pi^*) \geq \rho(\tilde{\pi}, \delta_{\tilde{\pi}}^*)$ for any other prior $\tilde{\pi}(\theta)$.

Thus, if there exists a Bayes rule $\delta_\pi^*(\mathbf{Y})$ w.r.t. a prior $\pi(\theta)$ such that its Bayes risk $\rho(\pi, \delta_\pi^*) = \int_\Theta R(\theta, \delta_\pi^*)\pi(\theta)d\theta$ (weighted average risk w.r.t. $\pi(\theta)$) equals to its maximal risk $\sup_{\theta \in \Theta} R(\theta, \delta_\pi^*)$ and, hence, the lower and upper bounds in (7.12) coincide, it is a minimax rule and $\pi(\theta)$ is a least favorable prior:

Theorem 7.4 Let δ_π^* be the Bayes rule corresponding to a (proper) prior $\pi(\theta)$ and $\rho(\pi, \delta_\pi^*) = \sup_{\theta \in \Theta} R(\theta, \delta_\pi^*)$. Then, δ_π^* is a minimax rule and $\pi(\theta)$ is a least favorable prior.

Proof. For any (not necessarily Bayes) rule $\delta(\mathbf{Y})$ we have

$$\sup_{\theta \in \Theta} R(\theta, \delta) \geq \int_\Theta R(\theta, \delta)\pi(\theta)d\theta = \rho(\pi, \delta) \geq \rho(\pi, \delta_\pi^*) = \sup_{\theta \in \Theta} R(\theta, \delta_\pi^*)$$

and, therefore δ_π^* is a minimax rule.

To prove that $\pi(\theta)$ is a least favorable prior, note that for any prior $\tilde{\pi}(\theta)$ and the corresponding Bayes rule $\delta_{\tilde{\pi}}^*(\mathbf{Y})$,

$$\rho(\tilde{\pi}, \delta_{\tilde{\pi}}^*) = \int_\Theta R(\theta, \delta_{\tilde{\pi}}^*)\tilde{\pi}(\theta)d\theta \leq \int_\Theta R(\theta, \delta_\pi^*)\tilde{\pi}(\theta)d\theta \leq \sup_{\theta \in \Theta} R(\theta, \delta_\pi^*) = \rho(\pi, \delta_\pi^*)$$

□

The following corollary provides an important particular case of Theorem 7.4 useful for deriving minimax rules:

Corollary 7.2 Let δ_π^* be a Bayes rule w.r.t. a proper prior $\pi(\theta)$. If its risk function $R(\theta, \delta_\pi^*)$ is constant for all $\theta \in \Theta$, then $\delta_\pi^*(\mathbf{Y})$ is a minimax rule and $\pi(\theta)$ is a least favorable prior.

Proof. If $R(\theta, \delta_\pi^*)$ is constant for any $\theta \in \Theta$, then obviously

$$R(\theta, \delta_\pi^*) = \sup_{\theta \in \Theta} R(\theta, \delta_\pi^*) = \int_\Theta R(\theta, \delta_\pi^*) \pi(\theta) d\theta = \rho(\pi, \delta_\pi^*)$$

and the proof follows from Theorem 7.4. □

Example 7.20 (Bernoulli trials—continuation of Example 7.13) Find a minimax estimator for p w.r.t. the quadratic loss in a series of Bernoulli trials $Y_1, ..., Y_n \sim \mathcal{B}(1, p)$. Consider a class of Bayes estimators of the form $\hat{p}_{\alpha,\beta} = \frac{\sum_{i=1}^n Y_i + \alpha}{n + \alpha + \beta}$, $\alpha, \beta > 0$ w.r.t. the Beta-priors $Beta(\alpha, \beta)$ from Example 7.18 and calculate their risk functions:

$$\begin{aligned} R(p, \hat{p}_{\alpha,\beta}) = MSE(\hat{p}_{\alpha,\beta}, p) &= \left(\frac{np + \alpha}{n + \alpha + \beta} - p \right)^2 + \frac{np(1-p)}{(n + \alpha + \beta)^2} \\ &= \frac{p^2((\alpha+\beta)^2 - n) + p(n - 2\alpha(\alpha+\beta)) + \alpha^2}{(n + \alpha + \beta)^2} \end{aligned}$$

In particular, for $\alpha = \beta = \frac{\sqrt{n}}{2}$ corresponding to the (proper) prior $Beta(\frac{\sqrt{n}}{2}, \frac{\sqrt{n}}{2})$, the resulting risk $R(p, \hat{p}_{\frac{\sqrt{n}}{2}, \frac{\sqrt{n}}{2}}) = \frac{1}{4(\sqrt{n}+1)^2}$ is constant (does not depend on p). Hence,

$$\hat{p}_{\frac{\sqrt{n}}{2}, \frac{\sqrt{n}}{2}} = \frac{\sum_{i=1}^n Y_i + \sqrt{n}/2}{n + \sqrt{n}} = \frac{n}{n + \sqrt{n}} \bar{Y} + \frac{\sqrt{n}}{n + \sqrt{n}} \frac{1}{2}$$

is a minimax estimator, while $Beta(\frac{\sqrt{n}}{2}, \frac{\sqrt{n}}{2})$ is a least favorable prior. This is an example, where the minimax principle yields a somewhat "unnatural" estimator: it is a weighted average of a standard estimator $\bar{Y}$ and $1/2$, and can be interpreted as the proportion of successes in a series of $n + \sqrt{n}$ Bernoulli trials with $\sqrt{n}$ artificially added trials with exactly half of them being successive. In Exercise 7.2 one will verify that the standard estimator $\hat{p} = \bar{Y}$ is minimax but w.r.t. the weighted quadratic loss $L(p, a) = \frac{(p-a)^2}{p(1-p)}$.

Example 7.21 (minimax test for two simple hypotheses) Consider again testing two simple hypotheses $H_0 : \theta = \theta_0$ vs. $H_1 : \theta = \theta_1$ w.r.t. 0-1 loss. In Example 7.15, we showed that the corresponding Bayes rule (test) is

$$\delta_{\pi_0}^*(\mathbf{y}) = \mathbb{I}\left\{ \frac{P(\theta = \theta_1 | \mathbf{y})}{P(\theta = \theta_0 | \mathbf{y})} \geq 1 \right\} = \mathbb{I}\left\{ \lambda(\mathbf{Y}) \geq \frac{\pi_0}{1 - \pi_0} \right\}$$

where π_0 is the prior probability for H_0 and $\lambda(\mathbf{Y}) = L(\theta_1; \mathbf{y})/L(\theta_0; \mathbf{y})$ is the likelihood ratio. From a frequentist point of view it corresponds to LRT with the critical value $\pi_0/(1 - \pi_0)$. Its risk is

$$R(\theta, \delta_{\pi_0}^*) = \begin{cases} \alpha, & \theta = \theta_0 \\ \beta, & \theta = \theta_1, \end{cases}$$

where $\alpha = P_{\theta_0}(\lambda(\mathbf{Y}) \geq \frac{\pi_0}{1 - \pi_0})$ and $\beta = P_{\theta_1}(\lambda(\mathbf{Y}) < \frac{\pi_0}{1 - \pi_0})$ are Type I and Type II error probabilities (see Example 7.5). Thus, if $\alpha = \beta$, the resulting risk is the same under both alternatives and by Corollary 7.2, the LRT that rejects the null if $\lambda(\mathbf{Y}) \geq C^*$ with the critical value C^* satisfying $P_{\theta_0}(\lambda(\mathbf{Y}) \geq C^*) = P_{\theta_1}(\lambda(\mathbf{Y}) < C^*)$, is a minimax test. The least favorable prior probabilities are $\pi_0 = \frac{C^*}{C^*+1}$ and $\pi_1 = 1 - \pi_0 = \frac{1}{C^*+1}$. These results can be extended for non-symmetric losses (see Exercise 7.5).

Theorem 7.4 and Corollary 7.2 can be applied only for *proper* least favorable priors. There are important examples of rules, however, that can be viewed as Bayes w.r.t. *improper* priors. For example, $\hat{\mu} = Y$ in estimating a normal mean from a single observation is a Bayes rule corresponding to an improper prior $\pi(\mu) \propto 1$ (see Example 7.16). Nevertheless, an improper prior can often be considered as a limiting distribution of a sequence of proper priors and the following theorem extends Theorem 7.4 for this case:

Theorem 7.5 Let $\delta^* = \delta^*(\mathbf{Y})$ be a rule with the maximal risk $\sup_{\theta \in \Theta} R(\theta, \delta^*) = \rho^*$. If there exists a sequence of (proper) priors $\{\pi_m(\theta)\}$ with corresponding Bayes risks $\rho(\pi_m, \delta^*_{\pi_m}) \underset{m \to \infty}{\to} \rho^*$, then δ^* is a minimax rule.

Proof. For any rule δ and for any m we have

$$\sup_{\theta \in \Theta} R(\theta, \delta) \geq \int_{\Theta} R(\theta, \delta) \pi_m(\theta) d\theta \geq \rho(\pi_m, \delta^*_{\pi_m})$$

Thus, as $m \to \infty$,

$$\sup_{\theta \in \Theta} R(\theta, \delta) \geq \rho^* = \sup_{\theta \in \Theta} R(\theta, \delta^*)$$

and, therefore, δ^* is a minimax rule.   □

Example 7.22 (continuation of Examples 7.6, 7.8) Consider again the estimation of an unknown normal mean μ from a single observation $Y \sim \mathcal{N}(\mu, 1)$ w.r.t. the quadratic loss. In Example 7.8, we showed that $\hat{\mu} = Y$ is a minimax estimator within the class of *affine* estimators (recall that $\delta_{0,1}(Y) = Y$). We now prove that it is minimax within *all* possible estimators of μ.

Although $R(\mu, Y) = 1$ and does not depend on μ, Corollary 7.2 cannot be applied since Y is a Bayes rule w.r.t. an improper prior $\pi(\mu) \propto 1$ (see above). Instead we exploit Theorem 7.5.

Consider a sequence of normal priors with infinitely increasing variances $\pi_m(\mu) = \mathcal{N}(0, \tau_m^2)$, where $\lim_{m \to \infty} \tau_m^2 = \infty$. The improper prior can be viewed as a limiting distribution of $\{\pi_m\}$. The corresponding Bayes estimators (posterior expectations) $\hat{\mu}_m = \frac{\tau_m^2}{\tau_m^2 + 1} Y$ and their Bayes risks

$$\rho(\pi_m, \hat{\mu}_m) = \frac{\tau_m^2}{\tau_m^2 + 1} \underset{m \to \infty}{\longrightarrow} 1 = R(\mu, Y)$$

for all μ (see (7.5) and (7.6) in Example 7.10). Theorem 7.5 implies then the minimaxity of $\hat{\mu} = Y$.

7.8 Exercises

Exercise 7.1 The annual number of earthquakes in a certain seismically active region is distributed Poisson with an unknown mean λ. Let $Y_1, ..., Y_n$ be the numbers of earthquakes in the area during the last n years (it is possible to assume that the Y_i are independent). Seismologists believe that $\lambda \sim Gamma(\alpha, \beta)$ and want to estimate λ w.r.t. the weighted quadratic loss $L(\lambda, a) = \frac{(\lambda - a)^2}{\lambda}$.

1. Find the corresponding Bayes estimator of λ. What is its risk? Is it admissible?

2. Find the Bayes estimator for the probability p that the following year will be "quiet" (there will be no earthquakes in the region) w.r.t. the weighted quadratic loss $L(p, a) = \frac{(p-a)^2}{p^2}$.

3. Are the above Bayes estimators for λ and p consistent?

Exercise 7.2 Continue Example 7.13 for estimating p in a series of Bernoulli trials $Y_1, ..., Y_n \sim \mathcal{B}(1, p)$ w.r.t. the weighted quadratic loss $L(p, a) = \frac{(p-a)^2}{p(1-p)}$. Calculate the risk of $\bar{Y}$ and show that it is minimax. Is it admissible?

Exercise 7.3 Let $Y_1, ..., Y_n \sim \exp(\theta)$.

1. In Example 6.8, we found that the noninformative Jeffreys prior for θ is $\pi(\theta) \propto 1/\theta$. What is the Bayes estimator for $\mu = EY = 1/\theta$ w.r.t. this prior and the weighted quadratic loss $L(\mu,a) = \frac{(a-\mu)^2}{\mu^2}$?

2. Calculate the risk of the Bayes estimator from the previous paragraph and compare it with that of the MLE of μ. Is the MLE admissible?

Exercise 7.4 Consider a single observation $Y \sim \mathcal{N}(\mu, \sigma^2)$, where $|\mu| \le a$ and σ is known. We want to estimate the unknown μ by a linear estimator of the form $\hat{\mu}_c = cY$, $c \in \mathbb{R}$.

1. Find the minimax linear estimator w.r.t. the quadratic loss. Is it admissible (within the family of linear estimators)? What is the minimax risk?

2. Is $\hat{\mu}_1 = Y$ an admissible estimator?

3. Show that the least favorable prior for μ is a degenerate two-point prior $\pi(a) = \pi(-a) = 0.5$.

4. Extend the previous paragraph and show that, in fact, any prior $\pi(\mu)$ with the second moment $E_\pi \mu^2 = a^2$ is least favorable.

5. Assume a noninformative uniform prior $\mathcal{U}(-a,a)$ on μ and find the resulting Bayes estimator w.r.t. the quadratic loss. Is it admissible?

6. Show that generally, $\hat{\mu}_c$ is admissible for $0 \le c < 1$ (within a family of linear estimators) and is inadmissible otherwise.

   (*Hint*: consider any prior on μ with support $(-a,a)$ and finite second moment, and find the corresponding Bayes linear estimator.)

Exercise 7.5 Continue Example 7.5 for testing two simple hypotheses $H_0 : \theta = \theta_0$ vs. $H_1 : \theta = \theta_1$ with the different losses K_0 and K_1 for erroneous rejection and acceptance of the null. Assume prior probabilities π_0 and $\pi_1 = 1 - \pi_0$ for H_0 and H_1 respectively.

1. Derive the resulting Bayes rule (test).

2. Interpret the Bayes test from the previous paragraph in terms of the frequentist approach. Is it LRT (MP test)? What is the critical value for the corresponding test statistic?

3. Show that any LRT is admissible w.r.t. $K_0 - K_1$ loss.

4. Let α and β be Type I and Type II errors probabilities respectively for the above test. Show that if the critical value for LRT is chosen such that $K_0 \alpha = K_1 \beta$, then it is also a minimax test .

5. Let $Y_1, ..., Y_{100} \sim \mathcal{N}(\mu, 25)$. Find the minimax test for testing $H_0 : \mu = 0$ vs. $H_1 : \mu = 1$ for $K_0 = 25$ and $K_1 = 10$.

Exercise 7.6 There is a device for blood type classification: A, B, AB or O. It measures a certain quantity Y, which has a shifted exponential distribution with the density $f_\tau(Y) = e^{-(y-\tau)}$, $y \ge \tau$, $\tau \ge 0$. If $0 \le \tau < 1$, the blood is of type AB; if $1 \le \tau < 2$, the blood is of type A; if $2 \le \tau < 3$, the blood is of type B; and if $\tau \ge 3$, the blood is of type O. It is known that in the population as a whole, $\tau \sim \exp(1)$.

The loss of misclassifying the blood is given in the following table:

		Classified As			
		AB	A	B	O
True	AB	0	1	1	2
Blood	A	1	0	2	2
Type	B	1	2	0	2
	O	3	3	3	0

A patient has been tested and $y = 4$ is observed. What is the Bayes action?

Exercise 7.7 Continue Exercise 6.7. In estimating IQ, it is deemed to be twice as harmful to underestimate as to overestimate and the following loss is felt appropriate:

$$L(\mu,a) = \begin{cases} 2(\mu - a) & , \mu \ge a \\ a - \mu & , \mu < a \end{cases}$$

Find the Bayes estimator for Genius' IQ w.r.t. this loss.

Exercise 7.8 Generalize the results of the previous Exercise 7.7. Let $Y_1, ..., Y_n$ be a random sample from a distribution with a density $f_\theta(y)$ and assume a prior $\pi(\theta)$ on θ. We want to estimate θ w.r.t. following loss:

$$L(\theta, a) = \begin{cases} c_1(\theta - a) & , \theta \geq a \\ c_2(a - \theta) & , \theta < a \end{cases}$$

Show that the resulting Bayes estimator $\hat{\theta}$ is $\Pi_{\theta|\mathbf{y}}^{-1}\left(\frac{c_1}{c_1 + c_2}\right)$—the $\left(\frac{c_1}{c_1 + c_2}\right) \cdot 100\%$-quantile of the posterior distribution $\theta|\mathbf{y}$.

Chapter 8

*Linear Models

8.1 Introduction

This chapter is somewhat different from the other chapters in this book. It does not deal with a particular concept or problem of statistical theory but with a particular type of statistical model. More specifically, we consider *linear regression models* (or simply linear models) that are undoubtedly the most important and used class of models. The chapter exemplifies how the various general results on statistical theory of the previous chapters can be applied to this more complex and realistic setup.

Linear models appear in regression analysis when one explores the impact of one or a group of variables called *independent* variables, *predictors*, *explanatory* variables, *inputs, factors, covariates*, etc., on another (quantitative) variable called a *dependent* variable, *response, output*, etc. How, for example, blood pressure depends on cholesterol level, type and dose of a drug, age, diet and other patient's characteristics. How one's income is related to education, professional experience, gender, region, and other socio-economical factors. These are just common examples where regression in general, and linear models in particular, play a key role.

A comprehensive study of linear models requires at least one full term course and is beyond the scope of this book. In this chapter, we provide just a brief introduction to explain their main concepts and utilize the theoretical tools from the previous chapters, for their analysis. The reader is assumed to have a basic background in linear algebra, and know the contents of Appendix A.4.16.

8.2 Definition and examples

Consider a standard regression setup. Suppose we have n observations $(\mathbf{x}_i, Y_i)$, $i = 1, \ldots, n$ over a set of m predictors $\mathbf{x} = (x_1, \ldots, x_m)$ and a response variable Y. A general (parametric) regression model assumes that

$$Y_i = g_\beta(\mathbf{x}_i) + \varepsilon_i, \qquad i = 1, \ldots, n, \tag{8.1}$$

where the response function $g_\beta(\cdot) : \mathbb{R}^m \to \mathbb{R}$ has a known parametric form and depends on $p \leq n$ unknown parameters $\beta = (\beta_1, \ldots, \beta_p)$, and the noise ε_i's are i.i.d. random variables, typically normal, with zero means and a finite (probably unknown) variance σ^2. The noise in (8.1) models either measurement errors or deviations from the true response function $g_\beta(\cdot)$ (e.g., due to the effect of other existing covariates not included in the model). The predictors $x_1, \ldots, x_m$ can be fixed or random and independent from the noise ε. In any case, our further analysis will be conditioned on their values. Note that the observed Y_i's in (8.1) are independent but typically not identically distributed, since each Y_i is associated with different values of explanatory variables.

In *linear* models, $g_\beta(\cdot)$ is considered to be linear in β, that is,

$$Y_i = \beta_1 x_{i1} + \beta_2 x_{i2} + \cdots + \beta_p x_{ip} + \varepsilon_i = \beta^\top \mathbf{x}_i + \varepsilon_i, \qquad i = 1, 2, \ldots, n; \ m = p$$

or, equivalently, in matrix form

$$\mathbf{Y} = \mathbb{X}\beta + \varepsilon, \tag{8.2}$$

DOI: 10.1201/9781003175407-8

139

where

$$\mathbf{Y} = \begin{pmatrix} Y_1 \\ \vdots \\ Y_n \end{pmatrix} \in \mathbb{R}^n, \quad \mathbb{X} = \begin{pmatrix} \mathbf{x}_1^\top \\ \vdots \\ \mathbf{x}_n^\top \end{pmatrix} = \begin{pmatrix} x_{11} & \cdots & x_{1p} \\ \vdots & & \\ x_{n1} & \cdots & x_{np} \end{pmatrix} \in \mathbb{R}^{n \times p}, \quad \beta = \begin{pmatrix} \beta_1 \\ \vdots \\ \beta_p \end{pmatrix} \in \mathbb{R}^p.$$

The matrix $\mathbb{X}$ is often called a *design matrix*. Typically, a regression model includes a constant term as one of predictors. In this case, $x_{i1} = 1$, $i = 1,...,n$ and β_1 is an *intercept*.

We emphasize that in linear models $g_\beta(\cdot)$ is a linear function of the unknown parameters (regression coefficients) β but not necessarily of the predictors. Thus, for example, a *polynomial* regression, where

$$Y_i = \beta_0 + \beta_1 x_i + ... + \beta_m x_i^m + \varepsilon_i, \qquad i = 1,...,n$$

is still a particular case of the general linear regression model (8.2) with $p = m + 1$ predictors $x_j = x^j$, $j = 0,...,m$.

Similarly, it is possible to include interactions between predictors within the linear model framework. Suppose there are two predictors (main effects) x_1 and x_2, and we want to include the interaction term and to consider the model

$$Y_i = \beta_0 + \beta_1 x_{1i} + \beta_2 x_{2i} + \beta_3 x_{1i} x_{2i} + \varepsilon_i, \qquad i = 1,...,n$$

Defining the interaction term as a new covariate $x_3 = x_1 x_2$, the above model becomes of the form (8.2) with the three predictors x_1, x_2, and x_3.

Generally, any regression model (8.1), where the response function is of the form $g_\beta(\mathbf{x}) = \sum_{j=1}^p \beta_j g_j(\mathbf{x})$ and the functions $g_j(\cdot) : \mathbb{R}^m \to \mathbb{R}$, $j = 1,...,p$ are known, is linear with p predictors $g_1(\mathbf{x}),...,g_p(\mathbf{x})$. Thus, without loss of generality, hereafter we will consider linear models of the type (8.2), where the predictors $x_1,...,x_p$ are not necessarily the original covariates but may be some (known) functions of them.

Furthermore, in a variety of applications, predictors may be *categorical* variables (factors). Automatic and manual transmission cars of the same model have different fuel consumptions. Transmission type is a two-valued categorical variable. In a medical experiment, the results of a test depend on whether a subject received the placebo, a low dose of a drug, or a higher dose. Thus, in this experiment, treatment type is a triple valued categorical variable. Categorical variables can be treated within the linear models framework by introducing indicator (dummy) variables. Suppose that z is a factor with p levels coded as $L_1,...,L_p$. For example, z is transmission type, where $L_1 = $ "*automatic*," $L_2 = $ "*manual*"; or a dose level of drug, where $L_1 = $ "*placebo*," $L_2 = $ "*low dose*," $L_3 = $ "*high dose*." Assume that when the categorical variable is equal L_j, the corresponding outputs are normal variates with the common mean β_j and variance σ^2. Given n observations $(z_1,...,z_n)$ on z, we can model them by defining p indicator variables $\mathbf{x}_1,...,\mathbf{x}_p$:

$$x_{ij} = \begin{cases} 1, & z_i = L_j \\ 0, & z_i \neq L_j \end{cases}, \quad i = 1,...,n; \, j = 1,...,p.$$

In terms of these p new dummy predictors the model becomes

$$Y_i = \beta_1 x_{i1} + \beta_2 x_{i2} + ... + \beta_3 x_{ip} + \varepsilon_i, \qquad i = 1,...,n,$$

where $\varepsilon \sim N(0, \sigma^2)$. This is still of the type (8.2), where β_j is the (unknown) mean value of Y for the j-th group where $z = L_j$. Without loss of generality we can assume that $z_i = L_1$ for the first n_1 observations $i = 1,...,n_1$, $z_i = L_2$ for the next n_2 observations $i = n_1 + 1,...,n_1 + n_2$, ..., $z_i = L_p$ for the last n_p observations, where evidently, $n_1 + n_2 + ... + n_p = n$. In this case, the design matrix $\mathbb{X}_{n \times p}$

is of the form

$$
\mathbb{X} = \begin{pmatrix}
1 & 0 & \cdots & 0 \\
\vdots & \vdots & \cdots & \vdots \\
1 & 0 & \cdots & 0 \\
0 & 1 & \cdots & 0 \\
\vdots & \vdots & \cdots & \vdots \\
0 & 1 & \cdots & 0 \\
\vdots & \vdots & \cdots & \vdots \\
0 & 0 & \cdots & 1 \\
\vdots & \vdots & \cdots & \vdots \\
0 & 0 & \cdots & 1
\end{pmatrix}. \tag{8.3}
$$

What are the main goals in the analysis of linear models? Possible aims can be:

1. Estimation of the regression coefficients β and the noise variance σ^2 to describe the relationship between the response and the predictors.

2. Inference about the regression coefficients β or part of them. Does a specific predictor or a group of predictors have a significant impact on the response? Are the effects of two predictors similar?

3. Prediction of the response for future observations. For example, fitting a regression model for the effect of various factors on electricity consumption based on past observations, an electric company would like to predict the expected consumption for today given the current values of covariates.

We consider these issues in the following sections.

8.3 Estimation of regression coefficients

The common method for estimating the unknown regression coefficients β in the linear model (8.2) is the method of *least squares* briefly introduced in Section 2.4. We consider now its application to linear models.

From Section 2.4, for a general linear model (8.2), the least squares estimator (LSE) $\hat{\beta}$ minimizes

$$
\|\mathbf{Y} - \mathbb{X}\beta\|^2 = \sum_{i=1}^{n}(Y_i - \sum_{j=1}^{p}\beta_j x_{ij})^2 \tag{8.4}
$$

over all $\beta \in \mathbb{R}^p$.

In what follows we always assume that the columns of $\mathbb{X}$ are linearly independent, that is, there is no linear combination of $\mathbf{x}_1, ..., \mathbf{x}_p$ that is a zero vector or, equivalently, $\mathbb{X}\mathbf{b} \neq \mathbf{0}$ for any $\mathbf{b} \neq \mathbf{0}$. Otherwise, part of the predictors are, in fact, linear combinations of others and the model (8.2) can be reduced to a smaller number of covariates. For linearly independent columns of $\mathbb{X}$ the matrix $\mathbb{X}^\top \mathbb{X}$ is invertible and the LSE $\hat{\beta}$ for β is given by the following theorem:

Theorem 8.1 The LSE of β in (8.2) is

$$
\hat{\beta} = (\mathbb{X}^\top \mathbb{X})^{-1} \mathbb{X}^\top \mathbf{Y}. \tag{8.5}
$$

Proof. We will show that $\hat{\beta} = (\mathbb{X}^\top \mathbb{X})^{-1} \mathbb{X}^\top \mathbf{Y}$ indeed minimizes (8.4) over all vectors $\beta \in \mathbb{R}^p$. For any $\beta \in \mathbb{R}^p$ we have

$$
\begin{aligned}
\|\mathbf{Y} - \mathbb{X}\beta\|^2 &= \|\mathbf{Y} - \mathbb{X}\hat{\beta} + \mathbb{X}(\hat{\beta} - \beta)\|^2 \\
&= \|\mathbf{Y} - \mathbb{X}\hat{\beta}\|^2 + 2(\mathbf{Y} - \mathbb{X}\hat{\beta})^\top \mathbb{X}(\hat{\beta} - \beta) + \|\mathbb{X}(\hat{\beta} - \beta)\|^2
\end{aligned} \tag{8.6}
$$

but

$$\mathbf{Y} - \mathbb{X}\hat{\beta} = \mathbf{Y} - \mathbb{X}(\mathbb{X}^\top \mathbb{X})^{-1}\mathbb{X}^\top \mathbf{Y} = \mathbf{Y} - P_\mathbb{X}\mathbf{Y} = (I - P_\mathbb{X})\mathbf{Y},$$

where $P_\mathbb{X} = \mathbb{X}(\mathbb{X}^\top \mathbb{X})^{-1}\mathbb{X}^\top$ is the projection matrix of the subspace spanned by the columns of $\mathbb{X}$ and $I - P_\mathbb{X}$ is the projection matrix on its orthogonal complement (see Section A.4.16). Thus, $P_\mathbb{X}\mathbf{Y}$ is the projection of $\mathbf{Y}$ on this subspace and, therefore, $\mathbf{Y} - \mathbb{X}\hat{\beta} = (I - P_\mathbb{X})\mathbf{Y}$ is its orthogonal complement, which is orthogonal to any vector in the span of the columns of $\mathbb{X}$ and, in particular, to $\mathbb{X}(\hat{\beta} - \beta)$. From (8.6) we then have

$$\|\mathbf{Y} - \mathbb{X}\beta\|^2 = \|\mathbf{Y} - \mathbb{X}\hat{\beta}\|^2 + \|\mathbb{X}(\hat{\beta} - \beta)\|^2. \tag{8.7}$$

Since the first term in the RHS of (8.7) does not involve β, while the second one is non-negative, the LHS is minimized w.r.t. β when the latter is zero, that is, for $\beta = \hat{\beta}$.   □

Note that the LSE $\hat{\beta} = (\mathbb{X}^\top \mathbb{X})^{-1}\mathbb{X}^\top \mathbf{Y}$ is a *linear* estimator, that is, it is a linear function of $\mathbf{Y}$. One should not be confused between a *linear model* and a *linear estimator*—they are two different notions! Thus, there may be linear estimators for nonlinear regression and nonlinear estimators for linear regression parameters.

As we have already mentioned in the proof of Theorem 8.1, geometrically, the method of least squares in (8.4) projects $\mathbf{Y}$ on the span generated by the columns of $\mathbb{X}$, where $P_\mathbb{X}\mathbf{Y} = \mathbb{X}(\mathbb{X}^\top \mathbb{X})^{-1}\mathbb{X}^\top \mathbf{Y} = \mathbb{X}\hat{\beta}$ is the resulting projection and the vector of LSE $\hat{\beta}$ is its coefficients.

The projection matrix $P_\mathbb{X} = \mathbb{X}(\mathbb{X}^\top \mathbb{X})^{-1}\mathbb{X}^\top$ and the corresponding matrix $I - P_\mathbb{X}$ have a series of specific properties. Some of them that will be used further in the chapter are given in Lemma 8.1 (see also Appendix A.4.16.2):

Lemma 8.1 Let $P_\mathbb{X} = \mathbb{X}(\mathbb{X}^\top \mathbb{X})^{-1}\mathbb{X}^\top$ be the projection matrix for the linear regression model (8.2). Then,

1. $P_\mathbb{X}$ and $I - P_\mathbb{X}$ are idempotent matrices, that is, $P_\mathbb{X}^2 = P_\mathbb{X}$ and $(I - P_\mathbb{X})^2 = I - P_\mathbb{X}$.
2. $P_\mathbb{X}$ and $I - P_\mathbb{X}$ are positive semi-definite matrices, that is, $\mathbf{a}^\top P_\mathbb{X}\mathbf{a} \geq 0$ and $\mathbf{a}^\top (I - P_\mathbb{X})\mathbf{a} \geq 0$ for any vector $\mathbf{a} \in \mathbb{R}^n$.
3. $rank(P_\mathbb{X}) = tr(P_\mathbb{X}) = p$ and $rank(I - P_\mathbb{X}) = tr(I - P_\mathbb{X}) = n - p$.

Proof. Simple straightforward calculus yields

$$P_\mathbb{X}^2 = \mathbb{X}(\mathbb{X}^\top \mathbb{X})^{-1}\mathbb{X}^\top \mathbb{X}(\mathbb{X}^\top \mathbb{X})^{-1}\mathbb{X}^\top = \mathbb{X}(\mathbb{X}^\top \mathbb{X})^{-1}\mathbb{X}^\top = P_\mathbb{X}.$$

Hence,

$$(I - P_\mathbb{X})^2 = I - 2P_\mathbb{X} + P_\mathbb{X}^2 = I - P_\mathbb{X}.$$

The second statement is an immediate consequence of the first one, since $P_\mathbb{X}^2$ and $(I - P_\mathbb{X})^2$ are clearly positive semi-definitive. To prove the last statement, recall from linear algebra, that since $P_\mathbb{X}$ is self-adjoint (symmetric) and idempotent, $rank(P_\mathbb{X}) = tr(P_\mathbb{X})$, while

$$tr(P_\mathbb{X}) = tr(\mathbb{X}(\mathbb{X}^\top \mathbb{X})^{-1}\mathbb{X}^\top) = tr(\mathbb{X}^\top \mathbb{X}(\mathbb{X}^\top \mathbb{X})^{-1}) = tr(I_p) = p.$$

Similarly, $rank(I - P_\mathbb{X}) = tr(I - P_\mathbb{X}) = n - p$.   □

What are the main statistical properties of the LSE $\hat{\beta}$? Assume that the linear model (8.2) is indeed a true underlying model for the data for some unknown β and σ^2. We then have

Theorem 8.2 Consider the LSE $\hat{\beta}$ in (8.5) of β in (8.2).

1. $\hat{\beta}$ is an unbiased estimator of β.
2. The covariance matrix $Var(\hat{\beta}) = \sigma^2(\mathbb{X}^\top \mathbb{X})^{-1}$. In particular, $Var(\hat{\beta}_j) = \sigma^2(\mathbb{X}^\top \mathbb{X})_{jj}^{-1}$.

Proof. Using the results on means and covariance matrices of random vectors (see Section A.3.9), straightforward calculus yields

$$
\begin{aligned}
E\hat{\beta} &= E\left((\mathbb{X}^{\top}\mathbb{X})^{-1}\mathbb{X}^{\top}\mathbf{Y}\right) = E\left((\mathbb{X}^{\top}\mathbb{X})^{-1}\mathbb{X}^{\top}(\mathbb{X}\beta + \varepsilon)\right) = \beta + E\left(\mathbb{X}^{\top}\mathbb{X})^{-1}\mathbb{X}^{\top}\varepsilon\right) \\
&= \beta + (\mathbb{X}^{\top}\mathbb{X})^{-1}\mathbb{X}^{\top}E\varepsilon = \beta
\end{aligned}
$$

and, therefore, $\hat{\beta}$ in an unbiased estimator of β.

Similarly, the covariance matrix

$$
\mathrm{Var}(\hat{\beta}) = \left((\mathbb{X}^{\top}\mathbb{X})^{-1}\mathbb{X}^{\top}\right)(\sigma^2 I)\left((\mathbb{X}^{\top}\mathbb{X})^{-1}\mathbb{X}^{\top}\right)^{\top} = \sigma^2(\mathbb{X}^{\top}\mathbb{X})^{-1}.
$$

The variance $Var(\hat{\beta}_j)$ is the j-th diagonal element $\sigma^2(\mathbb{X}^{\top}\mathbb{X})^{-1}_{jj}$ of $Var(\hat{\beta})$. $\qquad\square$

Is $\hat{\beta}_j$ an UMVUE for β_j? Without any assumptions on the noise distribution we can only claim that for all $j = 1,...,p$, $\hat{\beta}_j$ is the minimum variance unbiased estimator for β_j among all *linear* estimators, that is, the *best linear unbiased estimator* (BLUE):

Theorem 8.3 (Gauss–Markov theorem) The LSE $\hat{\beta}_j$ is the BLUE of β_j for all $j = 1,...,p$.

Proof. Consider any other linear unbiased estimator $\tilde{\beta}_j$ of β_j. Since $\tilde{\beta}_j$ is linear, $\tilde{\beta}_j = \mathbf{a}_j^{\top}\mathbf{Y}$ for some vector $\mathbf{a}_j \in \mathbb{R}^n$ that may depend on $\mathbb{X}$. Due to unbiasedness,

$$
E\tilde{\beta}_j = E(\mathbf{a}_j^{\top}\mathbf{Y}) = \mathbf{a}_j^{\top}E(\mathbf{Y}) = \mathbf{a}_j^{\top}\mathbb{X}\beta = \beta_j
$$

for all $\beta \in \mathbb{R}^p$. Hence, $\mathbf{a}_j$ should satisfy

$$
\mathbf{a}_j^{\top}\mathbb{X} = \mathbf{e}_j^{\top}, \tag{8.8}
$$

where $\mathbf{e}_j \in \mathbb{R}^p$ is the unit vector with one in the j-th coordinate and zero elsewhere.

Recall that $Var(\hat{\beta}_j) = \sigma^2(\mathbb{X}^{\top}\mathbb{X})^{-1}_{jj} = \sigma^2\mathbf{e}_j^{\top}(\mathbb{X}^{\top}\mathbb{X})^{-1}\mathbf{e}_j$ (see Theorem 8.2), while $Var(\tilde{\beta}_j) = Var(\mathbf{a}_j^{\top}\mathbf{Y}) = \sigma^2\mathbf{a}_j^{\top}\mathbf{a}_j$. Then, using (8.8),

$$
\begin{aligned}
Var(\tilde{\beta}_j) - Var(\hat{\beta}_j) &= \sigma^2(\mathbf{a}_j^{\top}\mathbf{a}_j - \mathbf{e}_j^{\top}(\mathbb{X}^{\top}\mathbb{X})^{-1}\mathbf{e}_j) \\
&= \sigma^2(\mathbf{a}_j^{\top}\mathbf{a}_j - \mathbf{a}_j^{\top}\mathbb{X}(\mathbb{X}^{\top}\mathbb{X})^{-1}\mathbb{X}^{\top}\mathbf{a}_j) \\
&= \sigma^2\mathbf{a}_j^{\top}(I - P_{\mathbb{X}})\mathbf{a}_j \geq 0,
\end{aligned}
$$

since the matrix $I - P_{\mathbb{X}}$ is the positive semi-definite (see Lemma 8.1). $\qquad\square$

So far we considered the ε_i to be i.i.d. with zero means and common variance σ^2 but did not make any assumptions on their distribution. Usually, however, in linear models the noise is assumed to be *normal* $\mathcal{N}(0,\sigma^2)$, that is, $\varepsilon \sim \mathcal{N}(\mathbf{0},\sigma^2 I)$. In this case the Y_i are also independent normal $\mathcal{N}(\beta^{\top}\mathbf{x}_i, \sigma^2)$ or $\mathbf{Y}$ is multivariate normal (see A.4.16), $\mathbf{Y} \sim \mathcal{N}(\mathbb{X}\beta, \sigma^2 I)$, and the likelihood function is

$$
L(\beta, \sigma; \mathbf{Y}) = \left(\frac{1}{\sqrt{2\pi}\sigma}\right)^n e^{-\frac{\|\mathbf{Y}-\mathbb{X}\beta\|^2}{2\sigma^2}}. \tag{8.9}
$$

Question 8.1 Show that:

1. The distribution of $\mathbf{Y}$ belongs to a $(p+1)$-parameter exponential family with the natural parameters $(\frac{\beta_1}{\sigma^2},...,\frac{\beta_p}{\sigma^2},\frac{1}{\sigma^2})$

2. $(\mathbb{X}^{\top}\mathbf{Y}, \mathbf{Y}^{\top}\mathbf{Y})$ is the sufficient statistic for (β, σ)

Under the normality assumption of the noise distribution we can say more about the properties of the LSE $\hat{\beta}$:

Theorem 8.4 Let $\hat{\beta}$ be the LSE of β in a linear model (8.2), where $\varepsilon \sim \mathcal{N}(\mathbf{0}, \sigma^2 I)$. Then,

1. $\hat{\beta} \sim \mathcal{N}(\beta, \sigma^2 (\mathbb{X}^\top \mathbb{X})^{-1})$

2. $\hat{\beta}$ is the MLE of β.

3. $\hat{\beta}_j$ is the UMVUE of β_j and achieves the Cramer–Rao lower bound.

Proof. The first statement follows immediately from the properties of multinormal random vectors (see Theorem A.1). The second statement has been established in Section 2.4 (see Question 2.4) even for a general (not necessarily linear) parametric regression. From Theorem 8.2 we already know that $\hat{\beta}$ is unbiased. To complete the last statement we show that $\hat{\beta}$ achieves the Cramer–Rao lower bound. We calculate the $(p+1) \times (p+1)$ Fisher information matrix (see Section 2.7.3):

$$I(\beta, \sigma) = -E \begin{pmatrix} \frac{\partial^2 l(\beta,\sigma;\mathbf{Y})}{\partial \beta \partial \beta^\top} & \frac{\partial^2 l(\beta,\sigma;\mathbf{Y})}{\partial \beta \partial \sigma} \\ \frac{\partial^2 l(\beta,\sigma;\mathbf{Y})}{\partial \sigma \partial \beta^\top} & \frac{\partial^2 l(\beta,\sigma;\mathbf{Y})}{\partial \sigma^2} \end{pmatrix},$$

where the log-likelihood $l(\beta, \sigma; \mathbf{Y}) = \ln L(\beta, \sigma; \mathbf{Y})$.

From (8.9) we have

$$l(\beta, \sigma; \mathbf{Y}) = -\frac{n}{2} \log(2\pi) - n \log(\sigma) - \frac{1}{2\sigma^2} \|\mathbf{Y} - \mathbb{X}\beta\|^2.$$

Differentiating $l(\beta, \sigma; \mathbf{Y})$ yields

$$\frac{\partial l(\beta,\sigma;\mathbf{Y})}{\partial \beta} = \frac{\mathbb{X}^\top \|\mathbf{Y} - \mathbb{X}\beta\|}{\sigma^2}, \quad \frac{\partial l(\beta,\sigma;\mathbf{Y})}{\partial \sigma} = -\frac{n}{\sigma} + \frac{1}{\sigma^3} \|\mathbf{Y} - \mathbb{X}\beta\|^2;$$

$$\frac{\partial^2 l(\beta,\sigma;\mathbf{Y})}{\partial \beta \partial \beta^\top} = -\frac{1}{\sigma^2}(\mathbb{X}^\top \mathbb{X}), \quad \frac{\partial^2 l(\beta,\sigma;\mathbf{Y})}{\partial \sigma^2} = \frac{n}{\sigma^2} - \frac{3\|\mathbf{Y} - \mathbb{X}\beta\|^2}{\sigma^4},$$

$$\frac{\partial^2 l(\beta,\sigma;\mathbf{Y})}{\partial \beta \partial \sigma} = -\frac{2\mathbb{X}^\top \|\mathbf{Y} - \mathbb{X}\beta\|}{\sigma^3}$$

and taking expectations we get

$$I(\beta, \sigma) = \begin{pmatrix} \frac{1}{\sigma^2} \mathbb{X}^\top \mathbb{X} & \mathbf{0}_p \\ \mathbf{0}_p^\top & \frac{2n}{\sigma^2} \end{pmatrix}.$$

Consider now the scalar-valued function $g(\beta, \sigma) = \beta_j$ whose $(p+1)$-dimensional gradient vector is $(0, ..., 0, 1, 0, ..., 0)^\top$. Applying Theorem 2.4, the Cramer–Rao lower bound for an unbiased estimate of β_j is then

$$(0, ..., 0, 1, 0, ..., 0) I(\beta, \sigma)^{-1} (0, ..., 0, 1, 0, ..., 0)^\top = \sigma^2 (\mathbb{X}^\top \mathbb{X})_{jj}^{-1} \tag{8.10}$$

which is exactly $Var(\hat{\beta}_j)$. $\qquad\square$

8.4   Residuals. Estimation of the variance

Although estimation of the regression coefficients β is usually of primary interest, estimation of the variance σ^2 in (8.2) is also relevant, in particular, for statistical inference on β.

Consider the vector of *predicted* values $\hat{\mathbf{Y}} = \mathbb{X}\hat{\beta}$ and the vector of *residuals* $\mathbf{e} = \mathbf{Y} - \hat{\mathbf{Y}}$. By simple calculus we have

$$\mathbf{e} = \mathbf{Y} - \hat{\mathbf{Y}} = (I - P_\mathbb{X})\mathbf{Y} = (I - P_\mathbb{X})(\mathbb{X}\beta + \varepsilon) = (I - P_\mathbb{X})\varepsilon. \tag{8.11}$$

The following Lemma 8.2 provides several useful properties of the vectors $\mathbf{e}$ and $\hat{\mathbf{Y}}$:

Lemma 8.2

1. $\mathbf{e}$ and $\hat{\mathbf{Y}}$ are orthogonal, i.e. $\mathbf{e}^\top \hat{\mathbf{Y}} = 0$

2. $||\mathbf{Y}||^2 = ||\hat{\mathbf{Y}}||^2 + ||\mathbf{e}||^2$

3. $\mathbf{e}$ is orthogonal to all columns of the design matrix $\mathbb{X}$, i.e. $\mathbb{X}^\top \mathbf{e} = \mathbf{0}$. In particular, if a linear model includes an intercept, $\sum_{i=1}^n e_i = 0$

4. If the noise ε in (8.2) is normal, $\mathbf{e}$ and $\hat{\mathbf{Y}}$ are statistically independent.

Proof. The proof becomes obvious once one realizes that $\hat{\mathbf{Y}} = P_{\mathbb{X}}\mathbf{Y}$ and that it is the projection of $\mathbf{Y}$ on the span of columns of $\mathbb{X}$, while the vector of residuals $\mathbf{e} = (I - P_{\mathbb{X}})\mathbf{Y}$ is its orthogonal complement (see Section 8.3). Thus, evidently, $\mathbf{e}^\top \hat{\mathbf{Y}} = 0$, while the second statement is essentially Pythagoras' theorem. Finally,

$$\mathbb{X}^\top \mathbf{e} = \mathbb{X}^\top (I - P_{\mathbb{X}})\mathbf{Y} = (\mathbb{X}^\top - \mathbb{X}^\top \mathbb{X}(\mathbb{X}^\top \mathbb{X})^{-1}\mathbb{X})\mathbf{Y} = \mathbf{0}.$$

In particular, if there is an intercept, then $\mathbf{x}_1 = \mathbf{1}_n = (1, ..., 1)^\top$ and, therefore, $\mathbf{x}_1^\top \mathbf{e} = \sum_{i=1}^n e_i = 0$.

The last assertion of the theorem follows from the independence of orthogonal projections of the multivariate normal distribution (see Theorem A.4.1). □

The quantity $||\mathbf{e}||^2 = \sum_{i=1}^n (Y_i - \hat{Y}_i)^2$ is often called the *residual sum of squares* (RSS). From (8.11) and Lemma 8.1 we then have

$$E(RSS) = E||(I - P_{\mathbb{X}})\varepsilon||^2 = \sigma^2 tr(I - P_{\mathbb{X}}) = \sigma^2(n - p)$$

and, therefore,

$$s^2 = \frac{RSS}{n - p} \tag{8.12}$$

is an *unbiased* estimator of σ^2.

Question 8.2 Is s^2 in (8.12) a linear estimator?

If, in addition, the noise distribution in (8.2) is normal, we have the following theorem:

Theorem 8.5 Consider the linear model (8.2), where $\varepsilon \sim \mathcal{N}(\mathbf{0}, \sigma^2 I)$. Then,

1. $s^2 = \frac{RSS}{n-p}$ is the UMVUE of σ^2.

2. The MLE of σ^2 is $\hat{\sigma}^2 = \frac{RSS}{n}$.

3. $RSS \sim \sigma^2 \chi_{n-p}^2$.

4. The MLE $\hat{\beta}$ in (8.5) and RSS are independent.

Proof. The unbiased estimator s^2 in (8.12) is a function of the sufficient statistic $(\mathbb{X}^\top \mathbf{Y}, \mathbf{Y}^\top \mathbf{Y})$ and following the arguments at the end of Section 2.7.4 for distributions from the exponential family, it is UMVUE.

In the proof of Theorem 8.4 we derived

$$\frac{\partial l(\beta, \sigma; \mathbf{Y})}{\partial \sigma} = -\frac{n}{\sigma} + \frac{1}{\sigma^3}||\mathbf{Y} - \mathbb{X}\beta||^2.$$

Equating $\frac{\partial l(\beta, \sigma; \mathbf{Y})}{\partial \sigma}$ to zero and substituting the MLE $\hat{\beta} = (\mathbb{X}^\top \mathbb{X})^{-1}\mathbb{X}^\top \mathbf{Y}$ (see Theorem 8.4) completes the proof for the MLE of σ^2.

Recall that $\mathbf{e} = (I - P_{\mathbb{X}})\varepsilon$, $rank(I - P_{\mathbb{X}}) = n - p$ (see Lemma 8.1) and $\varepsilon \sim \mathcal{N}(\mathbf{0}, \sigma^2 I)$. Thus, from the properties of normal vectors (see Appendix A.4.16), $RSS = \mathbf{e}^\top \mathbf{e} \sim \sigma^2 \chi_{n-p}^2$.

To prove the independence of $\hat{\beta}$ and RSS note that the former belongs to the span formed by the columns of $\mathbb{X}$, while RSS is the function of the vector of residuals $\mathbf{e} = (I - P_{\mathbb{X}})\mathbf{Y}$ lying in its orthogonal complement (see Section 8.3, and Theorem A.1). □

Question 8.3 Assume that $\varepsilon \sim \mathcal{N}(\mathbf{0}, \sigma^2 I)$ in (8.2). Show that the vector of residuals $\mathbf{e} \sim \mathcal{N}(\mathbf{0}, \sigma^2(I - P_{\mathbb{X}}))$ and the vector of predicted values $\hat{\mathbf{Y}} \sim \mathcal{N}(\mathbb{X}\beta, \sigma^2 P_{\mathbb{X}})$.

8.5   Examples

Consider several well-known statistical setups that can also be presented within the linear models framework.

8.5.1   Estimation of a normal mean

Suppose $Y_1, \ldots, Y_n$ are i.i.d. sample from $\mathcal{N}(\mu, \sigma^2)$. In terms of linear models (8.2), they correspond to a model with a single intercept term

$$Y_i = \beta + \varepsilon_i,$$

where $\beta = \mu$. The design matrix $\mathbb{X} = \mathbf{1}_n = (1, \ldots, 1)^\top$ and $\hat{\mu} = \hat{\beta} = (\mathbb{X}^\top \mathbb{X})^{-1} \mathbb{X}^\top \mathbf{Y} = n^{-1} \sum_{i=1}^n Y_i = \bar{\mathbf{Y}}$.

8.5.2   Comparison between the means of two independent normal samples with a common variance

Consider two independent normal random samples $U_1, \ldots, U_n \sim \mathcal{N}(\mu_1, \sigma^2)$ and $V_1, \ldots, V_m \sim \mathcal{N}(\mu_2, \sigma^2)$ with a common variance σ^2, where we want to estimate the means difference $\mu_2 - \mu_1$. One can think, for example, about $U_1, \ldots, U_n$ as test results for a control (placebo) group and $V_1, \ldots, V_m$ for a treatment (drug) group (see also Exercise 4.12). Without loss of generality we can assume that the first n observations belong to the control and the last m—to the treatment group. Define a vector $\mathbf{Y}_{n+m} = (U_1, \ldots, U_n, V_1, \ldots, V_m)^\top$ and an indicator variable $x : x_i = 0, \ i = 1, \ldots, n$ and $x_i = 1, \ i = n+1, \ldots, n+m$, and consider a linear regression with a response Y, a single predictor x and an intercept:

$$Y_i = \beta_1 + \beta_2 x_i + \varepsilon_i, \quad i = 1, \ldots, n+m.$$

Thus, $\beta_1 = \mu_1$ and $\beta_2 = \mu_2 - \mu_1$. The corresponding design matrix is

$$\mathbb{X} = \begin{pmatrix} \mathbf{1}_n & \mathbf{0}_n \\ \mathbf{1}_m & \mathbf{1}_m \end{pmatrix}.$$

Then, by straightforward calculus, $\hat{\beta}_1 = \bar{U}$ is the estimate of the mean within the placebo group $\hat{\mu}_1$, while $\hat{\beta}_2 = \bar{V} - \bar{U}$ estimates $\mu_2 - \mu_1$—the treatment effect w.r.t the placebo.

Question 8.4 Find the UMVUE s^2 in (8.12) for σ^2 and show that it is exactly the pooled variance estimator S_p^2 in the two-sample t-tests (see Exercise 4.12).

8.5.3   Simple linear regression

Consider a simple linear regression with a single predictor x and an intercept:

$$Y_i = \beta_1 + \beta_2 x_i + \varepsilon_i, \quad i = 1, \ldots, n$$

We have

$$\mathbb{X} = \begin{pmatrix} 1 & x_1 \\ 1 & x_2 \\ \vdots & \vdots \\ 1 & x_n \end{pmatrix}, \quad (\mathbb{X}^\top \mathbb{X}) = \begin{pmatrix} n & \sum_{i=1}^n x_i \\ \sum_{i=1}^n x_i & \sum_{i=1}^n x_i^2 \end{pmatrix}$$

$$(\mathbb{X}^\top \mathbb{X})^{-1} = \frac{1}{n \sum_{i=1}^n x_i^2 - \left(\sum_{i=1}^n x_i \right)^2} \begin{pmatrix} \sum_{i=1}^n x_i^2 & -\sum_{i=1}^n x_i \\ -\sum_{i=1}^n x_i & n \end{pmatrix}$$

$$\mathbb{X}^\top \mathbf{Y} = \begin{pmatrix} 1 & \cdots & 1 \\ x_1 & \cdots & x_n \end{pmatrix} \begin{pmatrix} Y_1 \\ \vdots \\ Y_n \end{pmatrix} = \begin{pmatrix} \sum_{i=1}^n Y_i \\ \sum_{i=1}^n x_i Y_i \end{pmatrix}$$

$$\hat{\beta} = \begin{pmatrix} \hat{\beta}_1 \\ \hat{\beta}_2 \end{pmatrix} = (\mathbb{X}^\top \mathbb{X})^{-1} \mathbb{X}^\top \mathbf{Y} = \frac{1}{n\sum_{i=1}^n x_i^2 - (\sum_{i=1}^n x_i)^2} \begin{pmatrix} \sum_{i=1}^n x_i^2 \sum_{i=1}^n Y_i - \sum_{i=1}^n x_i \sum_{i=1}^n x_i Y_i \\ n\sum_{i=1}^n x_i Y_i - \sum_{i=1}^n x_i \sum_{i=1}^n Y_i \end{pmatrix}.$$

The estimator for the slope β_2 is then

$$\hat{\beta}_2 = \frac{n\sum_{i=1}^n x_i Y_i - \sum_{i=1}^n x_i \sum_{i=1}^n Y_i}{n\sum_{i=1}^n x_i^2 - (\sum_{i=1}^n x_i)^2} = \frac{\sum_{i=1}^n (x_i - \bar{x}_n)(Y_i - \bar{Y})}{\sum_{i=1}^n (x_i - \bar{x}_n)^2}.$$

and for the intercept β_1

$$\hat{\beta}_1 = \frac{\sum_{i=1}^n x_i^2 \sum_{i=1}^n Y_i - \sum_{i=1}^n x_i \sum_{i=1}^n x_i Y_i}{n\sum_{i=1}^n x_i^2 - (\sum_{i=1}^n x_i)^2} = \bar{Y} - \hat{\beta}_2 \bar{x}.$$

The regression line therefore always passes through the midpoint $(\bar{x}, \bar{Y})$.

8.6 Goodness-of-fit: Multiple correlation coefficient

The properties of the LSE estimators discussed in the previous section were strongly based on the assumption that the true (unknown) response function $g_\beta(\cdot)$ in (8.1) is (or at least can be well-approximated by) a linear function of β. Can we somehow check that the linear model is indeed adequate for the data? Do the resulting $\hat{Y}_i$'s properly fit the observed Y_i's?

It seems natural to look at the residuals e_i's—the smaller the residuals, the better the model's fit. The goodness-of-fit of a model can be measured then by the $RSS = \|\mathbf{e}\|^2$. Note, however, that the residuals and RSS are not scale invariant and changing the scale of Y (say, from meters to kilometers) will change them accordingly. If the noise variance σ^2 was known, one could look at the ratio RSS/σ^2. But what does its value tell us about the goodness-of-fit of the model? Suppose that $RSS/\sigma^2 = 5$. Is it too large or still reasonable for adequacy of the linear model? For normal noise, under the assumption that the linear model is true, $RSS/\sigma^2 \sim \chi^2_{n-p}$ (see Theorem 8.5). Thus we can test the null hypothesis that the linear model properly fits the data at the level α by the corresponding χ^2-test and reject it iff $RSS/\sigma^2 > \chi^2_{n-p;\alpha}$. In practice, however, σ^2 is typically unknown and the above χ^2-test for the goodness-of-fit cannot be applied.

Instead we present another measure for the goodness-of-fit. Suppose that there is an intercept term in the model. Consider the sample variance $\frac{1}{n}\sum_{i=1}^n (Y_i - \bar{Y})^2$ of the response sample $\mathbf{Y}$. We have

$$\begin{aligned} \sum_{i=1}^n (Y_i - \bar{Y})^2 &= \sum_{i=1}^n \left((Y_i - \hat{Y}_i) + (\hat{Y}_i - \bar{Y}) \right)^2 \\ &= \sum_{i=1}^n (Y_i - \hat{Y}_i)^2 + 2\sum_{i=1}^n (Y_i - \hat{Y}_i)(\hat{Y}_i - \bar{Y}) + \sum_{i=1}^n (\hat{Y}_i - \bar{Y})^2, \end{aligned} \tag{8.13}$$

where the middle term $2\sum_{i=1}^n (Y_i - \hat{Y}_i)(\hat{Y}_i - \bar{Y}) = 2\sum_{i=1}^n e_i \hat{Y}_i - 2\bar{Y}\sum_{i=1}^n e_i = 0$ due to Lemma 8.2. The last term in the RHS is RSS that represents the deviations of the data from the fitted model. The proportion of the sample variance explained by the linear model can then be defined as

$$R^2 = \frac{\sum_{i=1}^n (Y_i - \hat{Y}_i)^2}{\sum_{i=1}^n (Y_i - \bar{Y})^2} = 1 - \frac{RSS}{\sum_{i=1}^n (Y_i - \bar{Y})^2}. \tag{8.14}$$

Evidently, $0 \le R^2 \le 1$ and is scale invariant. The closer R^2 is to one, the better the fit. $R^2 = 1$ corresponds to the ideal case when a linear model perfectly fits the data with zero residuals. The quantity R^2 defined in (8.14) is also known as the *multiple correlation coefficient* as is explained in Lemma 8.3:

Lemma 8.3 R^2 defined in (8.14) is the square of the sample correlation between the observed response vector $\mathbf{Y}$ and the vector of its predicted values $\hat{\mathbf{Y}} = \mathbb{X}\hat{\beta}$, where $\hat{\beta}$ is the LSE of β. More generally, R^2 is the square of the maximal sample correlation between $\mathbf{Y}$ and any vector of the form $\mathbb{X}\tilde{\beta}$, $\tilde{\beta} \in \mathbb{R}^p$.

Proof. Denote by corr_n and Var_n respectively sample correlations and sample variances. By Lemma 8.2 we then have

$$
\begin{aligned}
\text{corr}_n(\mathbf{Y}, \mathbb{X}\tilde{\beta}) &= \frac{(\mathbf{Y} - \bar{Y}\mathbf{1}_n)^\top \mathbb{X}\tilde{\beta}}{\sqrt{\text{Var}_n(\mathbf{Y})\,\text{Var}_n(\mathbb{X}\tilde{\beta})}} = \frac{((\mathbf{Y} - \mathbb{X}\hat{\beta}) + (\mathbb{X}\hat{\beta} - \bar{Y}\mathbf{1}_n))^\top \mathbb{X}\tilde{\beta}}{\sqrt{\text{Var}_n(\mathbf{Y})\,\text{Var}_n(\mathbb{X}\tilde{\beta})}} \\
&= \frac{(\mathbb{X}\hat{\beta} - \bar{Y}\mathbf{1}_n)^\top \mathbb{X}\tilde{\beta}}{\sqrt{\text{Var}_n(\mathbf{Y})\,\text{Var}_n(\mathbb{X}\tilde{\beta})}} = \sqrt{\frac{\text{Var}_n(\hat{\mathbf{Y}})}{\text{Var}_n(\mathbf{Y})}}\,\text{corr}_n(\mathbb{X}\hat{\beta}, \mathbb{X}\tilde{\beta}) \le \sqrt{R^2}
\end{aligned}
$$

with equality iff $\mathbb{X}\tilde{\beta} = \mathbb{X}\hat{\beta}$. $\qquad\square$

If there is no intercept in a model, $\sum_{i=1}^n (Y_i - \bar{Y})^2$ in (8.13) cannot be decomposed as $\sum_{i=1}^n (Y_i - \hat{Y}_i)^2 + \sum_{i=1}^n (\hat{Y}_i - \bar{Y})^2$ and the definition (8.14) of R^2 is not valid anymore. However, one still has $||\mathbf{Y}||^2 = ||\hat{\mathbf{Y}}||^2 + ||\mathbf{e}||^2$ (see Lemma 8.2) and can define $R^2 = ||\hat{\mathbf{Y}}||^2/||\mathbf{Y}||^2$ in this case. It is important to keep in mind that R^2 is defined differently when there is no intercept!

The multiple correlation coefficient is an important but not the only tool for checking the goodness-of-fit of a model. It allows one to check the adequacy of a linearity approximation, but says nothing about the appropriateness of assumptions on the noise distribution (normality, independence, common variance), where *analysis of residuals* plays a key role. We do not consider these topics in this book and refer the reader to numerous textbooks on regression.

8.7 Confidence intervals and regions for the coefficients

Confidence intervals and regions were discussed in Chapter 3. We now utilize its results for regression coefficients of a linear model (8.2). Unlike point estimation we have to assume something about the noise distribution to construct a confidence interval/region for the β_j's. We exploit the standard assumption that the noise is normal and $\varepsilon \sim \mathcal{N}(\mathbf{0}, \sigma^2 I_n)$.

Consider first a single coefficient β_j and construct the Wald confidence interval for β_j based on its MLE $\hat{\beta}_j$. Theorem 8.4 implies that $\hat{\beta}_j \sim \mathcal{N}(\beta_j, \sigma^2(\mathbb{X}^\top\mathbb{X})_{jj}^{-1})$. The noise variance σ^2 is usually also unknown but from Theorem 8.5 we know that $RSS/\sigma^2 \sim \chi^2_{n-p}$ and is independent of $\hat{\beta}_j$. Hence,

$$
\frac{(\hat{\beta}_j - \beta_j)/\sqrt{(\mathbb{X}^\top\mathbb{X})_{jj}^{-1}}}{\sqrt{RSS/(n-p)}} = \frac{\hat{\beta}_j - \beta_j}{s\sqrt{(\mathbb{X}^\top\mathbb{X})_{jj}^{-1}}} \sim t_{n-p},
$$

where $s^2 = \frac{RSS}{n-p}$ is the UMVUE for σ^2 (see (8.12)), and the resulting Wald $(1-\alpha)100\%$ confidence interval for β_j is therefore

$$
\hat{\beta}_j \pm t_{n-p,\alpha/2}\, s\sqrt{(\mathbb{X}^\top\mathbb{X})_{jj}^{-1}}. \tag{8.15}
$$

In linear models we are usually interested in a joint confidence region for a group of coefficients rather than in a confidence interval for a single β_j. As we have discussed in Section 3.5 one can use a conservative Bonferroni's method and construct $(1 - \alpha/p)100\%$ confidence intervals for each individual β_j, $j = 1, ..., p$. The resulting joint Bonferroni's $(1 - \alpha)100\%$ confidence region for the entire coefficients vector β is then a p-dimensional rectangle

$$
\mathscr{B}_B = \left\{ \beta : \hat{\beta}_j - t_{n-p,\alpha/(2p)}\, s\sqrt{(\mathbb{X}^\top\mathbb{X})_{jj}^{-1}} \le \beta_j \le \hat{\beta}_j + t_{n-p,\alpha/(2p)}\, s\sqrt{(\mathbb{X}^\top\mathbb{X})_{jj}^{-1}} \right\}
$$

(see Section 3.5).

However, similar to a single coefficient case, we can construct a Wald confidence region based on the joint distribution of the MLE $\hat{\beta}$ as a better alternative to a conservative Bonferroni's region. From Theorem 8.4, $\hat{\beta} \sim \mathcal{N}(\beta, \sigma^2(\mathbb{X}^\top\mathbb{X})^{-1})$ and, therefore, $\frac{1}{\sigma^2}(\hat{\beta} - \beta)^\top\mathbb{X}^\top\mathbb{X}(\hat{\beta} - \beta) \sim \chi_p^2$. Furthermore, $RSS/\sigma^2 \sim \chi_{n-p}^2$ and is independent of $\hat{\beta}$. Hence,

$$\frac{(\hat{\beta} - \beta)^\top\mathbb{X}^\top\mathbb{X}(\hat{\beta} - \beta)/p}{RSS/(n-p)} = \frac{(\hat{\beta} - \beta)^\top\mathbb{X}^\top\mathbb{X}(\hat{\beta} - \beta)/p}{s^2} \sim F_{p,n-p}$$

is a pivot (see Appendix A.4.15) and the corresponding $(1 - \alpha)100\%$ confidence region for β is a p-dimensional ellipsoid

$$\mathscr{B}_F = \{\beta : (\hat{\beta} - \beta)^\top\mathbb{X}^\top\mathbb{X}(\hat{\beta} - \beta) \leq p\, s^2 F_{p,n-p;\alpha}\}.$$

An F-based Wald confidence region for a subgroup of regression coefficients can be constructed in a similar way.

Question 8.5 Show that the F-based confidence interval for a *single* coefficient coincides with the t-based confidence interval (8.15).

8.8 Hypothesis testing in linear models

Fitting a model is usually only the first stage in the regression analysis. Statistical inference about the regression coefficients is essential. Does a specific predictor or a group of predictors have a significant effect on the response? Is it the same for several predictors? In this section we consider testing such types of hypotheses.

8.8.1 Testing significance of a single predictor

We start from testing a single predictor for significance. Does carrot consumption influence vision given that the other explanatory variables, e.g., age, gender, body mass index (BMI), sweet-potatoes consumption, remain the same?

Consider a general linear regression model (8.2) with p explanatory variables

$$\mathbf{Y}_i = \mathbf{x}_i^\top\beta + \varepsilon_i, \quad i = 1, \ldots, n, \quad \varepsilon_1 \sim N(0, \sigma^2)$$

and test

$$H_0 : \beta_j = 0 \ (\mathbf{x}_j \text{ has no effect on } Y) \quad vs. \quad H_1 : \beta_j \neq 0 \ (\text{has effect}).$$

We can use the duality between confidence intervals and hypothesis testing (see Section 4.4) to derive an α-level two-sided t-test from (8.15) rejecting the null hypothesis iff

$$\frac{|\hat{\beta}_j|}{s\sqrt{(\mathbb{X}^\top\mathbb{X})_{jj}^{-1}}} > t_{n-p,\alpha/2}.$$

Sometimes one is interested in testing one-sided hypotheses:

$$H_0 : \beta_j \leq 0 \quad vs. \quad H_1 : \beta_j > 0.$$

Does the drug have a positive effect or not? Does fuel consumption increase with a car's weight? The similar but one-sided t-test rejects the null hypothesis iff $\dfrac{\hat{\beta}_j}{s\sqrt{(\mathbb{X}^\top\mathbb{X})_{jj}^{-1}}} > t_{n-p,\alpha}$.

8.8.2   Testing significance of a group of predictors

It is often relevant to test the significance of an entire group of predictors rather than a single one. Do socio-economic factors such as parents' income, total number of children in the family, living area, and ethnic origin affect the results of school tests together with individual IQ, or are they determined solely by the latter? Similarly, suppose that a predictor is a categorical variable with m levels. In the linear model setup it is associated with m dummy predictors (see Section 8.2) and testing its effect on the response leads, therefore, to testing for significance of the corresponding m coefficients.

Suppose that there are p predictors $x_1, ..., x_p$ and we want to test the significance of a subset of r of them. Without loss of generality we can assume that they are the first r predictors $x_1, ..., x_r$:

$$H_0 : \beta_1 = \beta_2 = ... = \beta_r = 0, \quad 1 \le r \le p \tag{8.16}$$

against a general alternative. Under the null hypothesis, the original full model with p covariates is essentially reduced to a model with $p - r$ predictors—a *reduced* model.

Similarly to one parameter case, we can derive the Wald test. Let $\hat{\beta} = (\mathbb{X}^\top \mathbb{X})^{-1} \mathbb{X}^\top \mathbf{Y}$ be the MLE of β and $\hat{\beta}_{1:r} = (\hat{\beta}_1, ..., \hat{\beta}_r)^\top \in \mathbb{R}^r$ be the subvector of its first r components. From Theorem 8.4 we know that under the null hypothesis (8.16), $\hat{\beta}_{1:r} \overset{H_0}{\sim} \mathcal{N}(\mathbf{0}, \sigma^2 (\mathbb{X}_{1:r}^\top \mathbb{X}_{1:r})^{-1})$, where $\mathbb{X}_{1:r}$ is the $n \times r$ submatrix of $\mathbb{X}$ formed by its first r columns, and, therefore, $\frac{1}{\sigma^2} \hat{\beta}_{1:r}^\top \mathbb{X}_{1:r}^\top \mathbb{X}_{1:r} \hat{\beta}_{1:r} \overset{H_0}{\sim} \chi_r^2$. On the other hand, from Theorem 8.5, the residual sum of squares of the full model $RSS \sim \sigma^2 \chi_{n-p}^2$ and is independent of $\hat{\beta}_{1:r}$. Thus, $F = \frac{\hat{\beta}_{1:r}^\top \mathbb{X}_{1:r}^\top \mathbb{X}_{1:r} \hat{\beta}_{1:r}/r}{RSS/(n-p)} \overset{H_0}{\sim} F_{r,n-p}$ (see Section A.4.15) and the resulting α-level Wald F-test rejects the null hypothesis (8.16) iff

$$F = \frac{\hat{\beta}_{1:r}^\top \mathbb{X}_{1:r}^\top \mathbb{X}_{1:r} \hat{\beta}_{1:r}/r}{RSS/(n-p)} > F_{r,n-p;\alpha}. \tag{8.17}$$

In particular, for $r = 1$ (testing a single coefficient), the above F-test (8.17) is equivalent to the t-test from Section 8.8.1 (see Question 8.5).

An alternative to the Wald test would be a generalized likelihood ratio test (GLRT)—see Section 4.3.3. However, as we show below, for testing (8.16) in linear models the two tests coincide:

Proposition 8.1 *The Wald F-test (8.17) is also a GLRT for testing the null hypothesis (8.16).*

Proof. We construct the GLRT for testing (8.16) and verify that it is exactly the Wald F-test (8.17).

For the full model with p predictors, the MLEs for β and σ^2 are $\hat{\beta} = (\mathbb{X}^\top \mathbb{X})^{-1} \mathbb{X}^\top \mathbf{Y}$ and $\hat{\sigma}^2 = \frac{RSS}{n}$ respectively. For the reduced model with $p - r$ predictors $x_{r+1}, ..., x_p$ under the null hypothesis (8.16) the MLE for the corresponding $p - r$ coefficients is $\hat{\beta}_0 = (\mathbb{X}_{r+1:p}^\top \mathbb{X}_{r+1:p})^{-1} \mathbb{X}_{r+1:p}^\top \mathbf{Y}$, where $X_{r+1:p}$ is a $n \times (p-r)$ submatrix of $\mathbb{X}$ formed by its last $p - r$ columns. The MLE for the variance σ^2 under the null is $\hat{\sigma}_0^2 = \frac{RSS_0}{n}$, where RSS_0 is the residual sum of squares for the reduced model. Define $\Delta RSS = RSS_0 - RSS$.

Consider the corresponding GLRT (see Section 4.3.3). From (8.9) after a simple calculus the generalized likelihood ratio (GLR) is

$$\lambda(\mathbf{y}) = \frac{L(\hat{\beta}, \hat{\sigma}; \mathbf{y})}{L(\hat{\beta}_0, \hat{\sigma}_0; \mathbf{y})} = \left(\frac{\hat{\sigma}_0^2}{\hat{\sigma}^2} \right)^{\frac{n}{2}} = \left(\frac{RSS_0}{RSS} \right)^{\frac{n}{2}} = \left(1 + \frac{\Delta RSS}{RSS} \right)^{\frac{n}{2}}$$

$$= \left(1 + \frac{r}{n-p} \cdot \frac{\Delta RSS/r}{RSS/(n-p)} \right)^{\frac{n}{2}}$$

which is an increasing function of the statistic $\frac{\Delta RSS/r}{RSS/(n-p)}$.

In Section 8.8.3 we show that $\Delta RSS = \hat{\beta}_{1:r}^{\top} \mathbb{X}_{1:r}^{\top} \mathbb{X}_{1:r} \hat{\beta}_{1:r}$ and, thus, $\frac{\Delta RSS/r}{RSS/(n-p)}$ coincides with the F-statistic in (8.17). □

Proposition 8.1 shows that the F-test (8.17) can be written in the equivalent form rejecting the null hypothesis (8.16) iff

$$F = \frac{\Delta RSS/r}{RSS/(n-p)} > F_{r,n-p;\alpha}.$$

8.8.3 Testing a general linear hypothesis

So far we have considered testing significance of predictors. However, other types of hypotheses may also be of interest. Do several covariates have the same effect on the response? Is the impact of a certain predictor twice as strong as that of another one? Are $\beta_1 = 2$ and $\beta_2 = -3$? All these hypotheses can be viewed as particular cases of a general *linear hypothesis* of the form

$$H_0 : A\beta = \mathbf{b} \quad vs. \quad H_1 : \beta \text{ is unrestricted}, \tag{8.18}$$

where A is the $r \times p$ matrix of full rank ($1 \leq r \leq p$) and $\mathbf{b} \in \mathbb{R}^r$.

Question 8.6 Show that the following null hypotheses are particular cases of the general linear hypothesis (8.18) (derive the corresponding matrix A and vector $\mathbf{b}$):

1. $\beta_1 = ... = \beta_r = 0$
2. $\beta_1 = ... = \beta_r$
3. $\beta_1 = 2$, $\beta_2 = -3$
4. $\beta_1 = 1$, $\beta_3 = 2\beta_2$, $\beta_4 + 3\beta_6 = \beta_5$

Derive the GLRT for testing (8.18). We first need to find the MLE $\hat{\beta}_0$ and $\hat{\sigma}_0^2$ for β and σ^2 under the null hypothesis $A\beta = \mathbf{b}$. Maximizing the likelihood $L(\beta, \sigma; \mathbf{Y})$ in (8.9) w.r.t. β subject to the constraint $A\beta = \mathbf{b}$ is evidently equivalent to minimizing

$$\min_{\beta} ||\mathbf{Y} - \mathbb{X}\beta||^2 \text{ subject to } A\beta = \mathbf{b}. \tag{8.19}$$

The problem (8.19) can be solved by Lagrange multipliers, where the original constrained optimization is replaced by a nonconstrained one by adding an additional vector $\mathbf{w}$ of Lagrange multipliers:

$$\min_{\beta, \mathbf{w}} \left\{ ||\mathbf{Y} - \mathbb{X}\beta||^2 + (A\beta - \mathbf{b})^{\top} \mathbf{w} \right\}. \tag{8.20}$$

Differentiating (8.20) w.r.t β and $\mathbf{w}$ one has

$$\frac{\partial}{\partial \beta} = 2(\mathbb{X}^{\top}\mathbb{X})\beta - 2\mathbb{X}^{\top}\mathbf{Y} + A^{\top}\mathbf{w}.$$

and

$$\frac{\partial}{\partial \mathbf{w}} = A\beta - \mathbf{b}.$$

Equating the resulting expressions to zero, straightforward calculus yields

$$\begin{aligned}
\hat{\beta}_0 &= (\mathbb{X}^{\top}\mathbb{X})^{-1} \left(\mathbb{X}^{\top}\mathbf{Y} - A^{\top}(A(\mathbb{X}^{\top}\mathbb{X})^{-1}A^{\top})^{-1}(A(\mathbb{X}^{\top}\mathbb{X})^{-1}\mathbb{X}^{\top}\mathbf{Y} - \mathbf{b}) \right) \\
&= \hat{\beta} - (\mathbb{X}^{\top}\mathbb{X})^{-1}A^{\top}(A(\mathbb{X}^{\top}\mathbb{X})^{-1}A^{\top})^{-1}(A\hat{\beta} - \mathbf{b}),
\end{aligned} \tag{8.21}$$

where $\hat{\beta}$ is the unconstrained MLE (8.5) of β. Obviously, $\hat{\sigma}_0^2 = \frac{RSS_0}{n}$, where $RSS_0 = ||\mathbf{Y} - \mathbb{X}\hat{\beta}_0||^2$ is the residual sum of squares of the restricted model under the null hypothesis.

Substituting $\hat{\beta}_0$ from (8.21) after some matrix algebra one has

$$
\begin{aligned}
RSS_0 &= \|\mathbf{Y} - \mathbb{X}\hat{\beta}_0\|^2 = (\mathbf{Y} - \mathbb{X}\hat{\beta})^\top (\mathbf{Y} - \mathbb{X}\hat{\beta}) + (A\hat{\beta} - \mathbf{b})^\top (A(\mathbb{X}^\top \mathbb{X})^{-1} A^\top)^{-1} (A\hat{\beta} - \mathbf{b}) \\
&= RSS + (A\hat{\beta} - \mathbf{b})^\top (A(\mathbb{X}^\top \mathbb{X})^{-1} A^\top)^{-1} (A\hat{\beta} - \mathbf{b})
\end{aligned}
$$

and, hence,

$$
\Delta RSS = RSS_0 - RSS = (A\hat{\beta} - \mathbf{b})^\top (A(\mathbb{X}^\top \mathbb{X})^{-1} A^\top)^{-1} (A\hat{\beta} - \mathbf{b}).
$$

In particular, for the matrix $A = (I_r, 0_{r \times (p-r)})$ and vector $\mathbf{b} = \mathbf{0}$ corresponding to testing $\beta_1 = ... = \beta_r = 0$, $\Delta RSS = \hat{\beta}_{1:r}^\top \mathbb{X}_{1:r}^\top \mathbb{X}_{1:r} \hat{\beta}_{1:r}$ (see the proof of Proposition 8.1).
Repeating the calculus for the GLR in the proof of Proposition 8.1 one has

$$
\lambda(\mathbf{y}) = \frac{L(\hat{\beta}, \hat{\sigma}; \mathbf{y})}{L(\hat{\beta}_0, \hat{\sigma}_0; \mathbf{y})} = \left(\frac{\hat{\sigma}_0^2}{\hat{\sigma}^2} \right)^{\frac{n}{2}} = \left(1 + \frac{r}{n-p} \cdot \frac{\Delta RSS/r}{RSS/(n-p)} \right)^{\frac{n}{2}}
$$

which is an increasing function of the statistic

$$
F = \frac{\Delta RSS/r}{RSS/(n-p)} = \frac{(A\hat{\beta} - \mathbf{b})^\top (A(\mathbb{X}^\top \mathbb{X})^{-1} A^\top)^{-1} (A\hat{\beta} - \mathbf{b})/r}{RSS/(n-p)}. \tag{8.22}
$$

As we have already mentioned, $RSS \sim \sigma^2 \chi_{n-p}^2$. Note also that under the null hypothesis, $A\hat{\beta} \overset{H_0}{\sim}$ $\mathcal{N}(\mathbf{b}, \sigma^2 A(\mathbb{X}^\top \mathbb{X})^{-1} A^\top)$ and, therefore, $(A\hat{\beta} - \mathbf{b})^\top (A(\mathbb{X}^\top \mathbb{X})^{-1} A^\top)^{-1} (A\hat{\beta} - \mathbf{b}) \overset{H_0}{\sim} \sigma^2 \chi_r^2$. In addition, being the norms of projections of a multivariate normal random vector on orthogonal subspaces, ΔRSS and RSS are independent (see Theorem A.4.1). Thus, the F-statistic in (8.22) under the null hypothesis is an $F_{r,n-p}$ distributed pivot and the resulting α-level GLRT test for testing the general linear hypothesis (8.18) rejects the null iff

$$
F = \frac{(A\hat{\beta} - \mathbf{b})^\top (A(\mathbb{X}^\top \mathbb{X})^{-1} A^\top)^{-1} (A\hat{\beta} - \mathbf{b})/r}{RSS/(n-p)} > F_{r,n-p;\alpha}. \tag{8.23}
$$

8.9 Predictions

Prediction is one of the main goals in regression analysis. After fitting a model from the data we are typically interested to use it for predicting the response for particular values of covariates.

Consider again a linear model with p predictors $x_1, ..., x_p$ and the response Y based on n observations $(\mathbf{x}_i, Y_i)$:

$$
Y_i = \mathbf{x}_i^\top \beta + \varepsilon_i, \quad i = 1, ..., n
$$

where ε_i are i.i.d. with zero means and a common variance σ^2. The unknown regression coefficients β are estimated by the LSE $\hat{\beta} = (\mathbb{X}^\top \mathbb{X})^{-1} \mathbb{X}^\top \mathbf{Y}$ from the data and we want to predict the response Y_0 for a given $\mathbf{x}_0 = (x_{10}, ..., x_{p0})^\top$.

There may be two somewhat different possible questions: whether the interest is in predicting the *actual* response Y_0 or in the *expected* response $EY_0 = \mathbf{x}_0^\top \beta$. Suppose that an electrical engineer fits a regression model describing the daily electricity demand as a function of different variables, e.g., weather conditions and various economic factors. Provided with climate and economic forecasts for the forthcoming summer, he may be asked to predict accordingly the *average* daily demand during the summer. Another relevant question would be to predict the actual electricity consumption for a *single* summer day. Sometimes to distinguish between the two problems, prediction of EY_0 is called *prediction*, while prediction of Y_0 is known as *forecasting*.

The natural point predictor for both EY_0 and Y_0 is $\mathbf{x}_0^\top \hat{\beta}$ but the corresponding prediction intervals are different.

Consider first prediction of the expected response $EY_0 = \mathbf{x}_0^\top \beta$.

Theorem 8.6 Let $\widehat{EY_0} = \mathbf{x}_0^\top \hat{\beta}$, where $\hat{\beta}$ is the LSE (8.5).

1. $\mathbf{x}_0^\top \hat{\beta}$ is the BLUE of EY_0

2. If, in addition, the ε_i's are normal $\mathcal{N}(0, \sigma^2)$, $\mathbf{x}_0^\top \hat{\beta}$ is the MLE of EY_0 and the UMVUE that achieves the Cramer–Rao lower bound.

Proof. $\widehat{EY_0} = \mathbf{x}_0^\top \hat{\beta} = \mathbf{x}_0 (\mathbb{X}^\top \mathbb{X})^{-1} \mathbb{X}^\top \mathbf{Y}$ and thus $\widehat{EY_0}$ is a linear estimator. It is unbiased since $E(\mathbf{x}_0^\top \hat{\beta}) = \mathbf{x}_0^\top \beta = EY_0$. We now show that it is the BLUE. Consider any other linear estimator $\widetilde{EY_0} = \mathbf{a}^\top \mathbf{Y}$ of EY_0 for some $\mathbf{a} \in \mathbb{R}^n$. Then, $E(\mathbf{a}^\top \mathbf{Y}) = \mathbf{a}^\top \mathbb{X} \beta$ and due to unbiasedness, $\mathbf{a}$ should satisfy $\mathbf{a}^\top \mathbb{X} = \mathbf{x}_0^\top$.

As for the variances, from Theorem 8.2,

$$Var(\widetilde{EY_0}) = \mathbf{x}_0^\top Var(\hat{\beta}) \mathbf{x}_0^\top = \sigma^2 \mathbf{x}_0^\top (\mathbb{X}^\top \mathbb{X})^{-1} \mathbf{x}_0 = \sigma^2 \mathbf{a}^\top \mathbb{X} (\mathbb{X}^\top \mathbb{X})^{-1} \mathbb{X}^\top \mathbf{a} = \sigma^2 \mathbf{a}^\top P_{\mathbb{X}} \mathbf{a},$$

where $P_{\mathbb{X}}$ is the projection matrix on the span of columns $\mathbb{X}$ defined in Section 8.3, while $Var(\widetilde{EY_0}) = \sigma^2 \mathbf{a}^\top \mathbf{a}$. Thus,

$$Var(\widetilde{EY_0}) - Var(\widehat{EY_0}) = \sigma^2 \mathbf{a}^\top (I - P_{\mathbb{X}}) \mathbf{a} \geq 0$$

since the matrix $I - P_{\mathbb{X}}$ is positive semi-definite (see Lemma 8.1).

Under the additional assumption of noise normality, $\hat{\beta}$ is also the MLE of β (see Theorem 8.4) and therefore, $\widehat{EY_0} = \mathbf{x}_0^\top \hat{\beta}$ is the MLE of $\mathbf{x}_0^\top \beta = EY_0$.

Furthermore, from Theorem 2.4 and (8.10), the Cramer–Rao lower bound for unbiased estimation of $EY_0 = \mathbf{x}_0^\top \beta$ is

$$(x_{10}, ..., x_{p0}, 0) I(\beta, \sigma)^{-1} (x_{10}, ..., x_{p0}, 0)^\top = \sigma^2 \mathbf{x}_0^\top (\mathbb{X}^\top \mathbb{X})^{-1} \mathbf{x}_0$$

which is exactly $Var(\widehat{EY_0})$. □

We now derive the prediction confidence interval for EY_0 under the normality assumption for the noise distribution. Note that in this case $\mathbf{x}_0^\top \hat{\beta} \sim \mathcal{N}(\mathbf{x}_0^\top \beta, \sigma^2 \mathbf{x}_0^\top (\mathbb{X}^\top \mathbb{X})^{-1} \mathbf{x}_0)$. On the other hand, $RSS \sim \sigma^2 \chi_{n-p}^2$. Moreover, since $\hat{\beta}$ and RSS are independent (see Theorem 8.5), $\mathbf{x}_0^\top \hat{\beta}$ and RSS are also independent, $\dfrac{\mathbf{x}_0^\top \hat{\beta} - \mathbf{x}_0^\top \beta}{s \sqrt{\mathbf{x}_0^\top (\mathbb{X}^\top \mathbb{X})^{-1} \mathbf{x}_0}} \sim t_{n-p}$ and the resulting $100(1 - \alpha)\%$ confidence interval for $EY_0 = \mathbf{x}_0^\top \beta$ is, therefore,

$$\mathbf{x}_0^\top \hat{\beta} \pm t_{n-p, \alpha/2} \, s \sqrt{\mathbf{x}_0^\top (\mathbb{X}^\top \mathbb{X})^{-1} \mathbf{x}_0}. \tag{8.24}$$

Consider now the prediction of a new observation $Y_0 = \mathbf{x}_0^\top \beta + \varepsilon_0$, where $\varepsilon_0 \sim \mathcal{N}(0, \sigma^2)$ and is independent from $\varepsilon_1, ..., \varepsilon_n$. The point predictor of Y_0 is still $\mathbf{x}_0^\top \hat{\beta}$. To construct the prediction interval for Y_0, recall that $\mathbf{x}_0^\top \hat{\beta} \sim \mathcal{N}(\mathbf{x}_0^\top \beta, \sigma^2 \mathbf{x}_0^\top (\mathbb{X}^\top \mathbb{X})^{-1} \mathbf{x}_0)$ and $Y_0 \sim \mathcal{N}(\mathbf{x}_0^\top \beta, \sigma^2)$. In addition, $\mathbf{x}_0^\top \hat{\beta}$ and Y_0 are independent since $\hat{\beta}$ is a function of the past data $\mathbf{Y} = (Y_1, ..., Y_n)^\top$, while Y_0 is an independent new observation. By the same arguments, Y_0 is independent of the RSS. Thus, $Y_0 - \mathbf{x}_0^\top \hat{\beta} \sim \mathcal{N}(0, \sigma^2(1 + \mathbf{x}_0^\top (\mathbb{X}^\top \mathbb{X})^{-1} \mathbf{x}_0))$ and

$$\frac{Y_0 - \mathbf{x}_0^\top \hat{\beta}}{s(1 + \mathbf{x}_0^\top (\mathbb{X}^\top \mathbb{X})^{-1} \mathbf{x}_0)} \sim t_{n-p}$$

that yields the $100(1 - \alpha)\%$ prediction interval for Y_0

$$\mathbf{x}_0^\top \hat{\beta} \pm t_{n-p, \alpha/2} \, s \left(1 + \sqrt{\mathbf{x}_0^\top (\mathbb{X}^\top \mathbb{X})^{-1} \mathbf{x}_0} \right). \tag{8.25}$$

The prediction interval (8.25) for a new observation Y_0 is wider than (8.24) for EY_0 since in addition to an error in estimating $EY_0 = \mathbf{x}_0^\top \beta$ it also reflects an error due to the random noise ε_0.

8.10 Analysis of variance

Analysis of variance (ANOVA) is undoubtedly one of the most widely used statistical techniques. It is a general method in which the overall variance of the random variable of interest is partitioned into several components associated with different sources of variation. The variance in the results of medical experiments may be due to a series of factors, e.g., treatment type and severity of illness. The variance of the mean yield of cereal among different areas depends on types of soil and fertilizers, weather conditions, etc. In its simplest form, ANOVA analyzes the means of several populations. In this sense the name "analysis of variance" might be somewhat misleading since, in fact, ANOVA considers *variation in means*.

In this section we demonstrate how ANOVA models can also be treated within the linear models framework with categorical predictors.

8.10.1 One-way ANOVA

The simplest ANOVA model is *one-way* ANOVA that involves a single factor, e.g., the variance in the results of a medical test for various types of drugs.

The general one-way ANOVA model can be described as follows. We have observations Y_{jk} from K samples of sizes n_k from different populations (groups). The model is

$$Y_{kj} = \mu_k + \varepsilon_{kj}, \quad j = 1, \ldots, n_k, k = 1, \ldots, K, \tag{8.26}$$

where μ_k is the unknown mean of the k-th group. The assumptions on the noise may be

1. The ε_{kj} are uncorrelated with zero means.

2. The ε_{kj} are i.i.d. with zero means and the common variance σ^2.

3. The ε_{kj} are i.i.d. $\mathcal{N}(0, \sigma^2)$.

These are written in this order to stress the three levels of assumptions. Although we will use the last one, if the n_k's are not "too small" and the error distribution is tamed, the second assumption is essentially sufficient since the additional requirement of normality will be (approximately) satisfied due to the CLT (see Section 5.5). Finally, the resulting test statistics will be robust enough to cope with different variances among the groups.

The one way ANOVA model (8.26) is a linear regression model with a categorical explanatory variable with K levels (see Section 8.2) and, hence, fits the general linear model (8.2), where

$$
\begin{pmatrix} \mathbf{Y}_{11} \\ \vdots \\ \mathbf{Y}_{1n_1} \\ \mathbf{Y}_{21} \\ \vdots \\ \mathbf{Y}_{Kn_K} \end{pmatrix}
=
\begin{pmatrix} \mathbf{1}_{n_1} & 0_{n_1} & \cdots & 0_{n_1} \\ 0_{n_2} & \mathbf{1}_{n_2} & \cdots & 0_{n_2} \\ \vdots & & & \\ 0_{n_K} & 0_{n_K} & \cdots & \mathbf{1}_{n_K} \end{pmatrix}
\begin{pmatrix} \mu_1 \\ \mu_2 \\ \vdots \\ \mu_K \end{pmatrix}
+
\begin{pmatrix} \varepsilon_{11} \\ \vdots \\ \varepsilon_{1n_1} \\ \varepsilon_{21} \\ \vdots \\ \varepsilon_{Kn_K} \end{pmatrix}
$$

(see also (8.3)).

Let $N = \sum_{k=1}^{K} n_k$ be the total number of observations. Define the sample mean of the j-th sample $\bar{Y}_{k.} = \frac{1}{n_k} \sum_{j=1}^{n_k} Y_{kj}$ and the overall sample mean $\bar{Y}_{..} = \frac{1}{N} \sum_{k=1}^{K} \sum_{j=1}^{n_k} Y_{kj}$. Then,

$$
\mathbb{X}^\top \mathbb{X} = \begin{pmatrix} n_1 & 0 & 0 & \cdots & 0 \\ 0 & n_2 & 0 & \cdots & 0 \\ \vdots & & & & \\ 0 & 0 & 0 & \cdots & n_K \end{pmatrix}; \quad \mathbb{X}^\top \mathbf{Y} = \begin{pmatrix} n_1 \bar{Y}_{1.} \\ \vdots \\ n_K \bar{Y}_{K.} \end{pmatrix}
$$

and, therefore,

$$\begin{pmatrix} \hat{\mu}_1 \\ \vdots \\ \hat{\mu}_K \end{pmatrix} = \begin{pmatrix} \bar{\mathbf{Y}}_{1\cdot} \\ \vdots \\ \bar{\mathbf{Y}}_{K\cdot,} \end{pmatrix}.$$

The covariance matrix $Var(\hat{\mu})$ is

$$Var(\hat{\mu}) = \sigma^2 \begin{pmatrix} n_1^{-1} & 0 & 0 & \dots & 0 \\ 0 & n_2^{-1} & 0 & \dots & 0 \\ \vdots & & & & \\ 0 & 0 & 0 & \dots & n_K^{-1} \end{pmatrix}$$

(see Theorem 8.2) and, in particular, the $\hat{\mu}_k$'s are uncorrelated.

The first natural question is to test for the differences between the K groups:

$$H_0 : \mu_1 = \mu_2 = \cdots = \mu_K$$
$$H_1 : \mu_k \neq \mu_j \text{ for at least one pair } (k, j).$$

We can use the general F-test (GLRT) in (8.23) for the specific case of

$$A_{(k-1)\times k} = \begin{pmatrix} 1 & -1 & 0 & 0 & \dots & 0 & 0 \\ 0 & 1 & -1 & 0 & \dots & 0 & 0 \\ \vdots & & & & & & \\ 0 & 0 & 0 & 0 & \dots & 1 & -1 \end{pmatrix}$$

and $\mathbf{b}_{k-1} = \mathbf{0}$ (see Question 8.6). After a straightforward calculus the resulting F-statistic is

$$F = \frac{\sum_{k=1}^{K} \sum_{j=1}^{n_k} (\bar{\mathbf{Y}}_{k\cdot} - \bar{\mathbf{Y}}_{\cdot\cdot})^2/(K-1)}{\sum_{k=1}^{K} \sum_{j=1}^{n_k} (\mathbf{Y}_{kj} - \bar{\mathbf{Y}}_{k\cdot})^2/(N-K)} = \frac{\sum_{k=1}^{K} n_k (\bar{\mathbf{Y}}_{k\cdot} - \bar{\mathbf{Y}}_{\cdot\cdot})^2/(K-1)}{\sum_{k=1}^{K} \sum_{j=1}^{n_k} (\mathbf{Y}_{kj} - \bar{\mathbf{Y}}_{k\cdot})^2/(N-K)} \qquad (8.27)$$

and the null hypothesis of equality of group means is rejected at level α iff $F > F_{K-1,N-K;\alpha}$. The above F-test is an extension of the two-sample t-test for $K = 2$ (see Exercise 4.12) to $K > 2$.

The sum of squares $SSB = \sum_{k=1}^{K} n_k (\bar{\mathbf{Y}}_{k\cdot} - \bar{\mathbf{Y}}_{\cdot\cdot})^2$ in the numerator of the F-statistic (8.27) can be viewed as a measure of variation *between* the K samples, while the sum of squares $SSW = \sum_{k=1}^{K} \sum_{j=1}^{n_k} (\mathbf{Y}_{kj} - \bar{\mathbf{Y}}_{k\cdot})^2$ in the denominator measures *within* samples variation. Thus, the variability of the entire data $SST = \sum_{k=1}^{K} \sum_{j=1}^{n_k} (\mathbf{Y}_{kj} - \bar{\mathbf{Y}}_{\cdot\cdot})^2$ can be decomposed to

$$SST = SSB + SSW, \qquad (8.28)$$

that is, into variability between and within groups. Using this notation, the F-statistic (8.27)

$$F = \frac{SSB/(K-1)}{SSW/(N-K)}$$

is the ratio between SSB and SSW normalized by the corresponding degrees of freedom. The larger F is, the more dominating between groups variation relative to the within group variation in the total SST and the stronger the evidence for differences between the groups.

Question 8.7 Show that for the one-way ANOVA model, the ANOVA decomposition (8.28) is, in fact, a particular case of (8.13), where $RSS = SSW$.

8.10.2 Two-way ANOVA and beyond

Assume now that there is more than one component affecting the variation of the response. For example, in a medical experiment, severity of illness can matter in addition to the treatment type. This is an example of the *two*-way ANOVA model:

$$Y_{klj} = \mu_{kl} + \varepsilon_{klj}, \quad k = 1, ..., K; \; l = 1, ..., L; \; i = 1, ..., n_{kl}, \tag{8.29}$$

where μ_{kl} is the mean response of patients with the l-th level of severity of illness that received k-th type of treatment and the ε_{klj}'s are i.i.d. $N(0, \sigma^2)$.

Similarly to the one-way ANOVA, one can verify that the MLE/LSE of μ_{kj} is

$$\hat{\mu}_{kl} = \bar{Y}_{kl.} = \frac{1}{n_{kl}} \sum_{j=1}^{n_{kl}} Y_{klj}.$$

We can also write

$$\mu_{kl} = \mu_{..} + (\mu_{k.} - \mu_{..}) + (\mu_{.l} - \mu_{..}) + (\mu_{kl} - \mu_{k.} - \mu_{.l} + \mu_{..}),$$

where $\mu = \mu_{..}$ is the overall mean response, $\alpha_k = \mu_{k.} - \mu_{..}$ is the average main effect of the k-th treatment, $\beta_l = \mu_{.l} - \mu_{..}$ is the average main effect of the l-th level of severity of illness, and $\gamma_{kl} = \mu_{kl} - \mu_{k.} - \mu_{.l} + \mu_{..}$ is usually called the average *interaction* effect between the k-th type of treatment and the l-th level of severity of illness. In terms of these new $1 + K + L + KL$ parameters, the original two-way ANOVA model (8.29) with KL parameters is reparameterized as

$$Y_{klj} = \mu + \alpha_k + \beta_l + \gamma_{kj} + \varepsilon_{klj}$$

subject to the constraints

$$\sum_{k=1}^{K} \alpha_k = \sum_{l=1}^{L} \beta_l = 0$$

$$\sum_{k=1}^{K} \gamma_{kl} = 0, \quad l = 1, ..., L$$

$$\sum_{l=1}^{L} \gamma_{kl} = 0, \quad k = 1, ..., K$$

(note that the number of *independent* parameters remains KL).

We may be interested in testing the following null hypotheses:

1. $H_0 : \gamma = 0$ (there are no interactions between the treatment type and severity of illness—*additivity* of factors' effects)

2. $H_0 : \alpha = 0, \gamma = 0$ (there are no differences between treatments)

3. $H_0 : \beta = 0, \gamma = 0$ (effects of treatments do not depend on severity of illness)

All the above hypotheses are still particular cases of a general linear hypothesis (8.18) and the corresponding F-tests can be derived from (8.23). We omit the technical details and refer the interested reader to a variety of literature on ANOVA, e.g., Draper and Smith (1981, Chapter 9).

Chapter 9

*Nonparametric Estimation

9.1 Introduction

Suppose we observe a random sample $Y_1, \ldots, Y_n \sim f(y)$ with the unknown density $f(\cdot)$ and want to estimate it at a given point y. So far we assumed that $f(y)$ belongs to some known *parametric* family of distributions $\mathscr{F}_\theta, \theta \in \Theta$, e.g., normal or exponential, and the unknown quantity of interest is a parameter θ. Within the parametric framework, a natural approach would be to estimate θ by some (hopefully good) estimator $\hat\theta$, and estimate the density at y by $f_{\hat\theta}(y)$. In particular, if $\hat\theta$ is the MLE of θ, then $f_{\hat\theta}(y)$ is the MLE of $f_\theta(y)$.

Similarly, in Section 2.4 we considered a general regression model, where we observe a sample of pairs $(\mathbf{x}_i, Y_i)$, $i = 1, \ldots, n$ and $\mathbf{x}_i \in \mathbb{R}^m$. It is assumed that $Y_i = g_\theta(\mathbf{x}_i) + \varepsilon_i$, where the regression function $g_\theta(\cdot) : \mathbb{R}^m \to \mathbb{R}$ has the known parametric form with a vector of parameters $\theta \in \Theta \subseteq \mathbb{R}^p$ and ε_i's are i.i.d. random noise with zero mean and finite variance. The goal is to estimate $E_{\mathbf{x}}Y = g_\theta(\mathbf{x})$ for a given predictor vector $\mathbf{x}$. A reasonable approach would be to estimate θ by the least squares estimator and then to estimate $g_\theta(\mathbf{x})$ by $g_{\hat\theta}(\mathbf{x})$.

In both of these cases we had to estimate the value of some function (the density of Y or the predictor of $E_{\mathbf{x}}Y$) at a given point. We assumed that this function belongs to a parametric family, estimated the parameters and plugged them in to obtain the required estimate. However, in various situations such an assumption might be too restrictive and one should consider larger, *nonparametric* families of distributions with more general assumptions. For example, it can only be assumed that the function—the density or the regression function—is smooth and has, say, two bounded derivatives.

Generally speaking, we consider a random sample from a distribution that belongs to a family of distributions $\mathscr{F}$. Statistical methods that are based on the assumption that $\mathscr{F}$ can be parameterized by $\theta \in \Theta \subseteq \mathbb{R}^p$ are part of *parametric statistics*, while if $\mathscr{F}$ is a larger set and does not have such a parametrization, we talk about *nonparametric* statistics.

Our aim in this chapter is to discuss just a few ideas of nonparametric density estimation and nonparametric regression. We will consider only histogram, kernel, orthogonal series, and smoothing splines estimators although there exist other well-known methods, notably, wavelets, regression splines, local linear regression or k-nearest neighbors. We should remark that in a deep sense, all methods are asymptotically equivalent. We will consider estimation of real functions of univariate real variables with two bounded derivatives, and say almost nothing of what can be achieved under stronger or weaker constraints on the functions. Finally, all nonparametric methods depend on tuning parameters and we will briefly discuss some methods for their adaptive choice from the data.

9.2 The empirical distribution function and the histogram

Let Y be a random variable with cdf $F(y)$ and a density $f(y) = F'(y)$. Let $Y_1, \ldots, Y_n$ be a random sample from the same distribution. Finally, for any real y, let $T_i(y)$ be the indicator that $Y_i \le y$, that is, $T_i(y) = I\{Y_i \le y\}$. Evidently, $ET_i(y) = F(y)$; $T_1(y), \ldots, T_n(y)$ are i.i.d. Bernoulli $\mathscr{B}(1, F(y))$ and $\sum_{i=1}^n T_i(y) \sim \mathscr{B}(n, F(y))$. A natural estimator of $F(y)$ is then $\hat{F}_n(y) = \frac{1}{n} \sum_{i=1}^n T_i(y)$—the proportion

DOI: 10.1201/9781003175407-9

of observations in the sample not larger than y. We call $\hat{F}_n(y)$ the *empirical distribution function* (edf).

Question 9.1 Show that $\hat{F}_n(\cdot)$ is a non-decreasing step-function with jumps at points y_i. What are the jumps sizes?

Theorem 9.1 For any $y \in \mathbb{R}$, $\hat{F}_n(y) \xrightarrow{p} F(y)$.

The proof is an immediate consequence of the WLLN (see Theorem 5.2).

The simplest and intuitively obvious nonparametric estimator of the density $f(y)$ is the *histogram*. It can be considered as a way to present the edf. Split the support of f into intervals (bins) with fixed knots $a_0 < \ldots < a_k$ and let $a_{j-1} < y \leq a_j$. Then

$$f(y) = F'(y) \approx \frac{F(a_j) - F(a_{j-1})}{a_j - a_{j-1}} = \frac{\text{probability to be within the bin}}{\text{bin's width}} \tag{9.1}$$

and a natural estimator of $f(y)$ is

$$\hat{f}(y) = \frac{\hat{F}_n(a_j) - \hat{F}_n(a_{j-1})}{a_j - a_{j-1}} = \frac{\text{proportion of observations fallen within the bin}}{\text{bin's width}} \tag{9.2}$$

This idea leads us to the following definition:

Definition 9.1 (histogram) Let $a_0 < a_1 < \cdots < a_k$. The *histogram* with knots $a_0, \ldots, a_k$ is defined at point y by

$$\hat{f}_H(y) = \frac{\hat{F}_n(a_j) - \hat{F}_n(a_{j-1})}{a_j - a_{j-1}},$$

where $a_{j-1} < y \leq a_j$, and $\hat{f}_H(y) = 0$ for $y < a_0$ or $y > a_k$.

By its construction, the histogram $\hat{f}_H(y)$ is a piece-wise constant function with the same value within each bin.

The following proposition follows directly from the properties of the edf:

Proposition 9.1 Let $\hat{f}_H(y)$ be the histogram with fixed knots at $a_0, \ldots, a_k$. Then for any $a_{j-1} < y \leq a_j$,

$$n(a_j - a_{j-1})\hat{f}_H(y) \sim \mathscr{B}\big(n, F(a_j) - F(a_{j-1})\big).$$

Thus, in particular,

$$E\hat{f}_H(y) = \frac{F(a_j) - F(a_{j-1})}{a_j - a_{j-1}} \tag{9.3}$$

$$Var(\hat{f}_H(y)) = \frac{\big(F(a_j) - F(a_{j-1})\big)\big(1 - F(a_j) + F(a_{j-1})\big)}{n(a_j - a_{j-1})^2}.$$

We consider the asymptotic situation when the sample size n is large. Recall that the histogram should approximate $f(y)$. It is an unbiased estimator of the approximation $\frac{F(a_j) - F(a_{j-1})}{a_j - a_{j-1}}$ in (9.1) but is a biased estimator of $f(y)$ itself. The bias gets smaller as the bin becomes smaller. On the other hand, in a smaller bin there will be less observations and, thus, a larger variance (see Proposition 9.1). To be more precise:

Theorem 9.2 Suppose the density $f \in C^2$, where C^2 is the set of all bounded functions with two bounded and continuous derivatives. For any n, consider the histogram estimator with knots $a_{n,0} < \ldots < a_{n,k_n}$. Let y be a point in the bin $(a_{n,j-1}, a_{n,j}]$ with a half-width $h_n = \frac{a_{n,j} - a_{n,j-1}}{2}$, and $\delta_n(y) = h_n^{-1}\big(y - \frac{a_{n,j} + a_{n,j-1}}{2}\big) \in (-1, 1)$ be y's relative position in the bin. Then,

$$Var(\hat{f}_H(y)) = \frac{f(y)}{nh} + O\left(\frac{1}{n}\right)$$

$$E\big(\hat{f}_H(y) - f(y)\big) = 2\delta_n(y)f'(y)h_n + \left(\frac{1}{6} + \frac{\delta_n(y)^2}{2}\right)f''(y)h_n^2 + o(h_n^2).$$

Proof. The variance result follows immediately from the fact that $2h_n n \hat{f}_H(y) \sim \mathcal{B}(n, (F(a_{n,j}) - F(a_{n,j-1})))$ (see Proposition 9.1), while from (9.1), $F(a_{n,j}) - F(a_{n,j-1}) = 2h_n f(y) + o(h_n)$.

For the bias term, use Taylor's expansion:

$$
\begin{aligned}
E\hat{f}_H(y) - f(y) &= \frac{F(a_{n,j}) - F(a_{n,j-1})}{2h_n} - f(y) \\
&= \frac{1}{2h_n} \int_{a_{n,j-1}}^{a_{n,j}} f(t)dt - f(y) \\
&= \frac{1}{2h_n} \int_{y-(1+\delta_n(y))h_n}^{y+(1-\delta_n(y))h_n} f(t)dt - f(y) \\
&= \frac{1}{2} \int_{-(1+\delta_n(y))}^{1-\delta_n(y)} \big(f(y+sh_n) - f(y)\big)ds.
\end{aligned}
$$

Let $u(w) = \frac{1}{2} \int_{-(1+\delta_n(y))}^{1-\delta_n(y)} \big(f(y+wsh_n) - f(y)\big)ds$. Then, by Taylor's expansion, $u(1) = u(0) + u'(0) + \frac{1}{2}u''(w^*)$ for some $0 \le w^* \le 1$ and, therefore,

$$
\begin{aligned}
E\hat{f}_H(y) - f(y) &= u(1) \\
&= \frac{1}{2}f'(y)h_n \int_{-(1+\delta_n(y))}^{1-\delta_n(y)} s\,ds + \frac{1}{4}h_n^2 f''(y) \int_{-(1+\delta_n(y))}^{1-\delta_n(y)} s^2 ds + o(h_n^2) \\
&= 2\delta_n(y)f'(y)h_n + \left(\frac{1}{6} + \frac{\delta_n(y)^2}{2}\right)f''(y)h_n^2 + o(h_n^2).
\end{aligned}
$$

$\square$

Theorem 9.2 shows that for consistency in MSE of $\hat{f}_H(y)$, the bin's half-width h_n should decrease with n to reduce the bias but not "too fast" so that the number of observations in the bin ($\sim nh_n f(y)$) will increase to control the variance. In other words, h_n should tend to zero slower than n^{-1}.

What is the optimal number of bins k_n or, equivalently, what is the optimal bin's half-width h_n? From Theorem 9.2 it follows that at the center of a bin, where $\delta_n(y) = 0$,

$$
\begin{aligned}
MSE(\hat{f}_H(y), f(y)) &= Var(\hat{f}_H(y)) + bias^2(\hat{f}_H(y), f(y)) \\
&= \begin{cases} \frac{C_1}{nh_h} + C_2 h_n^4 + o\left(\frac{1}{n}\right) + o(h_n^4) & \delta_n(y) = 0 \\ \frac{C_1}{nh_h} + C_3 h_n^2 + o\left(\frac{1}{n}\right) + o(h_n^2) & \liminf \delta_n(y) > 0, \end{cases}
\end{aligned}
\tag{9.4}
$$

for some constant C_1, C_2 and C_3. Thus, the optimal value of h_n that balances the variance-bias trade-off in (9.4) in the center of the bins is $h_n = O(n^{-1/5})$ with the resulting MSE of $O(n^{-4/5})$. We will argue in Section 9.4 that, in fact, no better rate is possible under our minimal assumptions on the unknown density. We can compare it to the faster parametric rate of $O(n^{-1})$—we pay the unavoidable price for the weaker assumptions.

On the other hand, the estimation rate at the knot points, where $\delta_n(y) = 1$, is much slower since the bias is of $O(h_n)$. The optimal bin's width will be larger than needed for estimating at the center $- h_n = O(n^{-1/3})$ with the optimal MSE of only $O(n^{-2/3})$.

Example 9.2.1 1000 observations were drawn from a mixture of two beta distributions: Let $W_1, \ldots, W_n$ be an i.i.d. sample from $\mathcal{B}(1, 0.3)$, $Y_1', \ldots, Y_n'$ a sample from $Beta(8, 16)$ and $Y_1'', \ldots, Y_n''$ a sample from $Beta(10, 3)$. The observations are $Y_i = W_i Y_i' + (1 - W_i)Y_i''$, $i = 1, \ldots, n$. Figure 9.1 presents the true density and three histograms based on this sample, with different number of bins. We can see the importance of choosing the proper number of bins—if their number is too small (or, equivalently, bin's width is too large) there is over-smoothing, while if their number is too large (the bin's width is too small) there is a danger of over-fitting the data and having a too noisy histogram.

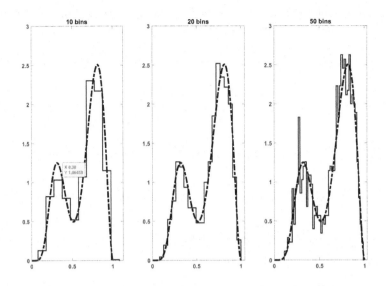

Figure 9.1: Histogram estimators of a mixture of two Beta densities with different numbers of bins based on 1000 observations (solid lines) and the true density (dashed line).

9.3   Kernel density estimation

The edf and the histogram are exact as a summary statistics. The histogram gives exactly the number of observations in each of the given bins. However, it has two main drawbacks as an estimator of the true density. First, as we saw in Section 9.2, the MSE is not uniform: it is much larger for points in edges of the bins than for points close to the bin center. The second problem with this estimator is esthetics. We assume that the density is smooth, but estimate it with a piece-wise constant function.

Is it possible to fix these two problems? Instead of using bins with fixed knots as in histrogrtam, consider a point of interest y as if it was the center of the bin. Namely, take a "window" with a bandwidth h around y, $y \pm h$, and estimate $\hat{f}(y)$ as

$$\hat{f}(y) = \frac{1}{n} \sum_{i=1}^{n} \frac{1}{2h} I\{y - h < y_i < y + h\} = \frac{\text{proportion of observations falling within the window}}{\text{window's width}}$$

(9.5)

similar to (9.2). The estimator $\hat{f}(y)$ in (9.5) can be viewed as a direct analogue of histogram estimator $\hat{f}_H(y)$ but with "moving" bins centered around the point of interest y. Unlike the histogram, we will see that its MSE rate is indeed uniform.

More generally, to get smoother estimators, we can consider fuzzy windows. Formally, let $K(t)$ be a *kernel* function such that $\int K(t)dt = 1$. Typically, but not necessarily, we consider kernels which are always positive and symmetric. Thus, the kernel is actually a probability density function of a zero mean random variable. A typical kernel used in practice is the normal density. A further common helpful restriction is to use a kernel with a finite support around zero. To summarize, we consider kernels which satisfy the following assumptions:

$$\int K(t)dt = 1;$$

(9.6)

$$\int t K(t)dt = 0;$$

(9.7)

$$\int t^2 K(t)dt < \infty;$$

(9.8)

$$K(t) = 0 \text{ for } |t| > 1; \tag{9.9}$$

$$K''(t) \text{ is continuous and bounded.} \tag{9.10}$$

Definition 9.2 (kernel density estimator) Let $Y_1, \ldots, Y_n \sim f(y)$. For a given kernel $K(\cdot)$ satisfying (9.6)–(9.10) and a bandwidth $h > 0$, the kernel density estimator of $f(y)$ is the function

$$\hat{f}_{K,h}(y) = \frac{1}{nh} \sum_{i=1}^{n} K\left(\frac{y - Y_i}{h}\right) \tag{9.11}$$

The estimator (9.5) is, in fact, a kernel estimator with the *uniform* kernel $K(t) = \frac{1}{2}$, $|t| \leq 1$ and 0 otherwise.

Question 9.2 Show that for a positive kernel $K(\cdot)$ with $\int K(t)dt = 1$, the kernel estimator $\hat{f}_{K,h}(y)$ is a proper density function, i.e. $\hat{f}_{K,h}(y) \geq 0$ for all y and $\int \hat{f}_{K,h}(y)dy = 1$.

The following theorem provides the $MSE(\hat{f}_{K,h}(y), f(y))$:

Theorem 9.3 Let $Y_1, \ldots, Y_n \sim f(y)$, where the unknown density $f \in C^2$. Let $K(\cdot)$ be a kernel satisfying (9.6)–(9.10) and the bandwidth $h_n \to 0$, but $nh_n \to \infty$. Then

1. $E\hat{f}_{K,h}(y) = f(y) + \frac{1}{2}C_1(K)f''(y)h_n^2 + o\left(h_n^2\right)$
2. $Var(\hat{f}_{K,h}(y)) \leq C_2(K)\frac{1}{nh_n} + O\left(\frac{1}{n}\right)$,

where the positive constants $C_1(K)$ and $C_2(K)$ depend on a particular kernel $K(\cdot)$.

Thus, $\hat{f}_{K,h}(y) \xrightarrow{MSE} f(y)$. If further, $nh_n^3 \to \infty$, then $\hat{f}'_{K,h}(y) \xrightarrow{MSE} f'(y)$.

In particular, Theorem 9.3 shows that the bias is larger at points where f is less smooth ($|f''|$ is large), while the variance is uniformly bounded.

Proof. Note that $\hat{f}_{K,h}(y)$ is the mean of n i.i.d. random variables, $h_n^{-1}K\left(\frac{y-Y_i}{h_n}\right)$, $i = 1, \ldots, n$. Hence we can use the properties of the mean. We consider first the expectation. By change of variables and condition (9.6):

$$
\begin{aligned}
E\hat{f}_{K,h}(y) &= \frac{1}{h_n}\int K\left(\frac{y-t}{h_n}\right)f(t)dt \\
&= \int K(s)f(y+sh_n)ds \\
&= f(y) + \int K(s)\left(f(y+sh_n) - f(y)\right)ds \\
&= u(1),
\end{aligned}
$$

where $u(w) = f(y) + \int K(s)\left(f(y+wsh_n) - f(y)\right)ds$. Expanding $u(w)$ in Taylor's series around $w = 0$: $u(w) = u(0) + u'(0) + \frac{1}{2}u''(w^*)$, we have

$$E\hat{f}_{K,h}(y) = f(y) + h_n f'(y)\int K(s)s\,ds + \frac{1}{2}h_n^2 f''(y)\int_{y-h_n}^{y+h_n} K(s)s^2 f''(y + w^* sh_n)ds$$

for some $0 \leq w^* \leq 1$, and from (9.7) and (9.10) it follows that $E\hat{f}_{K,h}(y) = f(y) + \frac{1}{2}C_1(K)f''(y)h_n^2 + o\left(h_n^2\right)$.

Similarly, for the variance

$$Var(\hat{f}_{K,h}(y)) = \frac{1}{nh_n^2}Var\left(K\left(\frac{y-Y_1}{h_n}\right)\right) = \frac{1}{nh_n^2}EK^2\left(\frac{y-Y_1}{h_n}\right) - \left(E\hat{f}_{K,h}(y)\right)^2.$$

But

$$E\frac{1}{h_n^2}K^2\left(\frac{y-Y_1}{h_n}\right) = \frac{1}{h_n^2}\int K^2\left(\frac{y-t}{h_n}\right)f(t)dt = \frac{1}{h_n}\int K^2(s)f(y+sh_n)ds$$

$$\leq \frac{\max_t K^2(t) \max_t f(t)}{h_n}.$$

Thus, if $h_n \to 0$, the second moment of each of the n terms converges to infinity, while its mean is $O(1)$. Therefore, the variance of each term is $O(h_n^{-1})$ and the variance of the mean is $O((nh_n)^{-1})$.

The results for the MSE-consistency follow. For the MSE-consistency of the derivative, note that $\hat{f}'_{K,h}(y)$ is a kernel estimator with the kernel $K'(\cdot)$:

$$\hat{f}'_{K,h}(y) \;=\; \frac{1}{nh_n^2} \sum_{i=1}^{n} K'\left(\frac{y - Y_i}{h_n}\right),$$

whose variance is given by

$$
\begin{aligned}
Var(\hat{f}'_{K,h}(y)) &= \frac{1}{nh_n^4} \int \left(K'\left(\frac{y-t}{h_n}\right)\right)^2 f(t)dt + O(n^{-1}) \\
&= \frac{1}{nh_n^3} \int \left(K'(s)\right)^2 f(y+sh_n)ds + O(n^{-1}) = O((nh_n^3)^{-1}) \to 0.
\end{aligned}
$$

$\square$

The optimal choice for a bandwidth h_n that minimizes $MSE(\hat{f}_{K,h}(y), f(y))$ is given in the following corollary, which is an immediate consequence of Theorem 9.3:

Corollary 9.1 Under the conditions of Theorem 9.3, h_n that minimizes $MSE(\hat{f}_{K,h}(y), f(y))$ is $h_n = O(n^{-1/5})$ with the resulting minimal $MSE(\hat{f}_{K,h_n}(y), f(y)) = O(n^{-4/5})$.

Thus, unlike histogram, a kernel estimator with $h_n = O(n^{-1/5})$ yields the MSE of $O(n^{-4/5})$ *uniformly* for all y. In Section 9.4 we will see that this is the best possible rate in the minimax sense. The choice of a kernel is asymptotically not important and affects only the constants. Theorem 9.3 also proves that a kernel estimator $\hat{f}_{K,h}(y)$ is a MSE-consistent estimator of $f(y)$ while its derivative is MSE-consistent estimator of $f'(y)$. However, the optimal choice of h_n for estimating the density itself is different from that for estimating its derivative.

Example 9.3.1 We apply a kernel density estimator with a normal kernel to the same data as in Figure 9.1 (see Figure 9.2). Again, we show three density estimators, each based on different bandwidths, having examples of over-smoothing and over-fitting estimators.

The MSE at a particular point is a local measure of goodness-of-fit at this point. An estimator may perform well at some points but worse at others. Thus, it may be useful to consider *integrated* MSE as a global criterion of estimator's accuracy:

Definition 9.3 (integrated mean squared error) The *integrated mean squared error* (IMSE) of the estimator $\hat{f}(x)$ is $IMSE(\hat{f}; f) = E \int (\hat{f}(y) - f(y))^2 dy$.

Assume that the density $f(y)$ has a finite support. It then follows from Theorem 9.3 that for a kernel estimator $\hat{f}_{K,h}(y)$ in (9.11) with the bandwidth $h_n = O(n^{-1/5})$, $IMSE(\hat{f}_{K,h_n}(y), f(y)) = O(n^{-4/5})$.[1]

As we have seen, the choice of a bandwidth h is essential for kernel estimation. Although we showed that the optimal h that minimizes IMSE is $h_{opt} = Cn^{-1/5}$, this result is mainly of theoretical interest since the constant C depends on the unknown f and its derivatives. We may want a rule for choosing h based on our assumptions and the data. Hence, instead of minimizing the IMSE that depends on the unknown f for choosing a bandwidth, we would like to find a criterion that doesn't depend on f. Consider $IMSE(\hat{f}_{K,h}, f)$:

$$
IMSE(\hat{f}_{K,h}, f) = E \int \left(\hat{f}_{K,h}(y) - f(y)\right)^2 dy = E \int \hat{f}_{K,h}(y)^2 dy - 2E \int \hat{f}_{K,h}(y)f(y)dy + \int f^2(y)dy
$$
$$(9.12)$$

---

[1] One should be more careful at boundary points, where Taylor's expansions used in the proof of Theorem 9.3 may not be applied and some additional boundary corrections are needed. We do not consider these issues here.

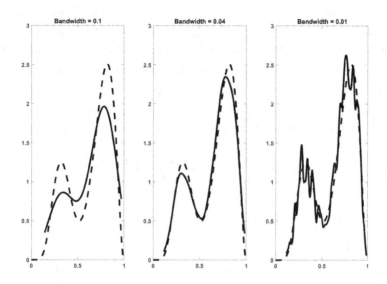

Figure 9.2: Kernel density estimators of a mixture of two Beta distributions with a normal kernel and different bandwidths based on 1000 observations (solid lines) and the true density (dashed line).

Note that the last term $\int f^2(y)dy$ in the right hand side of (9.12) does not depend on h and is, therefore, irrelevant for its choice. We would like then to find unbiased estimators $\hat{J}_1(h)$ and $\hat{J}_2(h)$ for the first two terms $J_1(h) = E \int \hat{f}_{K,h}(y)^2 dy$ and $J_2(h) = E \int \hat{f}_{K,h}(y) f(y) dy$ in (9.12) and choose h that minimizes $\hat{J}_1(h) - 2\hat{J}_2(h)$.

The unbiased estimator for $J_1(h)$ is trivial: $\hat{J}_1(h) = \int \hat{f}_{K,h}^2(y)dy$ and it remains to find the unbiased estimator for $J_2(h)$:

Proposition 9.2 $\hat{J}_2(h) = \frac{1}{n(n-1)h} \sum_{i=1}^{n} \sum_{j \neq i} K\left(\frac{Y_j - Y_i}{h}\right)$ is an unbiased estimator of $J_2(h)$.

Proof. Since $Y_1, \ldots, Y_n$ are i.i.d., we have

$$J_2(h) = E \int \hat{f}_{K,h}(y) f(y) dy = \int f(y) E \hat{f}_{K,h}(y) dy$$

$$= \frac{1}{h} \int f(y) E \left(\frac{1}{n} \sum_{i=1}^{n} K\left(\frac{y - Y_i}{h}\right)\right) dy = \frac{1}{h} \int f(y) E K\left(\frac{y - Y_1}{h}\right) dy$$

$$= \frac{1}{h} \int f(y) \int K\left(\frac{y - y_1}{h}\right) f(y_1) dy_1 dy$$

On the other hand, note that $\hat{J}_2$ is the average of identically distributed random variables $\xi_i = \frac{1}{(n-1)h} \sum_{j \neq i} K\left(\frac{Y_j - Y_i}{h}\right)$, $i = 1, \ldots, n$. Hence,

$$E\hat{J}_2 = E\xi_1 = \frac{1}{h} E \left(\frac{1}{n-1} \sum_{j \neq i} \int K\left(\frac{Y_j - y_1}{h}\right) f(y_1) dy_1\right)$$

$$= \frac{1}{h} \int f(y_1) \int K\left(\frac{y - y_1}{h}\right) f(y) dy dy_1$$

$\square$

The unbiased estimator $\hat{J}_2$ in Proposition 9.2 has, in fact, an interesting statistical meaning. Leave the first observation Y_1 out of the sample and estimate $f(y_1)$ by a kernel estimator using the remaining $n-1$ observations $Y_2, \ldots, Y_n$: $\hat{f}_{(-1)}(y_1) = \frac{1}{(n-1)h} \sum_{j=2}^{n} K(\frac{Y_1-Y_j}{h})$ which is exactly ξ_1 from the proof of Proposition 9.2. Return Y_1 to the sample, leave Y_2 out and estimate $f(y_2)$ from $Y_1, Y_3, \ldots, Y_n$ by $\hat{f}_{(-2)}(y_2) = \frac{1}{(n-1)h} \sum_{j \neq 2} K(\frac{Y_2-Y_j}{h}) = \xi_2$ and so on. At each step we leave out an observation Y_i and estimate $f(y_i)$ from the remaining $n-1$ observations Y_j, $j \neq i$ by $\hat{f}_{(-i)}(y_i) = \frac{1}{(n-1)h} \sum_{j \neq i} K(\frac{Y_i-Y_j}{h}) = \xi_i$. Such a procedure is known as a *leave-one-out cross-validation*. Thus, $\hat{J}_2(h) = \frac{1}{n} \sum_{i=1}^{n} \hat{f}_{(-i)}(y_i)$.

Following the proposed approach, we, therefore, choose the bandwidth adaptively from the data by minimizing

$$\int \hat{f}_{K,h}^2(y) dy - \frac{2}{n(n-1)h} \sum_{i=1}^{n} \sum_{j \neq i} K\left(\frac{Y_j - Y_i}{h}\right)$$

w.r.t. h.

9.4   The minimax rate

Corollary 9.1 shows that for any $f \in C^2$ the kernel estimator with a properly chosen bandwidth h_n of order $n^{-1/5}$, $n^{2/5}(\hat{f}_{K,h_n}(x) - f(x)) = O_p(1)$. Does there exist a better estimator that achieves a faster rate? Of course, the trivial estimator $\hat{f}(x) = f(x)$ is exact if the true density $f(x)$ is precisely $f(x)$ but it performs poorly under densities that are different from $f(x)$. Hence, we are interested in an estimator with the best possible error rate *uniformly* over some subset $\mathscr{C} \subset C^2$, i.e. we are looking for $\inf_{\hat{f}} \sup_{f \in \mathscr{C}} MSE(\hat{f}, f)$—the minimax estimator over $\mathscr{C}$ (see Section 7.3).

We are going to prove in this section that there is no estimator $\hat{f}(x)$ such that $n^{2/5}(\hat{f}(x) - f(x)) \xrightarrow{p} 0$ uniformly. Thus, the kernel estimator with the optimal choice of bandwidth $h_{opt} \sim n^{-1/5}$ is rate-optimal (in the minimax sense).

Theorem 9.4 Let $C^2(A)$ be the set of all densities with two continuous derivatives bounded by A and $Y_1, \ldots, Y_n$ be a random sample from a density $f \in C^2(A)$. Then there is a density $f_0 \in C^2(A)$ and a set of densities $\{f_1, f_2, \ldots\} \subset C^2(A)$ such that there is no estimator $\hat{f}$ satisfying $\hat{f}(y) - f_0(y) = o_p(n^{-2/5})$ if $Y_1, \ldots, Y_n$ are i.i.d. f_0 and, at the same time, $\hat{f}(y) - f_n(y) = o_p(n^{-2/5})$ if $Y_1, \ldots, Y_n$ are i.i.d. sampled from f_n.

Proof. There is one statistical problem in which we know exactly what should be done: testing between two simple hypotheses: the Neyman–Pearson lemma 4.1 proves that the LRT is a (unique) optimal test. We will use this to prove by contradiction that the convergence rate of the kernel estimator cannot be improved.

Suppose that $Y_1, \ldots, Y_n$ are sampled from some density $f \in C^2(A)$. Suppose further that $\hat{f}(x)$ is an estimator such that for any $f \in C^2(A)$ satisfies

$$\hat{f}(y) - f(y) = o_p(n^{-2/5}) \tag{9.13}$$

at any point y.

Consider $f_0 \in C^2(A)$ and a point y_0 such that $f_0(y_0) > 0$ and the density and its second derivative are bounded by $A/2$ at y_0. Let $f_n(y) = f_0(y) + h_n(y)$, where

$$h_n(x) = \begin{cases} a_n \cos^2(b_n(y-y_0)) & y_0 - \pi/2 < y \leq y_0 + \pi/2 \\ -a_n \cos^2(b_n(y-y_0)) & y_0 + \pi/2 < y < y_0 + 3\pi/2 \\ 0, & \text{otherwise,} \end{cases}$$

and

$$a_n = n^{-2/5}$$

$$b_n = \sqrt{\frac{A}{2}} n^{1/5}.$$

Note that $\int f_n(y)dy = 1$, and $f_n(y) > 0$ and $|f_n''(y)| < A$ for sufficiently large n.

Given a sample $Y_1, \ldots, Y_n \sim f$, test $H_0 : f = f_0$ vs. $H_1 : f = f_n$. If the estimator $\hat{f}$ is as good as it was assumed, we will have that under H_0, with probability converging to 1, $\hat{f}(y_0) < f(y_0) + n^{-2/5}/2$, while under H_1, $\hat{f}(y_0) > f(y_0) + n^{-2/5}/2$ with probability converging to 1. Use $\hat{T}_n = \hat{f}(y_0)$ as the test statistics and reject the null if $\hat{T}_n > f(y_0) + n^{-2/5}/2$. Thus, for some sequences c_n and $\varepsilon_n \to 0$,

$$P_{H_0}(\hat{T}_n \geq c_n) + P_{H_1}(\hat{T}_n < c_n) \leq \varepsilon_n, \tag{9.14}$$

that is, the sum of the probabilities of the two types of error is smaller than ε_n and thus converges to 0.

But we now show that this is impossible by the Neyman–Pearson lemma. According to the lemma, the optimal test statistic is the log-likelihood ratio

$$T_n = \sum_{i=1}^{n} \ell(Y_i), \quad \ell(Y_i) = \log\left(1 + \frac{h_n(Y_i)}{f_0(Y_i)}\right).$$

Consider the mean and variance of T_n under both hypotheses. Let $\Delta = E_{H_1}\ell(Y) - E_{H_0}\ell(Y)$, then

$$
\begin{aligned}
\Delta &= \int \ell(y) h_n(y) dy \\
&= \int \log\left(1 + \frac{h_n(y)}{f_0(y)}\right) h_n(y) dy \\
&= \int \frac{h_n^2(y)}{f_0(y)} dy + O\left(\int \frac{h_n^3(y)}{f_0^2(y)} dy\right) \\
&= \frac{a_n^2}{2b_n} + o\left(\frac{a_n^2}{2b_n}\right) = O(n^{-1}),
\end{aligned}
$$

where we used the Taylor's expansion $\log(1 + \xi) = \xi - \xi^2/2 + \xi^3/3 + O(\xi^4)$. On the other hand,

$$
\begin{aligned}
E(\ell^2(Y)) &= \int \log^2\left(1 + \frac{h_n(y)}{f_0(y)}\right) f(y)(y) dy \\
&= \int \frac{h_n^2(y)}{f_0(y)} dy + O\left(\int \frac{h_n^3(y)}{f_0^2(y)} dy\right) \\
&= \Delta + o(n^{-1}).
\end{aligned}
$$

Note the difference between the second moments of $\ell(Y)$ under H_0 and H_1 is of a smaller order, and the square of the first moment is of a lower order as well. Thus

$$
\begin{aligned}
\frac{ET_n}{\sqrt{Var(T_n)}} &= (1 + o(1)) \frac{n\Delta}{\sqrt{nE\ell^2(X)}} \\
&= (1 + o(1))\sqrt{n\Delta} \\
&= (1 + o(1))\sqrt{n\frac{a_n^2}{b_n}} \\
&= (1 + o(1))\left(\frac{2}{A}\right)^{1/4}.
\end{aligned}
$$

By the CLT, the (properly scaled) T_n is asymptotically normal, while the difference between its means under the two alternatives is the of same order as its standard deviation. In other words, the distributions of T_n under the two alternatives overlap, and for every c, $P_{H_0}(T_n \geq c) + P_{H_1}(T_n < c) > d > 0$ for some (fixed) $d > 0$. But by the Neyman–Pearson lemma, the sum of Type I and Type II errors of the LRT test is lower than the sum for any other test, in particular $\hat{T}_n$, contradicting (9.14). Thus, (9.13) cannot hold. $\square$

9.5   Nonparametric kernel regression

In this section we describe briefly how the ideas of histogram and kernel density estimation can also be applied to nonparametric regression problem.

Consider the model

$$Y_i = g(x_i) + \varepsilon_i, \quad i = 1, \dots, n, \tag{9.15}$$

where ε_i's are i.i.d. zero mean random noise with the variance σ^2. We don't assume any parametric form of the unknown g but only that it has two continuous derivatives, i.e. $g \in C^2$.

An analogue of the histogram for density estimation is the *regressogram* estimator $\hat{g}_R$: we divide the support of x into bins with fixed knots $a_0 < \dots < a_k$ and estimate $g(x)$, where $a_{j-1} < x \leq a_j$, by the average of all observations within the j-th bin, i.e.

$$\hat{g}_R(x) = \frac{1}{\#\{i : a_{j-1} < x_i \leq a_j\}} \sum_{i:a_{j-1}<x_i\leq a_j} Y_i$$

The regressogram $\hat{g}_R$ suffers from the same drawbacks as the histogram: it uses the fixed bins, doesn't utilize smoothness of g and estimate it by a piece-wise constant function.

Similarly to kernel density estimation, we consider instead a moving window around the point of interest x with a bandwidth h, $x \pm h$, and for a given kernel $K(\cdot)$, estimate $g(x)$ by a *weighted moving average* of observations within the window:

$$\hat{g}_{K,h}(x) = \frac{\sum_{i=1}^n K\left(\frac{x_i-x}{h}\right) Y_i}{\sum_{i=1}^n K\left(\frac{x_i-x}{h}\right)} \tag{9.16}$$

known as the *Nadaraya–Watson* kernel estimator.

Note that although the regression model (9.15) is nonparametric, the kernel estimator $\hat{g}_{K,h}$ in (9.16) is a *linear* estimator (a linear function of $\mathbf{Y}$) of the unknown g (see Section 8.3).

We have

$$E\hat{g}_{K,h}(x) = \frac{\sum_{i=1}^n K\left(\frac{x_i-x}{h}\right) g(x_i)}{\sum_{i=1}^n K\left(\frac{x_i-x}{h}\right)}$$

and

$$Var(\hat{g}_{K,h}) = \sigma^2 \frac{\sum_{i=1}^n K^2\left(\frac{x_i-x}{h}\right)}{\left(\sum_{i=1}^n K\left(\frac{x_i-x}{h}\right)\right)^2}$$

Going along the lines of the proof of Theorem 9.3 and using Taylor's expansions, one can show that for $g \in C^2[0,1]$, a kernel $K(\cdot)$ satisfying (9.6)–(9.10) and under mild conditions on the design, asymptotically, as $h_n \to 0$ but $nh_n \to \infty$,

$$IMSE(\hat{g}_{K,h}, g) = E \int (\hat{g}_{K,h}(x) - g(x))^2 dx = O(h_n^4) + O\left(\frac{1}{nh_n}\right)$$

(see, e.g., Bickel and Doksum, 2016, Section 11.2.2). Hence, similarly to kernel density estimation, the optimal bandwidth $h_n = O(n^{-1/5})$ with the resulting $IMSE(\hat{g}_{K,h_n}, g) = O(n^{-4/5})$.

9.6 Nonparametric estimation by orthonormal series

9.6.1 Orthonormal series

We present now another approach to nonparametric estimation based on an orthogonal series expansion of the unknown function f. Under some mild conditions on $\mathscr{F}$, there is sequence of functions $\psi_0, \psi_1, \ldots$ such that any function $f \in \mathscr{F}$ can be uniquely represented as $f(y) = \sum_{j=0}^{\infty} a_j \psi_j(y)$ with a sequence of coefficients a_j's. The set $\{\psi_j(y)\}_{j=0}^{\infty}$ is called a basis in $\mathscr{F}$. Instead of estimating f directly, we can use such a representation to estimate its coefficients a_js, replacing the original nonparametric estimation problem of a function by the estimation of an infinite number of real parameters. What are the possible benefits of such an approach? How can we estimate an infinite number of coefficients a_j's from a finite sample?

The idea is that if f can be well-approximated by a relatively small (finite) number of terms, i.e. if $f(y) \approx \sum_{j=0}^{k} a_j \psi_j(y)$ for some $k < n$, the approximation error of such series truncation would be "negligible" and we would need, in fact, to estimate only a finite number of coefficients $a_0, \ldots, a_k$. In this case we can reduce the nonparametric problem in the original functional domain to the (much simpler!) finite parametric problem in the coefficients domain and estimate f by a truncated series estimator $\hat{f}_k(y) = \sum_{j=0}^{k} \hat{a}_j \psi_j(y)$.

We will now formulate this idea more rigorously and develop some necessary basic tools. Assume that f has a finite support—without loss of generality, the unit interval $[0,1]$, and that f^2 is integrable on it, i.e. $\int_0^1 f^2(y)dy < \infty$. The space of all square integrable functions on $[0,1]$ is known as $L_2[0,1]$. We say that a set $\{\psi_j\}_{j=0}^{\infty}$ is a *basis* if for any $f \in L_2[0,1]$ there exists a unique sequence of coefficients $\{a_j\}_{j=0}^{\infty}$ such that f can be written as $f(y) = \sum_{j=0}^{\infty} a_j \psi_j(y)$ in the sense that the partial sums $f_k(y) = \sum_{j=0}^{k} a_j \psi_j(y)$ converge to f in the L_2-norm, i.e. $\lim_{k \to \infty} \int \left(f(y) - f_k(y)\right)^2 dy = 0$. The most well-known examples of bases are polynomials $1, y, y^2, \ldots$ and trigonometric Fourier series.

Often, it is helpful to work with an *orthonormal* basis, where, in addition, $\int_0^1 \psi_j^2(y)dy = 1$ for all $j \geq 0$ and $\int_0^1 \psi_j(y)\psi_k(y)dy = 0$ for all $j \neq k$. Thus, the polynomials are not an orthogonal basis, while a trigonometric Fourier series is. The main properties of function's coefficients in orthonormal series expansion are given in the following proposition:

Proposition 9.3 Let $\{\psi_j\}_{j=0}^{\infty}$ be an orthonormal basis in $L_2[0,1]$. Consider $f \in L_2[0,1]$ and let $\{a_j\}_{j=0}^{\infty}$ be its coefficients in this basis. Then,

1. $a_j = \int_0^1 f(y)\psi_j(y)dy$

2. Let $f_k(y) = \sum_{j=0}^{k} a_j \psi_j(y)$. Then, $\int_0^1 f_k^2(y)dy = \sum_{j=0}^{k} a_j^2$

3. $\int_0^1 f^2(y)dy = \sum_{j=0}^{\infty} a_j^2$ (Parseval identity)

Proof. We have $f(y) = \sum_{h=0}^{\infty} a_h \psi_h(y)$. Hence, due to orthonormality of the basis,

$$\int_0^1 f(y)\psi_j(y)dy = \sum_{h=0}^{\infty} a_j \int_0^1 \psi_h(y)\psi_j(y)dy = a_j \int_0^1 \psi_j^2(y)dy = a_j.$$

Similarly,

$$\int_0^1 f_k^2(y)dy = \sum_{j=0}^{k}\sum_{h=0}^{k} a_j a_h \int_0^1 \psi_j(y)\psi_h(y)dy = \sum_{j=0}^{k} a_j^2 \int_0^1 \psi_j^2(y) = \sum_{j=0}^{k} a_j^2.$$

Part 3 follows from Part 2 and the definition of the basis. □

Proposition 9.3 implies that for any chosen basis $\lim_{j \to \infty} a_j = 0$ for all $f \in L_2[0,1]$. We are interested in finite approximations, and thus the rate of convergence matters for being able to have good estimators. The $L_2[0,1]$ consists of "too different" types of functions (smooth and non-smooth) and the goal to find a good basis for the entire $L_2[0,1]$ is too ambitious. We need some extra restrictions on f and to choose a proper basis accordingly. For example, we may assume that $f \in C^2[0,1]$ as before. In the following subsections we will show that for such functions, (orthonormal) cosine series is a good choice for a sparse basis and will use it for nonparametric estimation. The cosine series are also relatively simple and we could present a relatively complete theory.

9.6.2   Cosine series

Consider the cosine series

$$\psi_0(y) = 1, \;\; \psi_j(y) = \sqrt{2}\cos(\pi j y), \; j \geq 1,$$

which is a variant of the well-known trigonometric Fourier series.

Theorem 9.5

1. The cosine series $\psi_0(y) = 1$, $\psi_j(y) = \sqrt{2}\cos(\pi j y)$, $j \geq 1$ is an orthonormal basis in $C^2[0,1]$.

2. Let $f \in C^2[0,1]$, $f'(0) = f'(1) = 0$, and $a_0 = \int_0^1 f(y)dy$, $a_j = \int_0^1 f(y)\cos(\pi j y)dy$, $j \geq 1$. Approximate f by a partial sum $f_k(y) = \sqrt{2}\sum_{j=0}^{k} a_j \cos(\pi j y)$. The integrated squared approximation error is

$$\int \big(f(y) - f_k(y)\big)^2 dy \leq \frac{1}{\pi^2 k^4}\int_0^1 \big(f''(y)\big)^2 dy$$

We need the following technical lemma for the proof:

Lemma 9.1 For $0 \leq t \leq 1$ and $k = 1,2,\ldots$:

$$H_k(t) \triangleq \frac{1}{2} + \sum_{j=1}^{k}\cos(\pi j t) \;=\; \frac{\sin\big(\pi(k+\frac{1}{2})t\big)}{2\sin\big(\frac{\pi}{2}t\big)},$$

where the RHS at $t = 0$ is understood as its limit as $t \to 0$, i.e., $k + \frac{1}{2}$.

Proof of Lemma 9.1. We prove the lemma by induction. If $k = 1$ the result holds since

$$\frac{\sin\big(\frac{3}{2}\pi t\big)}{2\sin\big(\frac{1}{2}\pi t\big)} - \frac{1}{2} \;=\; \frac{\sin\big(\frac{3}{2}\pi t\big) - \sin\big(\frac{1}{2}\pi t\big)}{2\sin\big(\frac{1}{2}\pi t\big)}$$

$$= \frac{2\sin\big(\frac{1}{2}\pi t\big)\cos(\pi t)}{2\sin\big(\frac{1}{2}\pi t\big)} = \cos(\pi t).$$

Suppose now that the identity holds for a given k, then

$$\sum_{j=0}^{k+1}\cos(\pi j t) \;=\; \frac{\sin\big(\pi(k+\frac{1}{2})t\big)}{2\sin\big(\frac{\pi}{2}t\big)} + \cos\big(\pi(k+1)t\big) - \frac{1}{2}$$

$$= \frac{\sin\big(\pi(k+\frac{1}{2})t\big) + 2\sin\big(\frac{\pi}{2}t\big)\cos\big(\pi(k+1)t\big)}{2\sin\big(\frac{\pi}{2}t\big)} - \frac{1}{2}$$

$$= \frac{\sin\big(\pi(k+\frac{1}{2})t\big) + \sin\big(\pi(k+\frac{3}{2})t\big) - \sin\big(\pi(k+\frac{1}{2})t\big)}{2\sin\big(\frac{\pi}{2}t\big)} - \frac{1}{2}.$$

$\square$

Proof of Theorem 9.5. For simplicity of exposition assume that $a_0 = 0$. To prove that the cosine series form a basis in $C^2[0,1]$, we need to show that $\lim_{k\to\infty}\int_0^1 \big(f(y) - f_k(y)\big)^2 dy = 0$ for any $f \in C^2[0,1]$. By straightforward calculus,

$$f_k(y) \;=\; \sqrt{2}\sum_{j=0}^{k} a_j \cos(\pi j y)$$

$$= 2 \sum_{j=1}^{k} \int f(t) \cos(\pi jt) dt \cos(\pi jy)$$

$$= 2 \sum_{j=1}^{k} \int \big(f(t) - f(y)\big) \cos(\pi jt) \cos(\pi jy) dt$$

$$= \int \big(f(t) - f(y)\big) \sum_{j=1}^{k} \big(\cos(\pi j(t-y)) + \cos(\pi j(t+y))\big) dt$$

$$= \int \big(f(t) - f(y)\big) \big(H_k(t-y) + H_k(t+y) - 1\big) dt$$

$$= f(y) + \int \big(f(t) - f(y)\big) \big(H_k(t-y) + H_k(t+y)\big) dt.$$

We break the last integral to two parts:

$$\left| \int_{y-\varepsilon}^{y+\varepsilon} \big(f(t) - f(y)\big) H_k(t-y) dt \right| \leq 2\varepsilon * \max_{|y-t| \leq \varepsilon} |f(t) - f(y)| \max_t |H_k(t)|$$

$$\leq 2n\varepsilon^2 \max_t |f'(t)|,$$

while the integrals over $(0, y-\varepsilon)$ and $(y+\varepsilon, 1)$ are dealt in similar ways and we consider the details of only one them:

$$\int_{y+\varepsilon}^{1} \big(f(t) - f(y)\big) H_k(t-y) dt$$

$$= \int_{y+\varepsilon}^{1} \frac{f(t) - f(y)}{2 \sin\big(\frac{\pi}{2}(t-y)\big)} \sin\big(\pi(n+\tfrac{1}{2})(t-y)\big) dt$$

$$= -\frac{1}{\pi(n+\frac{1}{2})} \frac{f(t) - f(y)}{2 \sin\big(\frac{\pi}{2}(t-y)\big)} \cos\big(\pi(n+\tfrac{1}{2})(t-y)\big) \Big|_{y+\varepsilon}^{1}$$

$$+ \frac{1}{\big(\pi(n+\frac{1}{2})\big)^2} \Big(\frac{f(t) - f(y)}{2 \sin\big(\frac{\pi}{2}(t-y)\big)}\Big)' \sin\big(\pi(n+\tfrac{1}{2})(t-y)\big) \Big|_{y+\varepsilon}^{1}$$

$$- \frac{1}{\big(\pi(n+\frac{1}{2})\big)^2} \int_{y+\varepsilon}^{1} \Big(\frac{f(t) - f(y)}{2 \sin\big(\frac{\pi}{2}(t-y)\big)}\Big)'' \sin\big(\pi(n+\tfrac{1}{2})(t-y)\big) dt$$

$$= O\Big(\frac{1}{n} + \frac{1}{n^2 \varepsilon}\Big),$$

since the first and second derivatives are bounded by ε^{-1}, while the first term $f(y+\varepsilon) - f(y) = O(\varepsilon)$. The integral $\int \big(f(t) - f(y)\big) H_k(t+y) dt$ is dealt in a similar way, but note that $|t+y| > \varepsilon$ always within the relevant range.

Thus, $\lim_{k \to \infty} \sup_t |f(t) - f_k(t)| = 0$ and, therefore, $\lim_{k \to \infty} \int_0^1 \big(f(t) - f_k(t)\big)^2 dt = 0$. Orthonormality of cosine series is verified by a straightforward integration.

To prove the approximation error, let $b_j = \sqrt{2} \int_0^1 f''(y) \cos(\pi jy) dy$ be the j-th cosine series coefficient of f''. By Proposition 9.3 and integration by parts one has

$$a_j = \sqrt{2} \int_0^1 f(y) \cos(\pi jy) dy$$

$$= \frac{\sqrt{2}}{(\pi j)^2} \int_0^1 f''(y) \cos(\pi jy) dy$$

$$= \frac{b_j}{\pi^2 j^2}.$$

and

$$\int_0^1 \left(f_k(y) - f(y)\right)^2 dt \;=\; \sum_{j=k+1}^{\infty} a_k^2$$

$$=\; \sum_{j=k+1}^{\infty} \frac{b_j^2}{\pi^4 j^4} \;\leq\; \frac{1}{\pi^4 k^4} \sum_{j=k+1}^{\infty} b_j^2$$

$$\leq\; \frac{1}{\pi^4 k^4} \sum_{j=1}^{\infty} b_j^2 = \frac{1}{\pi^4 k^4} \int_0^1 f''(y)^2 dy$$

$\square$

It follows from the proof that $f_k(y)$ converges also uniformly to $f(y)$ for all $0 \leq y \leq 1$: $\lim_{k \to \infty} \sup_{0 \leq y \leq 1} |f(y) - f_k(y)| = 0$.

We see from Theorem 9.5 that the integrated squared approximation error tends to zero as the number of terms in the truncated series k increases at a fast rate of k^{-4} that shows that the cosine series is a good (sparse) basis for functions in $C^2[0,1]$.

9.6.3 Nonparametric density estimation by cosine series

We apply now cosine series to nonparametric density estimation. Suppose $Y_1, \ldots, Y_n$ is a random sample from a density $f \in C^2[0,1]$. Following the proposed methodology, expand f in the cosine series as $f(y) = a_0 + \sqrt{2} \sum_{j=1}^{\infty} a_j \cos(\pi j y)$, where

$$a_0 = \int_0^1 f(y) dy = 1 \quad \text{and} \quad a_j = \sqrt{2} \int_0^1 f(y) \cos(\pi j y) dy = \sqrt{2} E \cos(\pi j Y), \; j \geq 1,$$

and approximate f by truncating the series: $f_k(y) = 1 + \sum_{j=1}^{k} a_j \cos(\pi j y)$ for a fixed $k < n$.

The coefficients a_j's can be easily estimated by

$$\hat{a}_j = \frac{\sqrt{2}}{n} \sum_{i=1}^{n} \cos(\pi j Y_i), \; j = 1, \ldots, k$$

and the resulting truncated cosine series estimator $\hat{f}_{C,k}$ is then

$$\hat{f}_{C,k}(y) = 1 + \sqrt{2} \sum_{j=1}^{k} \hat{a}_j \cos(\pi j y)$$

Question 9.3 Show that $\int_0^1 \hat{f}_{C,k}(y) dy = 1$.

How large should the optimal k be? The larger k, the smaller is the bias of $\hat{f}_{C,k}$ (the better is approximation) but the higher is the variance. Clearly, $\hat{a}_j$ is an unbiased estimator of a_j with the variance

$$Var(\hat{a}_j) = \frac{Var\left(\sqrt{2} \cos(\pi j Y)\right)}{n} \leq \frac{2}{n}.$$

Consider the $IMSE(\hat{f}_{C,k}, f) = E \int_0^1 \left(\hat{f}_{C,k}(y) - f(y)\right)^2 dy$. By Parseval identity and Theorem 9.5,

$$IMSE(\hat{f}_{C,k}, f) \;=\; E \sum_{j=1}^{k} (\hat{a}_j - a_j)^2 + \sum_{j=k+1}^{\infty} a_j^2$$

$$\leq\; \frac{k}{n} + \frac{C}{k^4},$$

Thus, choosing $k_{opt} \sim n^{1/5}$ yields the best possible (minimax) rate of $n^{-4/5}$ for estimating densities in $C^2[0,1]$ similar to kernel estimators.

9.6.4 *Nonparametric regression and orthonormal cosine series*

Consider now the nonparametric regression model with a univariate $x \in [0,1]$:

$$Y_i = g(x_i) + \varepsilon_i, \quad i = 1,\ldots,n, \tag{9.17}$$

where ε_i's are i.i.d. zero mean random noise with the variance σ^2. Again, we only assume that the unknown g has two continuous derivatives, i.e. $g \in C^2[0,1]$.

Let $\{\psi_j\}_{j=0}^{\infty}$ be a basis in $C^2[0,1]$ and expand g as

$$g(x) = \sum_{j=0}^{\infty} \beta_j \psi_j(x)$$

Similar to nonparametric density estimation, we cannot estimate an infinite number of coefficients β_j's of the expansion, so we truncate the series at some $k < n$ approximating $g(x)$ by $g_k(x) = \sum_{j=0}^{k} \beta_j \psi_j(x)$ and assuming that the resulting approximation error is negligible for a sufficiently large k. It leads to the following *parametric linear* regression model (see Chapter 8):

$$Y_i = \beta_0 \psi_0(x_i) + \beta_1 \psi_1(x_i) + \cdots + \beta_k \psi_k(x_i) + \varepsilon_i, \quad i = 1,\ldots,n \tag{9.18}$$

and the coefficients β_j's can be estimated by the least squares as

$$\hat{\beta} = (\Psi^T \Psi)^{-1} \Psi^T \mathbf{Y}, \tag{9.19}$$

where the matrix $\Psi \in \mathbb{R}^{n \times (k+1)}$ has entries $\Psi_{ij} = \psi_j(x_i)$, $i = 1,\ldots,n$; $j = 0,\ldots,k$ (see Theorem 8.1).

Question 9.4 Show that a truncated series estimator $\hat{g}_k(x) = \sum_{j=0}^{k} \hat{\beta}_j \psi_j(x)$ with the LSE $\hat{\beta}$ from (9.19), is a linear estimator of g.

For example, $\psi_j(x)$ could be polynomials x^j leading to polynomial regression discussed in Chapter 8, but this is not a good choice. Polynomial series might require high-order terms for good approximation. In addition, the polynomials x^j and x^k are highly collinear on the unit interval for large j and k that makes high-order polynomial regression unrobust.

Instead we will use cosine series which forms a sparse orthonormal basis in $C^2[0,1]$ (see Section 9.6.2). Thus, we consider the model

$$Y_i = \beta_0 + \beta_1 \sqrt{2}\cos(\pi x_i) + \cdots + \beta_k \sqrt{2}\cos(\pi k x_i) + \varepsilon_i, \quad i = 1,\ldots,n$$

or, equivalently, in the matrix form

$$\mathbf{Y} = \Psi\beta + \varepsilon,$$

where

$$\Psi = \begin{pmatrix} 1 & \sqrt{2}\cos(\pi x_1) & \ldots & \sqrt{2}\cos(\pi k x_1) \\ \vdots & \vdots & & \vdots \\ 1 & \sqrt{2}\cos(\pi x_n) & \ldots & \sqrt{2}\cos(\pi k x_n) \end{pmatrix}.$$

The resulting truncated cosine series estimator

$$\hat{g}_{C,k}(x) = \hat{\beta}_0 + \sqrt{2} \sum_{j=1}^{k} \hat{\beta}_j \cos(\pi j x), \tag{9.20}$$

where the LSE $\hat{\beta}$ is given in (9.19).

Suppose now that the design points $x_1, \ldots, x_n$ are equally spaced, where $x_i = \frac{i-1/2}{n}$. The design matrix Ψ is then

$$
\Psi \;=\; \begin{pmatrix} 1 & \sqrt{2}\cos(\pi(1-1/2)/n) & \cdots & \sqrt{2}\cos(\pi k(1-1/2)n) \\ \vdots & \vdots & & \vdots \\ 1 & \sqrt{2}\cos(\pi(n-1/2)/n) & \cdots & \sqrt{2}\cos(\pi k(n-1/2)/n) \end{pmatrix}.
$$

By a straightforward algebra one can verify that in this case the columns of Ψ are orthogonal and $\Psi^T\Psi = nI_n$.

Thus, $\hat{\beta} = \frac{1}{n}\Psi^T\mathbf{Y}$, i.e.

$$
\hat{\beta}_0 = \bar{Y} \quad \text{and} \quad \hat{\beta}_j = \frac{\sqrt{2}}{n}\sum_{i=1}^{n}\cos\left(\pi j \frac{i-1/2}{n}\right)Y_i, \quad j=1,\ldots,k,
$$

and, therefore,

$$
\begin{aligned}
E\hat{\beta}_0 &= \frac{1}{n}\sum_{i=1}^{n}g_k\left(\frac{i-1/2}{n}\right) \\
&= \int_0^1 g_k(x)dx + o(n^{-1}) \\
&= \beta_0 + o(n^{-1}),
\end{aligned}
$$

$$
\begin{aligned}
E\hat{\beta}_j &= \frac{\sqrt{2}}{n}\sum_{i=1}^{n}\cos\left(\pi j\frac{i-1/2}{n}\right)g_k\left(\frac{i-1/2}{n}\right) \\
&= \sqrt{2}\int_0^1 \cos(\pi jx)g_k(x)dx + o(n^{-1}) \\
&= \beta_j + o(n^{-1}), \quad j=1,\ldots,k,
\end{aligned}
$$

while from Theorem 8.2,

$$
\begin{aligned}
Var(\hat{\beta}_j) &= \sigma^2(\Psi^T\Psi)^{-1}_{jj} \\
&= \frac{\sigma^2}{n}
\end{aligned}
$$

for all $j = 0, \ldots, k$.

Apply Parseval identity to calculate the IMSE of $\hat{g}_{C,k}(x)$ from (9.20) in this case:

$$
\begin{aligned}
IMSE(\hat{g}_{C,k}, g) &= \sum_{j=0}^{k}E(\hat{\beta}_j - \beta_j)^2 + \sum_{j=k+1}^{\infty}\beta_j^2 \\
&= (k+1)\frac{\sigma^2}{n} + \sum_{j=k+1}^{\infty}\beta_j^2 + o(n^{-1}) \\
&\leq (k+1)\frac{\sigma^2}{n} + \frac{C}{k^4} + o(n^{-1}),
\end{aligned}
$$

where the last inequality follows from Theorem 9.5. Similar to density estimation, the optimal k is of the order $n^{1/5}$ yielding IMSE of $n^{-4/5}$.

Example 9.6.1 We consider now a sample of 1000 observations $(X_1, Y_1), \ldots, (X_{1000}, Y_{1000})$, where $X_i \sim U(0,1)$ and $Y_i = g(X_i) + \varepsilon_i$, $g(x) = 1000x^4(1-x)^4$ (see Figure 9.3). Again, we show three estimators of g, each based on different number of cosine functions, having examples of oversmoothing and over-fitting estimators.

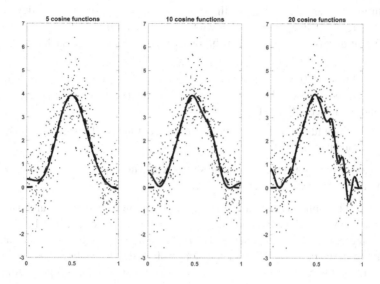

Figure 9.3: The data, cosine series estimators (solid lines) and the true regression function (dashed line).

9.7 Spline smoothing

Consider again a nonparametric regression model

$$Y_i = g(x_i) + \varepsilon_i, \quad i = 1, \dots, n$$

where as usual in this chapter we assume that $g(x)$ is smooth (in particular, $g \in C^2[0,1]$).

We are looking for a (nonparametric) estimator $\hat{g}$ which will fit the data but at the same time will be a smooth function. The goodness-of-fit can be evidently measured by the residual sum of squares (RSS) $\sum_{i=1}^{n}(Y_i - \hat{g}(x_i))^2$. How can we quantify the function's smoothness? Since we are talking about functions with two bounded derivatives, a natural measure of smoothness at a point x is $|\hat{g}''(x)|$ or $\hat{g}''(x)^2$. A global smoothness is then quantified by $\int (\hat{g}''(x))^2 dx$. Minimizing RSS necessarily implies an interpolating solution that passes through all data points with RSS=0 but is not smooth. On the other hand, the smoothest estimator with $\int (\hat{g}''(x))^2 dx = 0$ is a linear function that might be not appropriate for the data.

How can we merge both, often contradictory, requirements in the same estimator? Consider a *penalized* RSS criterion with a roughness penalty of the form

$$\hat{g}_{S,\lambda}(x) = \arg\min_{g \in C^2[0,1]} \sum_{i=1}^{n}(Y_i - g(x_i))^2 + \lambda \int (g''(x))^2 dx, \tag{9.21}$$

where the *smoothing parameter* $\lambda > 0$ controls the trade-off between fidelity to the data and smoothness. Thus, as $\lambda \to 0$, more and more attention is given to the fit and $\hat{g}$ converges to an interpolating curve (in fact, an interpolating spline as we will show later) implying over-fitting. At the same time, when $\lambda \to \infty$, most weight is given to smoothing and $\hat{g}$ is approaching the linear LSE that might yield a poor fit.

Finding a minimizer in (9.21) seems to be very difficult— it is a (nonparametric) minimization over the entire class of functions $C^2[0,1]^2$. Remarkably, the problem is not that hard and even has

---

[2]In fact, we can somewhat relax the requirement on the boundedness of the second derivative of g in (9.21) by its square integrability but, as we have mentioned, we will not discuss these issues.

an explicit solution. In fact, we will prove that $\hat{g}$ is a piece-wise polynomial of degree 3—a cubic spline and, therefore, such an approach is known as *spline smoothing*.

To simplify the discussion assume that the observations are ordered such that $0 < x_1 < x_2 < \cdots < x_n < 1$.

Definition 9.4 (cubic spline) A *cubic spline* with knots $x_1, \ldots, x_n$ is a continuous piece-wise cubic polynomial function, which is a polynomial of degree 3 on each of the intervals $(x_i, x_{i+1}]$, $i = 1, \ldots, n$ and has two continuous derivatives at x_i's.

Question 9.5 What would happen if we require *three* continuous derivatives at knots x_i's in the above definition of a cubic spline?

Definition 9.5 (natural cubic spline) A cubic spline with knots $x_1, \ldots, x_n$ is called a *natural cubic spline* if, in addition, it is linear on $(0, x_1)$ and $(x_n, 1)$.

Note that a natural cubic spline is an amalgam of $n-1$ polynomials of degree 3 and two polynomials of degree 1 at the boundaries. The total number of parameters is, therefore, $(n-1) \times 4 + 2 \times 2 = 4n$. However, there are 3 constraints on each knot x_i—the polynomials on the right and on the left of x_i and their two derivatives should have the same values due to continuity requirements. Thus, the number of free parameters is $4n - 3n = n$ and a natural smoothing spline is completely (and uniquely) defined by its values at the n knots $x_1, \ldots, x_n$.

We first prove that the solution of (9.21) should be a natural cubic spline reducing minimization over all functions in $C^2[0,1]$ to minimization over a space of n-parametric family of natural cubic splines. The proof will follow from the theorem below:

Theorem 9.6 The solution of

$$\min \int (g''(x))^2 dx \quad \text{s.t.} \quad g(x_i) = y_i, \ i = 1, \ldots, n$$

is a natural cubic spline with knots $x_1, \ldots, x_n$.

In other words, Theorem 9.6 states that the smoothest curve that interpolates the given n points (x_i, y_i) is an (interpolating) natural cubic spline with knots x_i's.

Proof. Let $g(x)$ be the natural cubic spline that satisfies $g(x_i) = y_i$ and $f(x)$ be any other interpolating function. Let $h(x) = f(x) - g(x)$. Evidently, $h(x_i) = 0$, $i = 1, \ldots, n$.

To simplify the notation, let $x_0 = 1$ and $x_{n+1} = 1$. By integration by parts we have

$$\int_0^1 h''(x)g''(x)dx = \sum_{i=0}^n \int_{x_i}^{x_{i+1}} h''(x)g''(x)dx$$

$$= \sum_{i=0}^n \left(h'(x_{i+1})g''(x_{i+1}) - h'(x_i)g''(x_i) - \int_{x_i}^{x_{i+1}} h'(x)g^{(3)}(x)dx \right)$$

But $g^{(3)}(x)$ is constant on $(x_i, x_{i+1}]$ and $h(x_i) = h(x_{i+1}) = 0$, hence the last integral is $g^{(3)}(x_{i+1})(h(i_j) - h(x_{i+1})) = 0$ and we are left with the telescopic sum:

$$\int_0^1 h''(x)g''(x)dx = \sum_{j=0}^n \left(h'(x_{i+1})g''(x_{i+1}) - h'(x_i)g''(x_i) \right)$$

$$= h'(1)g''(1) - h'(0)g''(0) = 0,$$

since $g''(0) = g''(1) = 0$ by definition of a natural cubic spline. Thus.

$$\int (f''(x))^2 dx = \int (g''(x) + h''(x))^2 dx$$

$$= \int (g''(x))^2 dx + 2 \int h''(x)g''(x)dx + \int (h''(x))^2 dx$$

$$= \int (g''(x))^2 dx + \int (h''(x))^2 dx$$

$$> \int (g''(x))^2 dx.$$

$\square$

Question 9.6 Show that Theorem 9.6 implies that the solution of (9.21) is necessarily a natural cubic spline with knots x_i's.

As we have argued above, a natural cubic spline $g(x)$ is completely defined by its values $g(x_i)$'s at n knots x_i's. In particular, we have the following lemma:

Lemma 9.2 Let $g(x)$ be a natural smoothing spline with knots $0 < x_1 \ldots < x_n < 1$ and $\mathbf{g} = (g(x_1), \ldots, g(x_n))^T$. Then,

$$\int_0^1 (g''(x))^2 dx = \mathbf{g}^T K \mathbf{g},$$

where the matrix $K = R^T H^{-1} R$, $R \in \mathbb{R}^{(n-2)\times n}$ is a matrix of second differences with elements $R_{i,i-1} = \frac{1}{x_i - x_{i-1}}$, $R_{i,i} = -\frac{1}{x_i - x_{i-1}} - \frac{1}{x_{i+1} - x_i}$, $R_{i,i+1} = \frac{1}{x_{i+1} - x_i}$ and zero otherwise, and $H \in \mathbb{R}^{(n-2)\times(n-2)}$ is a tri-diagonal matrix with elements $H_{i,i} = \frac{1}{3}(x_{i+1} - x_{i-1})$ and $H_{i,i+1} = H_{i+1,i} = \frac{1}{6}(x_{i+1} - x_i)$.

The tedious technical proof is omitted.

Summarizing all the above results, the solution $\hat{g}_{S,\lambda}$ of (9.21) can, therefore, be found by minimizing

$$\sum_{i=1}^n (Y_i - g(x_i))^2 + \lambda \int_0^1 (g''(x))^2 dx = ||\mathbf{Y} - \mathbf{g}||_2^2 + \lambda \mathbf{g}^T K \mathbf{g}$$

w.r.t. a vector $\mathbf{g} \in \mathbb{R}^n$. Differentiating w.r.t. $\mathbf{g}$ and equating to zero yields

$$-2(\mathbf{Y} - \mathbf{g}) + 2K\mathbf{g} = 0$$

and, therefore,

$$\hat{\mathbf{g}}_{S,\lambda} = (I_n + \lambda K)^{-1} \mathbf{Y} \qquad (9.22)$$

The formula (9.22) gives a simple way to calculate a smoothing spline $\hat{g}_{S,\lambda}$ at data points (knots) x_i's. To get an explicit expression for $\hat{g}_{S,\lambda}(x)$ at any point x we will use another representation for a natural cubic spline. Define a function

$$Q(x,u) = \int_0^1 (x-v)_+(u-v)_+ dv = \begin{cases} u\frac{x^2}{2} - \frac{x^3}{6} & , x \le u \\ x\frac{u^2}{2} - \frac{u^3}{6} & , x > u \end{cases} \qquad (9.23)$$

where recall that $\cdot_+ = \max(\cdot, 0)$. By straightforward calculus one can verify that as a function of x, $Q(\cdot, u)$ is a piece-wise (cubic-linear) polynomial function with a (single) knot u and two continuous derivatives.

Lemma 9.3 Let $g(x)$ be a natural cubic spline with knots $x_1, \ldots, x_n$. Then it can be uniquely represented as

$$g(x) = \alpha_0 + \alpha_1 x + \sum_{i=1}^n \beta_i Q(x, x_i), \qquad (9.24)$$

where $\sum_{i=1}^n \beta_i = 0$ and $\sum_{i=1}^n \beta_i x_i = 0$.

Moreover, $\int_0^1 (g''(x))^2 dx = \beta^T Q \beta$, where $Q_{ij} = Q(x_i, x_j)$, $i, j = 1, \ldots, n$.

Proof. By the construction, the right hand side of (9.24) is obviously a natural cubic spline with knots x_i's. As we have mentioned, a natural cubic spline is uniquely defined by $\mathbf{g} =$

$(g(x_1),\dots,g(x_n))^T$. Hence, the coefficients α and β are the (unique) solution of the system of $n+2$ linear equations $\mathbb{X}\alpha + Q\beta = \mathbf{g}$ and $\mathbb{X}^T\beta = \mathbf{0}_2$, where the design matrix

$$\mathbb{X} = \begin{pmatrix} 1 & x_1 \\ \vdots & \vdots \\ 1 & x_n \end{pmatrix}.$$

To prove the second statement of the lemma, use the representation (9.24) to get

$$g''(x) = \sum_{i=1}^n \beta_i Q''(x, x_i) = \sum_{i=1}^n \beta_i (x_i - x)_+$$

and, therefore, from (9.23)

$$
\begin{aligned}
\int_0^1 (g''(x))^2 dx &= \sum_{i=1}^n \sum_{j=1}^n \beta_i \beta_j \int_0^1 (x_i - x)_+ (x_j - x)_+ dx \\
&= \sum_{i=1}^n \sum_{j=1}^n \beta_i \beta_j Q_{ij} \\
&= \beta^T Q \beta
\end{aligned}
$$

$\square$

Using (9.24), re-write (9.21) as follows

$$\sum_{i=1}^n (Y_i - g(x_i))^2 + \lambda \int_0^1 (g''(x))^2 dx = ||\mathbf{Y} - \mathbb{X}\alpha - Q\beta||_2^2 + \lambda \beta^T Q \beta \tag{9.25}$$

and minimize it w.r.t. α and β. Differentiating (9.25) with respect to α and β and equating the derivatives to zero we have

$$-2\mathbb{X}^T(\mathbf{Y} - \mathbb{X}\alpha - Q\beta) = \mathbf{0}_2$$

and

$$-2Q(\mathbf{Y} - \mathbb{X}\alpha - Q\beta) + 2\lambda Q\beta = \mathbf{0}_n$$

Solving the above equations yields

$$\hat{\alpha} = (\mathbb{X}M^{-1}\mathbb{X})^{-1}\mathbb{X}^T M^{-1}\mathbf{Y} \quad \text{and} \quad \hat{\beta} = M^{-1}(\mathbf{Y} - \mathbb{X}\hat{\alpha}), \tag{9.26}$$

where $M = Q + \lambda I_n$. One can verify that $\hat{\beta}$ satisfies the required constraint $\mathbb{X}^T\hat{\beta} = \mathbf{0}_2$.

Finally, the solution of (9.21) is

$$\hat{g}_{S,\lambda}(x) = \hat{\alpha}_0 + \hat{\alpha}_1 x + \sum_{i=1}^n \hat{\beta}_i Q(x, x_i), \tag{9.27}$$

where the coefficients $\hat{\alpha}$ and $\hat{\beta}$ are given in (9.26), and the function $Q(\cdot, \cdot)$ is defined in (9.23).

Question 9.7 Show that the smoothing spline $\hat{g}_{S,\lambda}$ in (9.27) is a linear estimator.

Note that as a natural cubic spline, $\hat{g}_{S,\lambda}(x)$ implies linear extrapolation outside the data range, i.e. for $x < x_1$ and $x > x_n$.

To calculate the $IMSE(\hat{g}_{S,\lambda}, g)$ is not so trivial but nevertheless, it may be shown that for $\lambda_{opt} \sim n^{-1/4}$, $IMSE(\hat{g}_{S,\lambda_{opt}}, g)$ is of the optimal (minimax) rate $n^{-4/5}$.

Example 9.7.1 Data was generated from $Y = 100(X^3 - \frac{3}{2}X^2 + \frac{9}{16}X) + \varepsilon$, where $\varepsilon \sim N(0,1)$. Three natural cubic spline estimators with different λ's were calculated based on 500 observations (see Figure 9.4).

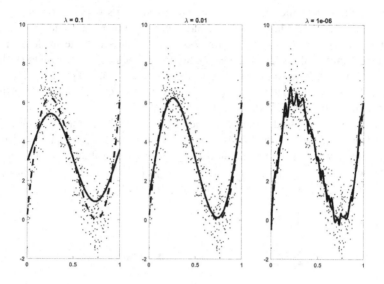

Figure 9.4: The data, smoothing spline estimators (solid lines) and the true regression function (dashed line).

9.8   Choice of the smoothing parameter

All nonparametric estimators considered in this chapter depend on a tuning (smoothing) parameter: the bandwidth in kernel estimation, the number of terms in truncated series approximation, the smoothing parameter in spline smoothing, etc. So far we considered a theoretical based choice of the smoothing parameter, and even that was considered only up to the rate. The actual choice should depend on the sample size, but also on the smoothness of the function to be estimated. The latter is clearly unknown. One possible method for choosing a bandwidth in kernel density estimation was considered in Section 9.3. In this section we discuss several approaches for choosing the smoothing parameter in nonparametric regression.

Consider a nonparametric regression model

$$Y_i = g(x_i) + \varepsilon_i, \ i = 1, \ldots, n,$$

where ε_i's are independent, zero mean random noise with a variance σ^2, and some family of estimators $\{\hat{g}_h\}$ of g that depend on a smoothing parameter h. Typically h is a positive real number (e.g., the kernel bandwidth or the smoothing parameter in spline smoothing) or an integer (the number of terms is truncated series approximation). We have some performance criterion of $\hat{g}_h$, e.g., the $IMSE(\hat{g}_h, g) = E \int (\hat{g}_h(x) - g(x))^2 dx$ or the average MSE, $AMSE(\hat{g}_h, g) = \frac{1}{n} E \sum_{i=1}^{n} (\hat{g}_h(x_i) - g(x_i))^2$, and ideally would seek the optimal value of h which minimizes this criterion. However, IMSE/AMSE depend on the unknown g. Instead we would like to replace them by some "good" estimators. A naive approach would be based on minimizing the observed prediction error (residual sum of squares) RSS_{obs}:

$$
\begin{aligned}
RSS_{obs}(h) &= \sum_{i=1}^{n} \left(Y_i - \hat{g}_h(x_i) \right)^2 \\
&= \sum_{i=1}^{n} \left(\hat{g}_h(x_i) - g(x_i) \right)^2 - 2 \sum_{i=1}^{n} \left(\hat{g}_h(x_i) - g(x_i) \right) \left(Y_i - g(x_i) \right) + \sum_{i=1}^{n} \left(Y_i - g(x_i) \right)^2 \\
&= \sum_{i=1}^{n} \left(\hat{g}_h(x_i) - g(x_i) \right)^2 - 2 \sum_{i=1}^{n} \left(\hat{g}_h(x_i) - g(x_i) \right) \varepsilon_i + \sum_{i=1}^{n} \varepsilon_i^2.
\end{aligned}
$$

The first term on the right hand side is of our interest since its expectation is $E \sum_{i=1}^{n} (\hat{g}_h(x_i) - g(x_i))^2 = n \cdot AMSE(\hat{g}_h, g)$. The last term has a non-zero mean $n\sigma^2$ but since it does not depend on the smoothing parameter, it is irrelevant for choosing h. The second terms looks like a sum of zero mean independent variables (since $E\varepsilon_i = 0$). So, ideally, it should be close to 0. However, this argument is far from being true since $\hat{g}_h(x_i)$ depends on the all sample $Y_1, \ldots, Y_n$, and, in particular it depends on ε_i! In fact, minimizing $RSS_{obs}(h)$ w.r.t. h would necessarily lead to over-fitting—note that any interpolating estimator has $RSS_{obs} = 0$!

We are looking for an unbiased estimator of the $AMSE(\hat{g}_h, g)$. The first approach is simple. Split the sample *randomly* into two unequal parts. The larger of them, $\mathscr{S}_{train} \subset \{1, \ldots, n\}$ of size n_{train}, is called the *training set*. It is used to calculate $\hat{g}_h(x)$ for different values of h in some set $\mathscr{H}$. The complementary set $\mathscr{S}_{test} = \mathscr{S}_{train}^c$ of size $n_{test} = n - n_{train}$ is the *test set*, and it is used to evaluate the goodness-of-fit of $\hat{g}_h$ for each $h \in \mathscr{H}$ by estimating the prediction error:

$$
\begin{aligned}
RSS_{test}(h) &= \sum_{i \in \mathscr{S}_{test}} (Y_i - \hat{g}_h(x_i))^2 \\
&= \sum_{i \in \mathscr{S}_{test}} (\hat{g}_h(x_i) - g(x_i))^2 - 2 \sum_{i \in \mathscr{S}_{test}} (\hat{g}_h(x_i) - g(x_i))\varepsilon_i + \sum_{i \in \mathscr{S}_{test}} \varepsilon_i^2.
\end{aligned}
$$

The main difference between $RSS_{test}(h)$ and $RSS_{obs}(h)$, is that $\hat{g}(x_i)$ in the second term for RSS_{test} depends only on Y_i's, $i \in \mathscr{S}_{train}$ and, therefore, is now independent of ε_i's, $i \in \mathscr{S}_{test}$. Hence, the second term does have a zero mean and, following our previous arguments, $RSS_{test}(h)/n_{test} + \sigma^2$ is, therefore, an *unbiased* estimator of the $AMSE(\hat{g}_h, g)$.

We then proceed to compute

$$
\hat{h} = \underset{h \in \mathscr{H}}{\arg\min} \, RSS_{test}(h)
$$

However, there is a price to pay for splitting the data and this price may be too high for small and even moderate samples: in order that RSS_{test} would be a good (with a small variance) estimator of the AMSE, the test set shouldn't be too small. On the other hand, it would leave less observations in the training set and, therefore, affect the accuracy of $\hat{g}_h$. Hence, the data set should be sufficiently large to make splitting efficient.

A common approach to overcome the efficiency problem while retaining the disentangling the test set from training test is *cross-validation* (CV) that uses different subsets of the entire data to test and train a model on different iterations. We consider the simplest *leave-one-out* cross-validation procedure (see also Section 9.3). Let $\hat{g}_{h,(-i)}(x)$ be the estimator based on all observations but the i-th one. Thus, $\hat{g}_{h,(-i)}(x_i)$ is independent of Y_i. Define

$$
CV(h) = \frac{1}{n} \sum_{i=1}^{n} (Y_i - \hat{g}_{h,(-i)}(x_i))^2
$$

and find $\hat{h}$ that minimizes it w.r.t h.

We now show that for Nadaraya–Watson, truncated series and smoothing spline estimators, CV admits quite simple expressions.

Example 9.1 (CV for Nadaraya–Watson kernel estimator) Define the matrix $A \in \mathbb{R}^{n \times n}$ with the entries $A_{ij} = \dfrac{K\left(\frac{x_j - x_i}{h}\right)}{\sum_{\ell=1}^{n} K\left(\frac{x_\ell - x_i}{h}\right)}$. In particular, $A_{ii} = \dfrac{K(0)}{\sum_{\ell=1}^{n} K\left(\frac{x_\ell - x_i}{h}\right)}$. From (9.16) we then have $\hat{\mathbf{g}} = \mathbf{A}\mathbf{y}$.

Hence,

$$
\hat{g}_{K,h,(-i)}(x_i) = \frac{\sum_{j \neq i} K\left(\frac{x_j - x_i}{h}\right) Y_j}{\sum_{j \neq i} K\left(\frac{x_j - x_i}{h}\right)}
$$

$$= \frac{\sum_{j=1}^{n} K\left(\frac{x_j - x_i}{h}\right) Y_j - K(0) Y_i}{\sum_{j=1}^{n} K\left(\frac{x_j - x_i}{h}\right) \left(1 - \frac{K(0)}{\sum_{\ell=1}^{n} K\left(\frac{x_\ell - x_i}{h}\right)}\right)}$$

$$= \frac{\hat{g}_{K,h}(x_i) - A_{ii} Y_i}{1 - A_{ii}}$$

and, therefore,

$$Y_i - \hat{g}_{K,h,(-i)}(x_i) = Y_i - \frac{\hat{g}_{K,h}(x_i) - A_{ii} Y_i}{1 - A_{ii}} = \frac{Y_i - \hat{g}_{K,h}(x_i)}{1 - A_{ii}}.$$

Thus, we don't need, in fact, to derive n kernel estimators $\hat{g}_{K,h,(-i)}$, $i = 1, \dots, n$ from $n - 1$ observations $j \neq i$ each to calculate $CV(h)$ but to obtain it directly from a single kernel estimator $\hat{g}_{K,h}$ based on the entire sample:

$$CV(h) = \frac{1}{n} \sum_{i=1}^{n} \left(\frac{Y_i - \hat{g}_{K,h}(x_i)}{1 - A_{ii}}\right)^2 \tag{9.28}$$

and minimize w.r.t. h.

Example 9.2 (CV for truncated series estimator) Consider a truncated series estimator $\hat{g}_k(x_i) = \sum_{j=0}^{k} \hat{\beta}_j \psi_j(x_i)$ or, in the matrix form,

$$\hat{\mathbf{g}}_k = \Psi \hat{\beta} = H \mathbf{Y},$$

where $\hat{\beta}$ is the LSE of β from (9.19) and $H = \Psi(\Psi^T \Psi)^{-1} \Psi^T$ is the projection matrix on the $(k+1)$ columns of the design matrix Ψ.

Consider the LSE $\hat{\beta}_{(-i)} = (\Psi_{(-i)}^T \Psi_{(-i)})^{-1} \Psi_{(-i)}^T \mathbf{Y}_{(-i)}$ based on $n - 1$ observations (x_j, Y_j), $j \neq i$. We have

$$\Psi_{(-i)}^T \Psi_{(-i)} = \sum_{j \neq i} \psi_{j \cdot}^T \psi_{j \cdot} = \sum_{j=1}^{n} \psi_{j \cdot}^T \psi_{j \cdot} - \psi_{i \cdot}^T \psi_{i \cdot}.$$

and, therefore,

$$(\Psi_{(-i)}^T \Psi_{(-i)})^{-1} = (\Psi^T \Psi)^{-1} + \frac{(\Psi^T \Psi)^{-1} \psi_{i \cdot}^T \psi_{i \cdot} (\Psi^T \Psi)^{-1}}{1 - \psi_{i \cdot} (\Psi^T \Psi)^{-1} \psi_{i \cdot}^T} = (\Psi^T \Psi)^{-1} + \frac{(\Psi^T \Psi)^{-1} \psi_{i \cdot}^T \psi_{i \cdot} (\Psi^T \Psi)^{-1}}{1 - H_{ii}}$$

Similarly,

$$\Psi_{(-i)}^T \mathbf{Y}_{(-i)} = \sum_{j \neq i} \psi_{j \cdot}^T Y_j = \sum_{j=1}^{n} \psi_{j \cdot}^T Y_j - \psi_{i \cdot}^T Y_i = \Psi^T \mathbf{Y} - \psi_{i \cdot}^T Y_i$$

Hence, after some algebra,

$$\hat{\beta}_{(-i)} = \left((\Psi^T \Psi)^{-1} + \frac{(\Psi^T \Psi)^{-1} \psi_{i \cdot}^T \psi_{i \cdot} (\Psi^T \Psi)^{-1}}{1 - H_{ii}}\right)(\Psi^T \mathbf{Y} - \psi_{i \cdot}^T Y_i)$$

$$= \hat{\beta} + \frac{(\Psi^T \Psi)^{-1} \psi_{i \cdot}^T \psi_{i \cdot} \hat{\beta}}{1 - H_{ii}} - \frac{(\Psi^T \Psi) \psi_{i \cdot}^T H_{ii} Y_i}{1 - H_{ii}} - (\Psi^T \Psi)^{-1} \psi_{i \cdot}^T Y_i$$

$$= \hat{\beta} - \frac{(\Psi^T \Psi)^{-1} \psi_{i \cdot}^T (Y_i - \psi_{i \cdot} \hat{\beta}_i)}{1 - H_{ii}} = \hat{\beta} - \frac{(\Psi^T \Psi)^{-1} \psi_{i \cdot}^T (Y_i - \hat{g}_k(x_i))}{1 - H_{ii}}$$

and

$$Y_i - \hat{g}_{k,(-i)}(x_i) = Y_i - \psi_{i \cdot} \hat{\beta}_{(-i)}$$

$$= Y_i - \hat{g}_k(x_i) + \frac{H_{ii}(Y_i - \hat{g}_k(x_i))}{1 - H_{ii}}$$

$$= \frac{Y_i - \hat{g}_k(x_i)}{1 - H_{ii}}$$

Thus,

$$CV(k) = \frac{1}{n}\sum_{i=1}^{n}\left(Y_i - \hat{g}_{k,(-i)}(x_i)\right)^2 = \frac{1}{n}\sum_{i=1}^{n}\left(\frac{Y_i - \hat{g}_k(x_i)}{1 - H_{ii}(k)}\right)^2 \tag{9.29}$$

and

$$\hat{k} = \arg\min_{0 \le k \le n-1}\sum_{i=1}^{n}\left(Y_i - \hat{g}_{k,(-i)}(x_i)\right)^2 = \sum_{i=1}^{n}\left(\frac{Y_i - \hat{g}_k(x_i)}{1 - H_{ii}(k)}\right)^2$$

In particular, the above criterion can be used for choosing the order of the polynomial regression or the number of terms in the orthonormal cosine series estimator $\hat{g}_{C,k}$ in (9.20).

Example 9.3 (CV for spline smoothing) Consider a smoothing spline $\hat{g}_{S,\lambda}$ from (9.21) and a smoothing spline $\hat{g}_{S,\lambda,(-i)}$ based on $n-1$ observations with (x_i, Y_i) left out of the sample. From the definition, evidently,

$$\begin{aligned}
\hat{g}_{S,\lambda,(-i)}(x) &= \arg\min_{g\in C^2[0,1]}\sum_{j\neq i}(Y_j - g(x_j))^2 + \lambda\int(g''(x))^2 dx \\
&= \arg\min_{g\in C^2[0,1]}\sum_{j\neq i}(Y_j - g(x_j))^2 + (\hat{g}_{S,\lambda,(-i)}(x_i) - g(x_i))^2 + \lambda\int(g''(x))^2 dx,
\end{aligned}$$

that is, $\hat{g}_{S,\lambda,(-i)}$ is also a smoothing spline that solves (9.21) for n observations: (x_j, Y_j), $j \neq i$ and the extra observation $(x_i, \hat{g}_{S,\lambda,(-i)}(x_i))$. Hence, from (9.22),

$$\begin{aligned}
\hat{g}_{S,\lambda,(-i)}(x_i) &= \sum_{j\neq i}A_{ij}(\lambda)Y_j + A_{ii}(\lambda)\hat{g}_{S,\lambda,(-i)}(x_i) = \sum_{j=1}^{n}A_{ij}(\lambda)Y_j + A_{ii}(\lambda)(\hat{g}_{S,\lambda,(-i)}(x_i) - Y_i) \\
&= \hat{g}_{S,\lambda}(x_i) + A_{ii}(\lambda)(\hat{g}_{S,\lambda,(-i)}(x_i) - Y_i),
\end{aligned}$$

where the matrix $A(\lambda) = (I_n + \lambda K)^{-1}$ and the matrix K is given in Lemma 9.2.

Thus,

$$Y_i - \hat{g}_{S,\lambda,(-i)}(x_i) = Y_i - \hat{g}_{S,\lambda}(x_i) + A_{ii}(\lambda)(Y_i - \hat{g}_{S,\lambda,(-i)}(x_i))$$

implying

$$Y_i - \hat{g}_{S,\lambda,(-i)}(x_i)) = \frac{Y_i - \hat{g}_{S,\lambda}(x_i)}{1 - A_{ii}(\lambda)}$$

Therefore,

$$CV(\lambda) = \frac{1}{n}\sum_{i=1}^{n}\left(\frac{Y_i - \hat{g}_{S,\lambda}(x_i)}{1 - A_{ii}(\lambda)}\right)^2 \tag{9.30}$$

and $\hat{\lambda}$ is chosen by minimizing (9.30) w.r.t. λ.

From (9.28), (9.29) and (9.30) one realizes that CV for Nadaraya–Watson kernel, truncated series and spline smoothing estimators can be obtained explicitly from a single estimator $\hat{g}_h$ based on the entire sample and are all of the form

$$CV(h) = \frac{1}{n}\sum_{i=1}^{n}\left(\frac{Y_i - \hat{g}_h(x_i)}{1 - A_{ii}(h)}\right)^2 \tag{9.31}$$

for the corresponding matrices $A(h)$.

Another approach to find an unbiased estimator of AMSE is *Stein's unbiased risk estimation* (SURE). Assume that $\varepsilon \sim N(0, \sigma^2)$. The idea is based on the following lemma:

Lemma 9.8.1 (Stein's lemma) *Let $Y \sim \mathcal{N}(\mu, \sigma^2)$ and $u(\cdot)$ is a differentiable function such that $E|u'(Y)| < \infty$. Then,*

$$E\big(u(Y)(Y - \mu)\big) = \sigma^2 E u'(Y).$$

Proof. Note that

$$\frac{d}{dy}\varphi\Big(\frac{y-\mu}{\sigma}\Big) = -\frac{y-\mu}{\sigma^2}\varphi\Big(\frac{y-\mu}{\sigma}\Big),$$

$\square$

where $\varphi(y)$ is the standard normal density. Hence, the proof follows by integration by parts:

$$
\begin{aligned}
E\big(u(Y)(Y-\mu)\big) &= \sigma^2\int u(y)\frac{y-\mu}{\sigma^2}\frac{1}{\sigma}\varphi\Big(\frac{y-\mu}{\sigma}\Big)dy \\
&= \sigma^2\int u'(y)\frac{1}{\sigma}\varphi\Big(\frac{y-\mu}{\sigma}\Big)dy = \sigma^2 Eu'(Y).
\end{aligned}
$$

Stein's lemma allows us to obtain an unbiased estimator of the AMSE:

Theorem 9.7 Suppose $Y_i = g(x_i) + \varepsilon_i$, where $\varepsilon_1,\ldots,\varepsilon_n$ are i.i.d. $\mathcal{N}(0,\sigma^2)$ and $\hat{g}_h(x) = \hat{g}_h(x;Y_1,\ldots,Y_n)$ some estimator of $g(x)$ differentiable w.r.t. Y_i. Define

$$SURE(h) = \frac{1}{n}\sum_{i=1}^{n}(Y_i - \hat{g}_h(x_i)) + \frac{2\sigma^2}{n}\sum_{i=1}^{n}\frac{\partial \hat{g}_h(x_i)}{\partial Y_i} - \sigma^2 \tag{9.32}$$

$SURE(h)$ is an unbiased estimator of the $AMSE(\hat{g}_h,g) = \frac{1}{n}E\sum_{i=1}^{n}\big(\hat{g}_h(x_i) - g(x_i)\big)^2$.
The smoothing parameter h is then chosen by minimizing $SURE(h)$ w.r.t. h.

Proof. We have

$$\frac{1}{n}\sum_{i=1}^{n}\big(\hat{g}_h(x_i) - g(x_i)\big)^2 \tag{9.33}$$

$$= \frac{1}{n}\sum_{i=1}^{n}\big(\hat{g}_h(x_i) - Y_i\big)^2 + 2\frac{1}{n}\sum_{i=1}^{n}\big(\hat{g}_h(x_i) - Y_i\big)\big(Y_i - g(x_i)\big) + \frac{1}{n}\sum_{i=1}^{n}\big(Y_i - g(x_i)\big)^2.$$

The expectation of the left hand side is exactly $AMSE(\hat{g}_h,g)$. On the other hand, for the right hand side, the expectation of the third term is σ^2, while for the second term by Lemma 9.8.1,

$$
\begin{aligned}
E\Big(\frac{2}{n}\sum_{i=1}^{n}\big(\hat{g}_h(x_i) - Y_i\big)\big(Y_i - g(x_i)\big)\Big) &= \frac{2\sigma^2}{n}E\Big(\sum_{i=1}^{n}\frac{\partial\big(\hat{g}_h(x_i) - Y_i\big)}{\partial Y_i}\Big) \\
&= \frac{2\sigma^2}{n}E\Big(\sum_{i=1}^{n}\frac{\partial \hat{g}_h(x_i)}{\partial Y_i}\Big) - 2\sigma^2.
\end{aligned}
$$

$\square$

Example 9.4 (SURE for a general linear estimator) Recall that kernel estimators, truncated series and smoothing splines are examples of linear estimators. Consider a general linear estimator $\hat{g}_h$ depending on a smoothing parameter h, where $\hat{g}_h = A(h)\mathbf{Y}$ for some matrix $A(h)$. In this case $\hat{g}_h(x_i) = \sum_{j=1}^{n}A_{ij}(h)Y_j$ and $\frac{\partial}{\partial Y_i}\hat{g}_h(x_i) = A_{ii}(h)$. Hence, by (9.32)

$$
\begin{aligned}
SURE(h) &= \frac{1}{n}||(I - A(h)\mathbf{Y}||^2 + \frac{2\sigma^2}{n}\sum_{i=1}^{n}A_{ii}(h) - \sigma^2 \\
&= \frac{1}{n}RSS(h) + \frac{2\sigma^2}{n}trA(h) - \sigma^2
\end{aligned}
$$

and we choose

$$\hat{h} = \arg\min_{h\in\mathscr{H}} RSS(h) + 2\sigma^2 trA(h)$$

Such a criterion may be viewed as a penalized RSS with a penalty on h to avoid over-fitting.

In particular, for Nadaraya–Watson kernel estimator in (9.16), $A_{ij}(h) = \frac{K((x_i-x_j)/h)}{\sum_{i=1}^n K((x_\ell-x_j)/h)}$, $i,j = 1,\ldots,n$. Thus,

$$SURE(h) = \frac{1}{n}RSS(h) + \frac{2\sigma^2}{n}\sum_{i=1}^n \frac{K(0)}{\sum_{j=1}^n K\left(\frac{x_j-x_i}{h}\right)} - \sigma^2$$

and

$$\hat{h} = \arg\min_{h\in\mathscr{H}} RSS(h) + 2\sigma^2 K(0)\sum_{i=1}^n \frac{1}{\sum_{j=1}^n K\left(\frac{x_j-x_i}{h}\right)}$$

For the truncated series estimator $\hat{g}_k$, $A(k)$ is the projection matrix on the $(k+1)$ columns of design matrix Ψ and, therefore, $trA(k) = k+1$ (see Chapter 8). Hence,

$$SURE(k) = \frac{1}{n}RSS(k) + \frac{\sigma^2}{n}(2(k+1) - n)$$

and

$$\hat{k} = \arg\min_{0\le k\le n-1} RSS(k) + 2\sigma^2 k$$

which is a penalized RSS criterion with a complexity penalty on the number of terms in the truncated series.

Similarly, for the smoothing spline $\hat{g}_{S,\lambda}$ in (9.21), $A(\lambda) = (I_n + \lambda K)^{-1}$ (see (9.22)), where the matrix K is given in Lemma 9.2,

$$\hat{\lambda} = \arg\min_{\lambda} RSS(\lambda) + 2\sigma^2 tr(I_n + \lambda K)^{-1}$$

Example 9.8.1 We consider the data presented in Figure 9.5. The optimal bandwidth according to the SURE was used to construct the kernel estimator.

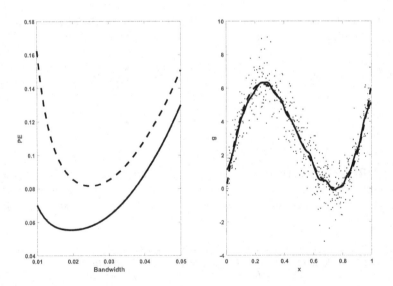

Figure 9.5: Unbiased estimation of the risk. The left panel presents $SURE(h)$ (solid line) and $CV(h)$ (dashed line) as a function of a bandwidth h. The right panel exhibits the data, the true regression function (dashed line) and Nadaraya–Watson kernel estimator with a normal kernel and the bandwidth selected by SURE (solid line).

We conclude this section with some heuristics discussion of CV and SURE criteria on the example of a general linear estimator $\hat{\mathbf{g}}_h = A(h)\mathbf{Y}$. Recall that

$$h_{SURE} = \arg\min_{h \in \mathcal{H}} RSS(h) + 2\sigma^2 trA(h), \tag{9.34}$$

while for Nadaraya–Watson kernel, truncated series and spline smoothing estimators from (9.31) we have

$$h_{CV} = \arg\min_{h \in \mathcal{H}} \sum_{i=1}^{n} \left(\frac{Y_i - \hat{g}_h(x_i)}{1 - A_{ii}(h)} \right)^2 \tag{9.35}$$

for the corresponding matrices $A(h)$.

Approximate $A_{ii}(h)$'s in the denominator in (9.35) by their average value $trA(h)/n$:

$$CV(h) = \sum_{i=1}^{n} \left(\frac{Y_i - \hat{g}_h(x_i)}{1 - A_{ii}} \right)^2 \approx \frac{RSS(h)}{\left(1 - \frac{trA(h)}{n} \right)^2}$$

The above approximation of CV is commonly called a *generalized cross-validation* (GCV).

Then, assuming $trA(h) \ll n$ for large n,

$$\frac{RSS(h)}{\left(1 - \frac{trA(h)}{n} \right)^2} \quad \sim \quad \frac{RSS(h)}{1 - 2\frac{trA(h)}{n}}$$

$$\sim \quad RSS(h) \left(1 + 2\frac{trA(h)}{n} \right)$$

$$\sim \quad RSS(h) + 2\sigma^2 trA(h),$$

where we used $\hat{\sigma}^2 = RSS(h)/n \sim \sigma^2$.

Thus, we see that at least for these examples both SURE and CV criteria (9.34) and (9.35) are asymptotically close and usually lead to similar $\hat{h}$.

Appendix A

Probabilistic Review

A.1 Introduction

In this appendix we provide a basic review on probability needed for the book. By no means should it be seen as an (even a brief!) introduction to probability theory. It is assumed that the student took such a course as a prerequisite. The presentation is quite elementary and intuitive. In particular, we avoid using the arguments from measure theory. We do not always try to be rigorous from the mathematical point of view and introduce the proofs only if we consider them enlightening.

A.2 Basic probabilistic laws

Probability theory deals with experiments and their outcomes. In an abstract way an experiment is defined by three objects $(\Omega, \mathscr{A}, P)$ known as a *probability space*, where *a sample space* Ω is the collection of all end points or outcomes of the experiment. Probability theory does not assign probability to single outcomes as such but to events. An *event A* is a subset of Ω, such that given the outcome of the experiment one knows whether it belongs to A or not. A *set of events* $\mathscr{A}$ is the collection of all possible events that may be considered. In a way, $\mathscr{A}$ is the collection of all possible yes/no questions that can be answered after the completion of the experiment. Finally, P assigns *probability* to every event.

Technically, a set of events $\mathscr{A}$ should satisfy some conditions. In particular, the empty set $\emptyset$ and Ω itself should belong to $\mathscr{A}$, and $\mathscr{A}$ should be closed to some set operations:

1. If $A \in \mathscr{A}$, then its *complement* $A^c \in \mathscr{A}$, where A^c contains all elements in Ω not belonging to A. This is natural demand, since if we know whether A happened or not, we necessarily know whether A^c happened or not as well.
2. If $A_1, A_2, \cdots \in \mathscr{A}$, then $A_1 \cup A_2 \cup \cdots \in \mathscr{A}$ (the event "at least one of $A_1, A_2, \ldots$ has happened" belongs to $\mathscr{A}$).
3. If $A_1, A_2, \cdots \in \mathscr{A}$, then $A_1 \cap A_2 \cap \cdots \in \mathscr{A}$ (the event "all of $A_1, A_2, \ldots$ have happened" belongs to $\mathscr{A}$).

The probability function assigns to each event $A \in \mathscr{A}$ a real number $P(A)$ called the *probability* of A satisfying the following conditions:

1. $0 \le P(A) \le 1$.
2. Something should necessarily happen: $P(\Omega) = 1$ and $P(\emptyset) = 0$.
3. $P(A^c) = 1 - P(A)$ for any $A \in \mathscr{A}$.
4. If $A_1, A_2, \ldots$ are disjoint events, that is, at most only one of them can happen and $P(A_i \cap A_j) = \emptyset$ for all $i \ne j$, then $P(A_1 \cup A_2 \cup \ldots) = \sum_{i=1}^{\infty} P(A_i)$.
5. Since $(A_1 \cup A_2)^c = A_1^c \cap A_2^c$, then $P(A_1 \cap A_2) = 1 - P\big((A_1 \cap A_2)^c\big) = 1 - P(A_1^c \cup A_2^c)$.

Two events are *independent* if $P(A \cap B) = P(A)P(B)$. Note that $A \cap B^c$ and $A \cap B$ are disjoint and their union is A. Hence if A and B are independent then also A and B^c are independent:

$$P(A \cap B^c) = P(A) - P(A \cap B) = P(A) - P(A)P(B) = P(A)\big(1 - P(B)\big) = P(A)P(B^c).$$

DOI: 10.1201/9781003175407-A

The *conditional probability* of an event B given an event A (with non-zero probability) is

$$P(B \mid A) = \frac{P(A \cap B)}{P(A)}.$$

It tells us what the probability is that we should assign to the event B if we already know that the event A occurred. If A and B are independent, then $P(A \mid B) = P(A)$ and the probability we assign to A does not depend on whether B happened or not.

The *law of total probability* states that if $A_1, A_2, \ldots$ is a *partition* of Ω, that is, $\bigcup_{i=1}^{\infty} A_i = \Omega$ and $A_i \cap A_j = \emptyset$ for all $i \neq j$, then

$$P(B) = P\left(\bigcup_{i=1}^{\infty} \{A_i \cap B\}\right) = \sum_{i=1}^{\infty} P(\{A_i \cap B\}).$$

The celebrated *Bayes Theorem* is the direct result of a double application of the definition of conditional probability:

$$P(A \mid B) = \frac{P(A \cap B)}{P(B)} = \frac{P(A)P(B \mid A)}{P(B)}.$$

In particular, if $A_1, A_2, \ldots$ is a partition of Ω,

$$P(A_i \mid B) = \frac{P(A_i)P(B \mid A_i)}{\sum_j P(A_j)P(B \mid A_j)}.$$

A.3   Random variables

Usually we deal with random variables. A random variable is defined as a real-valued function defined on the sample space Ω (in fact, it should satisfy some conditions but we do not discuss measure theory here!). Typically we have the triplet $(\Omega, \mathscr{A}, P)$ in the back of our mind, but deal directly with random variables, which, simply, yield a numerical value observed at the end of the experiment. Random variables are traditionally denoted by capital letters from the end of the Latin alphabet, e.g. X, Y, U, V. Note that on the same probability space we may define more than one random variable. Thus, for a randomly picked student, we can measure his height, weight and IQ.

Random variables are characterized by the set of their possible values and the corresponding probabilities. The random variables we consider in this book are of two types. A *discrete* random variable can take only a countable number of values. For example, the number of dots on a die, the number of girls in a family with 3 siblings or the number of calls to a service center per hour. The number of possible values can be finite (6 and 4 respectively in the two first examples above) or infinite (the last example). A discrete random variable X is defined by its *probability function* $p_i = P(X = x_i)$, $i = 1, 2, \ldots$, where $x_1, x_2, \ldots$ are possible values of X. Obviously, $\sum_i p_i = 1$.

A (absolutely) *continuous* random variable may assume any value within an entire (finite or infinite) interval, e.g., weight or temperature. For a continuous random variable, the probability of any particular value is zero and the notion of a probability function becomes therefore useless in this case. Instead, one can define a *cumulative distribution function* (cdf) $F(x) \triangleq P(X \leq x)$. For most "common" continuous random variables, a cdf $F(\cdot)$ is differentiable and hence $F(x) = \int_{-\infty}^{x} f(y) dy$, where $f = F'$ is called a *density function*, or simply a *density*. As we have noted above, for a continuous random variable we can only speak about the probability that its value will be within a certain range. Thus, $P(a < X \leq b) = F(b) - F(a) = \int_a^b f(x) dx$, for any $a < b$. Clearly, the density $f(\cdot)$ is a non-negative function, and the area under its graph (the integral of f over the entire region) is one—it is just the probability that the random variable gets any value between $-\infty$ to ∞. To be more rigorous, a density is not well defined and can be changed arbitrarily in any countable set of points, but there exists its unique continuous version (or even right-continuous). Unlike the cdf, a value of $f(x)$ itself has no probabilistic meaning. In particular, it may be larger than one.

Evidently, a cdf $F(\cdot)$ can be defined for a discrete random variable as well. In this case, it is a stepwise function. It is easy to see that for any random variable $F(\cdot)$ possess the following properties:

1. It is a monotone non-decreasing function.
2. It increases from 0 to 1: $\lim_{x \to -\infty} F(x) = 0$ and $\lim_{x \to \infty} F(x) = 1$.
3. It is right-continuous: $\lim_{t \searrow 0} F(x+t) = F(x)$.

A.3.1 Expected value and the variance

It is usual to characterize a random variable by a few parameters, that although do not single out its entire distribution, gives one nevertheless some idea about its main properties.

One of the most important characteristics of a distribution of a random variable is its *expected value* (*expectation* or *mean*) EX defined as

$$EX = \begin{cases} \sum_i x_i p_i, & \text{if } X \text{ is discrete} \\ \int x f(x) dx, & \text{if } X \text{ is continuous} \end{cases} \tag{A.1}$$

If we consider the plot of a probability function for a discrete random variable or a density for a continuous one, the expected value is the x coordinate of its "center of mass" in the sense it is the center of the distribution. However, such a picture may be very misleading. Consider for simplicity a discrete random variable, where a tiny probability p_i is assigned to a very large value x_i. The value of EX may be totally distorted since $x_i p_i$ might be large. It is easy to see that if $P(a \leq X \leq b) = 1$ (i.e., roughly speaking, X cannot be smaller than a and larger than b), where a can be infinitely negative and b can be infinitely positive, then $a \leq EX \leq b$. More generally, we have:

Lemma A.3.1 *If two random variables X and Y are defined on the same probability space and $P(X \leq Y) = 1$, then $EX \leq EY$.*

It is important to note that a random variable may have an infinite expectation or even may not have a well defined (finite or infinite) expectation. This would happen if the sum or the integral in the definition (A.1) of EX fails to converge.

The expectation alone may not tell the complete story about a random variable. The spread (dispersion) of the values of a random variable around the expectation also matters. The standard measure of dispersion is a *variance* defined as the mean squared deviation of a random variable from its mean:

$$\text{Var}(X) = E(X - EX)^2 = \begin{cases} \sum (x_i - EX)^2 p_i, & \text{if } X \text{ is discrete} \\ \int (x - EX)^2 f(x) dx, & \text{if } X \text{ is continuous} \end{cases}$$

In both cases the variance can be calculated as

$$\text{Var}(X) = E(X^2) - (EX)^2,$$

which is typically more convenient for computations.

The larger the dispersion of a random variable around its expectation, the larger the variance. The variance, however is not measured on the same scale as a random variable itself and its expectation. Thus, if X is the annual salary of a graduate of a given high school, the mean salary is given in US Dollars (say, 70,000 USD), but its variance may be something like 100 million USD^2, which is hard to evaluate. A *standard deviation* $SD(X) = \sqrt{\text{Var}(X)}$ brings the measure of dispersion back to the original scale of the distribution (10,000 USD in the example).

There exist other measures of the center and dispersion of the distribution of a random variable. Thus, a *median* $Med(X)$ of a random variable X is defined as the point x where $F(x) = 0.5$ (the definition should be corrected for discrete random variables if there is no unique value satisfying $F(x){=}0.5$). It is a more robust measure than the mean and small changes in the distribution does not affect it strongly. More generally, a *q-quantile* $\xi_q(X)$ is defined as the solution of $F(\xi_q(X)) = 1 - q$ [1]. In particular, $Med(X) = \xi_{0.5}(X)$. An *interquartile range* $IQR = \xi_{1/4}(X) - \xi_{3/4}(X)$ is another measure for the dispersion.

A.3.2 Chebyshev's and Markov's inequalities

Chebyshev's inequality provides a more rigorous probabilistic statement on the deviation of a random variable X from its mean in terms of the variance:

$$P(|X - EX| \leq a) \geq 1 - \frac{\text{Var}(X)}{a^2}.$$

for any distribution of X (of course, the existence of the mean and the variance is assumed). For example, with a probability of at least 8/9, the deviation from the mean is within three standard deviations.

Chebyshev's inequality is a particular case of *Markov's inequality* stating that for any non-negative random variable Y and $b > 0$

$$P(Y \geq b) \leq \frac{EY}{b}$$

Indeed, applying Markov's inequality for $Y = (X - EX)^2$ and $b = a^2$ one immediately has

$$P(|X - EX| \geq a) = P\left((X - EX)^2 \geq a^2\right) \leq \frac{E(X - EX)^2}{a^2} = \frac{\text{Var}(X)}{a^2}$$

A.3.3 Expectation of functions and the Jensen's inequality

Given a random variable X we are often interested in its functions, e.g. X^2, $1/X$, e^X, etc. For any function $h(\cdot)$, $h(X)$ is a random variable (at least if we forget about measurability issues). To calculate the expectation $Eh(X)$ one can derive the distribution of $h(X)$ (see Section A.3.8) and then use the definition (A.1). However, the following direct calculation is typically easier:

$$Eh(X) = \begin{cases} \sum h(x_i) f_i, & \text{if } X \text{ is discrete} \\ \int h(x) f(x) dx, & \text{if } X \text{ is continuous} \end{cases}$$

In particular, for $h(x) = (X - EX)^2$, $Eh(x) = \text{Var}(X)$. For linear functions of the form $h(X) = a + bX$ one has

$$E(a + bX) = a + bEX. \tag{A.2}$$

It is also easy to check that

$$\text{Var}(a + bX) = b^2 \text{Var}(X).$$

Another important example is the class of convex functions. A function $h(\cdot)$ is convex if for every $0 \leq \alpha \leq 1$ and for every x_1 and x_2 we have $h(\alpha x_1 + (1 - \alpha)x_2) \leq \alpha h(x_1) + (1 - \alpha)h(x_2)$. Examples of convex functions include x^2, e^x, etc. Geometrically, convexity of $h(x)$ means that for any x_0 there always exists a b such that the line $h(x_0) + b(x - x_0)$ lies below $h(x)$, that is, $h(x) \geq h(x_0) + b(x - x_0)$ for any x. In particular, $h(X) \geq h(EX) + b(X - EX)$ for some b. It then follows from Lemma A.3.1 and (A.2) that

$$Eh(X) \geq h(EX).$$

Thus, the expectation of a convex function of a random variable is always not less than the same function of its expectation. This is known as *Jensen's inequality*.

---

[1] There is no unique conventional definition of a quantile and in some books a q-quantile $\xi_q(X)$ is defined as the solution of $F(\xi_q(X)) = q$

A.3.4 Joint distribution

In statistics we typically deal simultaneously with several random variables. Thus, in medical tests a series of measurements are taken from a patient, e.g., blood pressure, pulse rate, cholesterol level, etc. For random variables defined on the same probability space, their *joint distribution* assigns probabilities of events defined in terms of all of them. What is the probability that a randomly chosen patient has a systolic blood pressure above 140 and a pulse rate between 90 and 100? All the concepts introduced above for a single random variable can be directly extended to the multivariate case. For simplicity of exposition we consider the case of two random variables but its generalization is straightforward (see also Section A.3.9).

Let X and Y be two discrete random variables with possible values $x_1, \ldots, x_n$ and $y_1, \ldots, y_m$ respectively. Their joint distribution is characterized by the joint probability function $p_{ij} = P(X = x_i, Y = y_j)$, $i = 1, \ldots, n$ $j = 1, \ldots, m$. The joint cdf is $F_{XY}(x, y) = P(X \leq x, Y \leq y)$. For continuous X and Y, we define the joint density $f_{XY}(x, y)$ satisfying $F_{XY}(x, y) = \int_{-\infty}^{x} \int_{-\infty}^{y} f_{XY}(x, y) dx dy$.

Let $A \subset \mathbb{R}^2$. The joint probability is then

$$P((X, Y) \in A) = \sum_{(i,j):(x_i, y_j) \in A} p_{ij} \qquad \text{or}$$

$$P((X, Y) \in A) = \int_{A} f_{XY}(x, y) dx dy$$

for discrete and continuous cases respectively.

Given the joint distribution of X and Y one can always obtain their *marginal* distributions. Thus, for the continuous case, if $f_{XY}(x, y)$ is the joint density of X and Y, their marginal densities $f_X(x)$ and $f_Y(y)$ are $f_X(x) = \int_{-\infty}^{\infty} f_{XY}(x, y) dy$ and $f_Y(y) = \int_{-\infty}^{\infty} f_{XY}(x, y) dx$.

Two random variables X and Y are called *independent* if their joint density is a product of the marginal densities $f_{X,Y}(x, y) = f_X(x) \cdot f_Y(y)$ if they are continuous with the obvious modification in terms of probability functions for the discrete case.

A.3.5 Covariance, correlation, and the Cauchy–Schwarz inequality

While the means and variances characterize the marginal distributions, the *covariance* is a characterization of the joint distribution of two random variables. For two random variable X and Y defined on the same probability space, the covariance $\text{cov}(X, Y)$ (sometimes also called a joint variance) is defined as

$$\text{cov}(X, Y) = \text{E}\Big((X - \text{E}X)(Y - \text{E}Y)\Big) = \text{E}(XY) - \text{E}(X)\text{E}(Y). \tag{A.3}$$

Obviously, $\text{cov}(X, X) = \text{Var}(X)$.

A direct consequence of the celebrated *Cauchy–Schwarz inequality* is that

$$|\text{cov}(X, Y)| \leq \sqrt{\text{Var}X \cdot \text{Var}Y}. \tag{A.4}$$

for any two random variables X and Y.

If it is hard to grasp the meaning of a particular value of the variance, it is even harder for the covariance. The Cauchy–Schwarz inequality above suggest a normalized version of the covariance called the *correlation coefficient* $\rho(X, Y)$ between two random variables X and Y:

$$\rho(X, Y) = \frac{\text{cov}(X, Y)}{\sqrt{\text{Var}(X) \cdot \text{Var}(Y)}}.$$

Unlike the covariance, $\rho(X, Y)$ is given in absolute unit and $|\rho(X, Y)| \leq 1$. Moreover, the equality holds if and only if Y is an affine function of X, that is, $Y = a + bX$ for some a and b with probability

1. $\rho(X,Y) = 1$ if and only if $b > 0$ and $\rho(X,Y) = -1$ if $b < 0$. The correlation coefficient therefore is a measure of "linear dependency" between X and Y. If $\rho(X,Y) = 0$, X and Y are called *uncorrelated*. Independent random variables are necessarily uncorrelated. However, uncorrelated random variables may still be highly dependent (though not linearly).

The covariance and correlation coefficient have a simple behavior under linear transformations:

$$\mathrm{cov}(a+bX, c+dY) = bd\,\mathrm{cov}(X,Y)$$
$$\rho(a+bX, c+dY) = \mathrm{sgn}\,(bd)\rho(X,Y).$$

Hence, the correlation coefficient is invariant to linear transformations up to its sign.

A.3.6   Expectation and variance of a sum of random variables

It is easy to verify that for any random variables $X_1, \ldots, X_n$ the expectation of their sum is the sum of their expectations:

$$\mathrm{E}\left(\sum_{i=1}^{n} X_i\right) = \sum_{i=1}^{n} \mathrm{E} X_i.$$

The situation is more complicated with the variance of sum of random variables. For two random variables,

$$\mathrm{Var}(X+Y) = \mathrm{Var}(X) + \mathrm{Var}(Y) + 2\,\mathrm{cov}(X,Y),$$

while in the general case

$$\mathrm{Var}(\sum_{i=1}^{n} X_i) = \sum_{i=1}^{n} \mathrm{Var}(X_i) + 2\sum_{i=1}^{n-1}\sum_{j=i+1}^{n} \mathrm{cov}(X_i, X_j).$$

However, if $X_1, \ldots, X_n$ are uncorrelated (in particular, independent),

$$\mathrm{Var}\left(\sum_{i=1}^{n} X_i\right) = \sum_{i=1}^{n} \mathrm{Var}(X_i). \tag{A.5}$$

A.3.7   Conditional distribution and Bayes Theorem

Suppose X and Y are two continuous random variable with a joint density $f_{XY}(x,y)$. We may ask about the density of Y given $X = x$. What is the distribution (density) of blood pressure among patients with a pulse rate of 90? This is a somewhat problematic question, because we condition on event with probability zero! However, the same intuition behind conditional probability from Section A.2 is still mostly valid (if used carefully). To be more precise, we would ask about the distribution of Y given $x - \varepsilon \leq X \leq x + \varepsilon$ and then let ε tend to zero. The resulting *conditional distribution* has the density $f_{Y|X}(\cdot)$ given by

$$f_{Y|X}(y|x) = \frac{f_{XY}(x,y)}{f_X(x)}.$$

According to Bayes Theorem,

$$f_{X|Y}(x|y) = \frac{f_X(x)f_{Y|X}(y|x)}{\int f_X(x')f_{Y|X}(y|x')dx'}.$$

A.3.8 Distributions of functions of random variables

We have already mentioned that any function $h(X)$ of a random variable is also a random variable. What is the distribution of $Y = h(X)$ if we know the distribution of the original random variable X? The answer is simple in the discrete case and follows directly from the definition:

$$P(Y = y_j) = \sum_{i:\, h(x_i)=y_j} P(X = x_i).$$

The situation is somewhat more complicated for continuous random variables. Suppose for simplicity that $h(x)$ is an increasing function. Then there exists an inverse function $h^{-1}(y)$, which is also increasing, and we have

$$F_Y(y) = P(Y \le y) = P(h^{-1}(Y) \le h^{-1}(y)) = P(X \le h^{-1}(y)) = F_X\left(h^{-1}(y)\right).$$

Taking derivatives of both sides (assume that h^{-1} is differentiable) and using the chain rule yield the density $f_Y(y)$:

$$f_Y(y) = (h^{-1}(y))' f_X\left(h^{-1}(y)\right).$$

Similarly, allowing $h(x)$ to be also a decreasing function, we have

$$f_Y(y) = |(h^{-1}(y))'| f_X\left(h^{-1}(y)\right). \tag{A.6}$$

An important case is where $Y = h(X) = a + bX$, or $X = h^{-1}(Y) = (Y - a)/b$. Applying (A.6) we have:

$$f_Y(y) = \frac{1}{|b|} f_X\left(\frac{y-a}{b}\right). \tag{A.7}$$

In particular, if $a = -EX/SD(X)$ and $b = -SD(X)$ so that $Y = \frac{X-\mu}{SD(X)}$, then $EY = 0$, $SD(Y) = 1$ and Y is called the *standardized* version of X.

The minimum and maximum are other important functions of a series of random variables. Suppose that $X_1, X_2, \ldots, X_n$ are independent variables with cdf's $F_1, F_2, \ldots, F_n$ respectively. The cdf $F_M(\cdot)$ of $\max(X_1, \ldots, X_n)$ is then given by

$$\begin{aligned} F_M(x) &= P(\max\{X_1, \ldots, X_n\} \le x) = P(X_1 \le x, X_2 \le x, \ldots, X_n \le x) = P\left(\bigcap_{i=1}^{n}\{X_i \le x\}\right) \\ &= \prod_{i=1}^{n} P(X_i \le x) = \prod_{i=1}^{n} F_i(x). \end{aligned}$$

In particular, if $X_1, \ldots, X_n$ are i.i.d. with the common cdf F,

$$F_M(x) = F^n(x).$$

The distribution of the minimum can be found in a similar way by considering the complements. Thus, if $X_1, \ldots, X_n$ are i.i.d. with cdf $F(\cdot)$, the cdf $F_m(\cdot)$ of $\min(X_1, \ldots, X_n)$ is

$$F_m(x) = 1 - \left(1 - F(x)\right)^n. \tag{A.8}$$

A.3.9 Random vectors

A random X vector in $\mathbb{R}^p$ is just a collection of p random variables $X_1, \ldots, X_p$ with some joint distribution. One can define the mean (expectation) of X as a vector of the means of $X_1, \ldots, X_p$:

$$EX = E\begin{pmatrix} X_1 \\ \vdots \\ X_p \end{pmatrix} = \begin{pmatrix} EX_1 \\ \vdots \\ EX_p \end{pmatrix}.$$

The variance-covariance matrix $\mathrm{Var}(X)$ (or simply variance or covariance matrix) of a random vector X is the self-adjoint positive semi-definite $p \times p$ matrix

$$\mathrm{Var}(X) = \begin{pmatrix} \mathrm{Var}(X_1) & \mathrm{cov}(X_1,X_2) & \mathrm{cov}(X_1,X_3) & \ldots & \mathrm{cov}(X_1,X_p) \\ \mathrm{cov}(X_2,X_1) & \mathrm{Var}(X_2) & \mathrm{cov}(X_2,X_3) & \ldots & \mathrm{cov}(X_2,X_p) \\ \mathrm{cov}(X_3,X_1) & \mathrm{cov}(X_3,X_2) & \mathrm{Var}(X_3) & \ldots & \mathrm{cov}(X_3,X_p) \\ \vdots & \vdots & \vdots & \vdots & \vdots \\ \mathrm{cov}(X_p,X_1) & \mathrm{cov}(X_p,X_2) & \mathrm{cov}(X_p,X_3) & \ldots & \mathrm{Var}(X_p) \end{pmatrix}.$$

Similarly to the univariate case:

$$\mathrm{Var}(X) = \mathrm{E}\big((X - \mathrm{E}X)(X - \mathrm{E}X)^\top\big) = \mathrm{E}(XX^\top) - (\mathrm{E}X)(\mathrm{E}X)^\top,$$

where the mean of a random matrix is the matrix of the means and $(\mathrm{E}X)^\top$ is the adjoint of $\mathrm{E}X$.

For a fixed matrix A, the standard rules for means, variances and covariances of a linear transformation can still be applied. Namely,

$$\begin{aligned} \mathrm{E}(AX) &= A\mathrm{E}X \\ \mathrm{Var}(AX) &= A\,\mathrm{Var}(X)A^\top \end{aligned} \tag{A.9}$$

Assume that a p-dimensional random vector X has a joint density $f_X(x)$. For any event S

$$P(X \in S) = \underset{x \in S}{\int \cdots \int} f_X(x)dx_1 \ldots dx_p.$$

Let X be a p-dimensional random vector with a joint density $f_X(\cdot)$ and consider its one-to-one transformation $Y = h(X)$, where $h(\cdot) = (h_1(\cdot),\ldots,h_p(\cdot)) : \mathbb{R}^p \to \mathbb{R}^p$, that is, $Y_1 = h_1(X_1,\ldots,X_p),\ldots,Y_p = h_p(X_1,\ldots,X_p)$. Since $h(\cdot)$ is a one-to-one transformation, there exists an inverse transformation $h^{-1} = (h_1^{-1},\ldots,h_p^{-1})$ such that $X = h^{-1}(Y)$. To derive the multivariate analogue of (A.6) for the joint density $f_Y(y)$, recall that the *Jacobian* $J_{h^{-1}}(y)$ of $h^{-1}(y)$ is defined as the determinant

$$J_{h^{-1}}(y) = \begin{vmatrix} \dfrac{\partial h_1^{-1}(y)}{\partial y_1} & \cdots & \dfrac{\partial h_1^{-1}(y)}{\partial y_p} \\ \vdots & \vdots & \vdots \\ \dfrac{\partial h_p^{-1}(y)}{\partial y_1} & \cdots & \dfrac{\partial h_p^{-1}(y)}{\partial y_p} \end{vmatrix}$$

Then,

$$f_Y(y) = |J_{h^{-1}}(y)|\, f_X(h^{-1}(y)) \tag{A.10}$$

As an example of (A.10) consider the distribution of a sum of two independent random variables. Let X and Y be two independent random variables with densities $f_X(\cdot)$ and $f_Y(\cdot)$ and find the density of $X + Y$. Define two random variables $U = X + Y$ and $V = X$. Hence, $X = V$, $Y = U - V$ and the corresponding Jacobian is

$$J(u,v) = \begin{vmatrix} 0 & 1 \\ 1 & -1 \end{vmatrix} = -1$$

Since X and Y are independent, $f_{XY} = f_X \cdot f_Y$ and applying (A.10) one obtains the joint density f_{UV} of U and V as

$$f_{UV}(u,v) = f_X(v) \cdot f_Y(u - v)$$

The marginal density of $U = X + Y$ is then

$$f_U(u) = \int f_{UV}(u,v)dv = \int f_X(v)f_Y(u - v)dv = f_X * f_Y,$$

where $f_X * f_Y$ is a *convolution* of f_X and f_Y.

Consider now a linear transformation $\mu + AX$ of a random vector X with a joint density $f_X(x)$, where μ is a fixed p-dimensional vector and A is a fixed non-singular $p \times p$ matrix. The joint density $f_Y(y)$ of Y is then

$$f_Y(y) = |det(A^{-1})|\, f_X(A^{-1}(y-\mu)) = |det(A)|^{-1} f_X(A^{-1}(y-\mu)), \tag{A.11}$$

which is a multivariate analogue of (A.7).

A.4 Special families of distributions

In this section we consider a few important families of random variables. A family of random variables is what statistics is all about. We do not know the exact distribution of the model, but typically we have good reason to assume that the distribution belongs to a family parameterized by a few unknown parameters. In the sequel, we present several most important and common parametric families of distributions.

A.4.1 Bernoulli and binomial distributions

The simplest random variable is a *Bernoulli* random variable X, which is a result of an experiment with only two possible outcomes: "success" or "failure" (Bernoulli trial). For example, it can be a result of a coin tossing, where getting "head," say, is considered a success, or of a basketball game, or the gender of a randomly selected student. Denoting "success" by 1 and "failure" by 0, the distribution of X has a Bernoulli distribution $\mathscr{B}(1, p)$, where

$$X = \begin{cases} 1, & \text{with probability p} \\ 0, & \text{with probability 1-p} \end{cases}$$

Trivially, $EX = p$ and $\text{Var}(X) = p(1-p)$.

Assume now that we perform a series of i.i.d. n Bernoulli trials and let X be the total number of successes. Obviously, $X = \sum_{i=1}^{n} X_i$, where a Bernoulli random variable $X_i \sim \mathscr{B}(1,p)$ is the result of the i-th trial. X can take the values of $0, 1, \ldots, n$ and by simple combinatorial arguments,

$$p_i = P(X = i) = \binom{n}{i} p^i (1-p)^{n-i}, \; i = 0, \ldots, n$$

This is known as the *binomial distribution* $\mathscr{B}(n, p)$, which is a generalization of a Bernoulli random variable for $n > 1$. By the properties of a sum of random variables (see Section A.3.6),

$$EX = np,$$
$$\text{Var}(X) = np(1-p).$$

Note that if $X_1 \sim \mathscr{B}(n_1, p)$ is the number of successes in a series of n_1 Bernoulli trials with probability of success p, and $X_2 \sim \mathscr{B}(n_2, p)$ is the number of successes in another independent series of n_2 trials with the same probability of success, the overall number of successes $X_1 + X_2$ in $n_1 + n_2$ trials is also binomial $\mathscr{B}(n_1 + n_2, p)$.

A.4.2 Geometric and negative binomial distributions

Consider a sequence $X_1, X_2, \ldots$ of i.i.d. Bernoulli trials and let X be the number of trials until the first success (including the last successful trial). Since the common probability of success in a single trial is p, due to independency we have

$$p_i = P(X = i) = (1-p)^{i-1} p, \; i = 1, 2, \ldots \tag{A.12}$$

The probabilities p_i in (A.12) decrease geometrically as i increases and such distribution is called the *geometric Geom(p)* [2].

The cdf $F(\cdot)$ of geometric distribution $Geom(p)$ is

$$F(x) = \sum_{j=1}^{x}(1-p)^{j-1}p = p\frac{1-(1-p)p^x}{1-(1-p)} = 1-(1-p)^x, \ \ x = 1,2,\ldots.$$

It is easy to check that

$$\mathrm{E}X = \frac{1}{p}$$

$$\mathrm{Var}(X) = \frac{1-p}{p}.$$

The interesting property of the geometric distribution is its *memoryless*: the probability that we will have to wait for another x trials for the first success does not depend on how many trials t we have waited so far, that is, $P(X > x+t \mid X > t) = P(X > x)$ for any t. Indeed,

$$P(X > x+t \mid X > t) = \frac{P(X > x+t, X > t)}{P(X > t)} = \frac{P(X > x+t)}{P(X > t)} = \frac{1-F(x+t)}{1-F(t)} = (1-p)^x.$$

The *negative binomial* distribution $NB(r,p)$ is similar to the geometric, but we now count the number of Bernoulli trials until the r-th success. It can thus be presented as the sum of r independent geometric random variables $Geom(p)$ and we have

$$P(X = i) = \binom{i-1}{r-1}(1-p)^{i-r}p^r, \ \ i = r, r+1, \ldots$$

$$\mathrm{E}X = \frac{r}{p}$$

$$\mathrm{Var}(X) = \frac{r(1-p)}{p}.$$

A.4.3   *Hypergeometric distribution*

Suppose there is a list of people composed of d registered Democrats and r registered Republicans. A sample of n names are drawn at random from the list. Let X be the number of Democrats in the sample. What is the distribution of X? As asked, this question cannot be answered since it was not stated how the random sample was taken.

For better illustration, imagine the situation where we have a deck with $r+d$ cards. Each card has one of the names and his affiliation on it—d cards are marked as "Democrats" and r as "Republicans." We shuffle the deck, draw one card, check its affiliation, return it to the deck and repeat the process again n times. This is known as sampling with replacement. In this case we are essentially talking about a series of n independent Bernoulli trials with a probability of picking a Democrat $\frac{d}{d+r}$ and, therefore, $X \sim \mathscr{B}\left(n, \frac{d}{d+r}\right)$. Note however that in such sampling scheme the same card (person) may be randomly selected more than once, and we may not want this to happen.

Alternatively, we can pick a random card, record its affiliation and put it aside. We then select another card out of the $r+d-1$ remaining cards, put it aside as well, and continue doing this n times. This is sampling without replacement. Affiliation of a chosen card can still be viewed as an outcome of a Bernoulli trial $\mathscr{B}\left(1, \frac{d}{d+r}\right)$ but the trials are now not independent. What is the distribution of X in the this case? First note that logically $\max\{0, n-r\} \le X \le \min\{n, d\}$. Furthermore, there are

---

[2]Sometimes, the geometric distribution is defined as the distribution of the number of failures until the first success X'. Obviously, $X' = X-1$ and $P(X' = i) = (1-p)^i p$, $i = 0, 1, \ldots$.

$\binom{d}{i}\binom{r}{n-i}$ ways to obtain exactly i Democrats and $n-i$ Republicans in sampling n cards without replacement (the order is evidently not important). The total number of all possible samples of size n without replacement is $\binom{d+r}{n}$. Thus,

$$p_i = P(X = i) = \frac{\binom{d}{i}\binom{r}{n-i}}{\binom{d+r}{n}}.$$

This is known as the *hypergeometric* distribution $\mathscr{H}(d,r,n)$.

To compute its mean and variance we do not need to use the above formula for the p_i's. As we have discussed above, $X = \sum_{i=1}^{n} X_i$, where $X_i \sim \mathscr{B}\left(1, \frac{d}{d+r}\right)$. Hence,

$$\mathrm{E}X = \sum_{i=1}^{n} \mathrm{E}X_i = n\,\frac{d}{d+r}$$

exactly as for sampling with replacement.

However, the dependency between the X_i's matters for calculating the variance. Consider the first two successive outcomes X_1 and X_2. It is easy to derive their covariance. By (A.3),

$$
\begin{aligned}
\mathrm{cov}(X_1,X_2) &= \mathrm{E}(X_1 X_2) - \mathrm{E}(X_1)\mathrm{E}(X_2) \\
&= P(\text{both } X_1 \text{ and } X_2 \text{ are Democrats}) - P(X_1 \text{ is Democrat}) \cdot P(X_2 \text{ is Democrat}) \\
&= P(X_1 \text{ is Democrat}) \cdot P(X_2 \text{ is Democrat} \mid X_1 \text{ is Democrat}) \\
&\qquad - P(X_1 \text{ is Democrat}) \cdot P(X_2 \text{ is Democrat}) \\
&= \frac{d}{d+r}\frac{d-1}{d+r-1} - \left(\frac{d}{d+r}\right)^2 = -\frac{dr}{(d+r)^2(d+r-1)}.
\end{aligned}
$$

By the parallel description of the sampling, which is totally equivalent, it is intuitively clear that all variances $\mathrm{Var}(X_i)$ and covariances $\mathrm{cov}(X_i, X_j)$ are the same (recall that we do not always present the rigorous arguments in this chapter) and hence by (A.5):

$$
\begin{aligned}
\mathrm{Var}(X) &= n\,\mathrm{Var}(X_1) + n(n-1)\,\mathrm{cov}(X_1,X_2) = \frac{ndr}{(d+r)^2} - n(n-1)\frac{dr}{(d+r)^2(d+r-1)} \\
&= \frac{ndr}{(d+r)^2}\frac{d+r-n}{d+r-1}.
\end{aligned}
$$

If the sample size n is much smaller than the population size $d+r$, the last term is close to one, and the variance of the hypergeometric distribution $\mathscr{H}(d,r,n)$ is close to that of the corresponding binomial distribution $\mathscr{B}\left(n, \frac{d}{d+r}\right)$. It is intuitively clear in this case that the difference between sampling with and without replacement is hardly noticeable. However, if the sample size is large relatively to the population size, the former variance is much smaller. In particular, in the extreme case, where the sample is the entire population, $\mathrm{Var}(X) = 0$! In fact, it is not surprising since in this case we know exactly that $X = d$.

A.4.4   Poisson distribution

The *Poisson* distribution, denoted by $\mathrm{Pois}(\lambda)$, is a discrete distribution on the non-negative integers. A random variable is called Poisson if for some $\lambda > 0$

$$p_i = P(X = i) = e^{-\lambda}\frac{\lambda^i}{i!}, \qquad i = 0, 1, \dots.$$

Poisson random variables are used to describe rare events, for example, the number of independent calls to an emergency center between midnight and 1am of a particular day, the number of flight cancelations, the annual number of earthquakes in a region, etc.

In fact, the Poisson distribution can be viewed as an approximation to the binomial $\mathscr{B}(n,p)$ distribution, when n is large (e.g., the population of a city) and p is small (the probability that a citizen will have a stroke at a given date and hour), while np is moderate (e.g., the mean number of strokes in the city in a given hour is, say, 1.3). We prove that now. Suppose we consider a sequence of binomial random variables $X_n \sim \mathscr{B}(n, p_n)$, where $p_n \to 0$ but $np_n \to \lambda$. Recall that $\lim_{n \to \infty}(1 + a_n)^n = \exp(\lim_{n \to \infty} na_n)$. Hence,

$$
\begin{aligned}
P(X_n = i) &= \binom{n}{i} p_n^i (1 - p_n)^{n-i} = \frac{n(n-1)\ldots(n-i+1)}{i!} p_n^i (1 - p_n)^{n-i} \\
&= \frac{1}{i!} \frac{n}{n(1-p_n)} \frac{(n-1)}{n(1-p_n)} \cdots \frac{(n-i+1)}{n(1-p_n)} (np_n)^i (1-p_n)^n \\
&\to \frac{1}{i!} \lambda^i e^{-\lambda}.
\end{aligned}
$$

The mean and variance of the Poisson random variables can be obtained then as the limits of $EX_n = np_n$ and $\text{Var} X_n = np_n(1 - p_n)$:

$$
EX = \lambda
$$
$$
\text{Var} X = \lambda.
$$

Similarly to the binomial, the Poisson distribution is closed to addition: if $X_1 \sim Pois(\lambda_1)$, $X_2 \sim Pois(\lambda_2)$ and they are independent, then $X_1 + X_2 \sim Pois(\lambda_1 + \lambda_2)$.

It can also be shown that if the number of events per time unit follows the Poisson distribution $Pois(\lambda)$, the times between events' arrivals are independent and identically distributed exponential random variables $\exp(\lambda)$ (see Section A.4.6).

A.4.5 Uniform distribution

A (continuous) uniform distribution $\mathscr{U}(a,b)$ has a constant density on the entire interval (a,b), where

$$
f(x) = \begin{cases} \frac{1}{b-a}, & a \leq x \leq b \\ 0, & \text{otherwise} \end{cases}
$$
$$
F(x) = \begin{cases} 0, & x < a \\ \frac{x-a}{b-a}, & a \leq x \leq b \\ 1, & x > b \end{cases}
$$

Due to the symmetry of the density function, the expectation of the uniform random variable $X \sim \mathscr{U}(a,b)$ is the midpoint of the interval (a,b), i.e. $EX = \frac{a+b}{2}$, while a simple calculus yields $\text{Var}(X) = \frac{(b-a)^2}{12}$.

A.4.6 Exponential distribution

A positive random variable X has an exponential distribution $\exp(\theta)$ with a parameter θ if its density and cdf are

$$
f(x) = \theta e^{-\theta x}, \quad x > 0
$$
$$
F(x) = 1 - e^{-\theta x}.
$$

Simple integration by parts yields

$$
EX = \theta^{-1}
$$
$$
\text{Var} X = \theta^{-2}.
$$

The exponential distribution is the continuous counterpart of the geometric distribution in being memoryless:

$$P(X > x+t | X > t) = \frac{P(X > x+t)}{P(X > t)} = \frac{1 - F(x+t)}{1 - F(t)} = e^{-\theta x} = 1 - F(x)$$

for any $t > 0$.

The exponential distribution is used to model the lifetime of an entity that does not wear out with age, and thus memoryless. This may be a good model for lifetimes of various devices and animals (those who mainly do not get to "old age"). For example, this would be the case if there are many independent components, each with a lifetime which has a non-zero density near 0, and the device fails when the first of these components stops functioning. Let $T_1, \ldots, T_n$ be the lifetimes of the components, then $X = \min\{T_1, \ldots, T_n\}$, and by (A.8) for small t (on the scale of the lifetime of an individual component) we have

$$P(X > t) = \prod_{i=1}^{n} P(X_i > t) \approx \prod_{i=1}^{n} (1 - f_i(0)t) \approx e^{-t\left(\sum_{i=1}^{n} f_i(0)\right)} = 1 - F(t),$$

where $F(\cdot)$ is the cdf of an exponential random variable $\exp(\sum_{i=1}^{n} f_i(0))$.

In addition, if the number of events occurring per time unit is $Pois(\lambda)$, the time between events' arrivals is $\exp(\lambda)$ (see also Section A.4.4).

The two closely related distributions to the exponential are the double exponential and the shifted exponential distributions. A *double exponential* (also called the *Laplace*) distribution has the density

$$f(x) = \frac{\theta}{2} e^{-\theta|x|}, \ x \in \mathbb{R}, \ \theta > 0$$

Evidently, if X has a double exponential distribution with parameter θ, then $|X| \sim \exp(\theta)$. The mean $EX = 0$ and the variance is

$$\mathrm{Var}\, X = \int x^2 \frac{\theta}{2} e^{-\theta|x|} dx = \int_0^\infty x^2 \theta e^{-\theta x} dx = \frac{2}{\theta^2}.$$

The density of a *shifted exponential* distributed random variable X is

$$f(x) = \theta e^{-\theta(x-\tau)}, \ x \geq \tau, \ \theta > 0$$

Hence, $Y = X - \tau \sim \exp(\theta)$ immediately implying $EX = \tau + EY = \tau + \frac{1}{\theta}$ and $\mathrm{Var}(X) = \mathrm{Var}(Y) = \frac{1}{\theta^2}$.

A.4.7 Weibull distribution

Let $Y \sim \exp(\theta)$ and $Y = X^\alpha$, where $\alpha > 0$. The cdf of X is then given by

$$F_X(x) = 1 - P(X > x) = 1 - P(X^\alpha > x^\alpha) = 1 - P(Y > x^\alpha) = 1 - e^{-\lambda x^\alpha}.$$

Taking the derivatives of both sides implies

$$f(x) = \alpha \lambda x^{\alpha-1} e^{-\lambda x^\alpha}$$

which is known as the *Weibull* distribution $\mathcal{W}(\theta, \alpha)$. The Weibull distribution plays a key role in survival analysis.

A.4.8   Gamma-distribution

A *Gamma*-distribution $Gamma(\alpha,\beta)$ with a *shape* parameter $\alpha > 0$ and a *scale* parameter $\beta > 0$ has the density

$$f(x) = \frac{\beta^\alpha}{\Gamma(\alpha)} x^{\alpha-1} e^{-\beta x}, \quad x > 0,$$

where $\Gamma(\alpha) = \int_0^\infty t^{\alpha-1} e^{-t} dt$ is the so-called Gamma function. The Gamma function has the remarkable property $\Gamma(\alpha) = (\alpha-1)\Gamma(\alpha-1)$. Thus, for a natural α, $\Gamma(\alpha) = (\alpha-1)!$. In particular, $\exp(\theta) = Gamma(1,\theta)$.

The mean and the variance of $Gamma(\alpha,\beta)$ are

$$EX = \frac{\alpha}{\beta},$$

$$Var(X) = \frac{\alpha}{\beta^2}.$$

An important property of the Gamma-distribution is that the sum of n independent Gamma-variables $Gamma(\alpha_i,\beta)$, $i = 1,...,n$ with the same scale parameter β also has a Gamma-distribution $Gamma(\sum_{i=1}^n \alpha_i, \beta)$ with the same scale parameter.

A.4.9   Beta-distribution

A Beta-distribution $Beta(\alpha,\beta)$ with parameters $\alpha > 0$ and $\beta > 0$ is a continuous distribution on the unit interval $(0,1)$ with the density

$$f(x) = \frac{\Gamma(\alpha+\beta)}{\Gamma(\alpha)\Gamma(\beta)} x^{\alpha-1}(1-x)^{\beta-1}, \quad 0 < x < 1$$

Its moments are given by

$$EX = \frac{\alpha}{\alpha+\beta}$$

$$Var X = \frac{\alpha\beta}{(\alpha+\beta)^2(\alpha+\beta+1)}.$$

If $U \sim Gamma(\alpha,\theta)$ and $V \sim Gamma(\beta,\theta)$ are two independent Gamma-variables with the same scale parameter θ, then the ratio $U/(U+V) \sim Beta(\alpha,\beta)$.

A.4.10   Cauchy distribution

The Cauchy distribution $Cauchy(\theta)$ is the primary example of a distribution with a density but no mean or variance. Its density is given by [3]

$$f(x) = \frac{1}{\pi(1+(x-\theta)^2)}, \quad x \in \mathbb{R}$$

It is easy to see that the Cauchy distribution has an undefined expected value and variance, but $Med(X) = \theta$.

The Cauchy distribution can be generated by a "real situation." Let Z_1 and Z_2 be two i.i.d. normals $\mathcal{N}(0,\sigma^2)$. Then, their ratio $X = Z_1/Z_2 \sim Cauchy(0)$. Indeed, clearly $X < 0$ with probability 0.5. Let $x > 0$. Then, $X < x$ if X is negative, $0 < Z_1 < xZ_2$, or $0 > Z_1 > xZ_2$. These events are mutually exclusive. Hence

---

[3] It is possible to consider a more general Cauchy distribution with an additional scale parameter σ: $f(x) = \frac{\sigma}{\pi(\sigma^2+(x-\theta)^2)}$.

$$P(X < x) = P(Z_1 \le xZ_2) = 0.5 + 2 \int_0^\infty \int_0^{xz_2} \frac{1}{\sqrt{2\pi}} e^{-z_1^2/2} \frac{1}{\sqrt{2\pi}} e^{-z_2^2/2} dz_1 dz_2.$$

Taking derivatives of both sides:

$$
\begin{aligned}
f(x) &= \frac{1}{\pi} \int_0^\infty z_2 e^{-x^2 z_2^2/2 - z_2^2/2} dz_2 = \frac{1}{\pi} \int_0^\infty z_2 e^{-(1+x^2)z_2^2/2} dz_2 = \frac{1}{\pi} \frac{1}{1+x^2} \int_0^\infty z e^{-z^2/2} dz \\
&= \frac{1}{\pi} \frac{1}{1+x^2}.
\end{aligned}
$$

which is the density of *Cauchy*(0).

A.4.11 Normal distribution

The *normal* distribution (sometimes also called *Gaussian*) is undoubtedly the most well-known and important. The density of the normal distribution $\mathcal{N}(\mu, \sigma^2)$ is

$$f(x) = \frac{1}{\sqrt{2\pi}\sigma} e^{-\frac{(x-\mu)^2}{2\sigma^2}}, \quad x \in \mathbb{R}$$

where $EX = \mu$ and $\mathrm{Var}(X) = \sigma^2$.

If $X \sim \mathcal{N}(\mu, \sigma^2)$, then the standardized random variable $Z = \frac{X-\mu}{\sigma} \sim \mathcal{N}(0, 1)$ (see (A.7)), which is called the *standard* normal distribution. The cdf $F_X(\cdot)$ of a normal distribution $\mathcal{N}(\mu, \sigma^2)$ does not have a closed form but can be presented in the terms of the cdf $\Phi(\cdot)$ of a standard normal distribution:

$$F_X(x) = P(X \le x) = P\left(\frac{X-\mu}{\sigma} \le \frac{x-\mu}{\sigma}\right) = \Phi\left(\frac{x-\mu}{\sigma}\right)$$

The normal distribution plays a fundamental role in probability and statistics. One of the main reasons is the celebrated *central limit theorem* (CLT) which, under mild conditions, implies asymptotic normal approximations for a wide variety of distributions in large sample setups. See Section 5.5.

A.4.12 Log-normal distribution

A strictly positive random variable X has a log-normal distribution if $\ln X \sim \mathcal{N}(\mu, \sigma^2)$. From (A.6) the density $f_X(\cdot)$ is given by

$$f_X(x) = \frac{1}{x\sqrt{2\pi}\sigma} e^{-\frac{(\ln x - \mu)^2}{2\sigma^2}}, \quad x \ge 0.$$

Since $X = e^{\mu + \sigma Z}$, where $Z \sim \mathcal{N}(0, 1)$, we have

$$EX = Ee^{\mu + \sigma Z} = \int e^{\mu + \sigma z} \frac{1}{\sqrt{2\pi}} e^{-z^2/2} = e^{\mu + \sigma^2/2} \int \frac{1}{\sqrt{2\pi}} e^{-(z-\sigma)^2/2} = e^{\mu + \sigma^2/2},$$

since the integrand is the density of $\mathcal{N}(\sigma, 1)$.

Similarly, $EX^2 = e^{2\mu + 2\sigma^2}$ and, therefore,

$$\mathrm{Var}(X) = EX^2 - (EX)^2 = e^{2\mu + 2\sigma^2} - e^{2\mu + \sigma^2} = e^{2\mu + \sigma^2}\left(e^{\sigma^2} - 1\right).$$

A.4.13 χ^2 distribution

Let Z_1 be standard normal random variable $\mathcal{N}(0,1)$ and consider $X = Z^2$. From (A.6), the density f_X is

$$f_1(x) = \frac{1}{\sqrt{2\pi x}}\, e^{-x/2}.$$

This is the density for $Gamma(0.5, 0.5)$. It is also known as the χ^2-distribution with one degree of freedom χ_1^2.

More generally, let $Z_1, ..., Z_n$ be i.i.d. standard normals $\mathcal{N}(0,1)$ and consider $X_n = \sum_{i=1}^n Z_i^2$. Since each $Z_i^2 \sim Gamma(0.5, 0.5)$ and the Gamma-distribution is closed to the sum of independent variables with the same scale parameter (see Section A.4.8), $X_n \sim Gamma(\frac{n}{2}, 0.5)$ with the density

$$f_{X_n}(x) = \frac{x^{n/2-1}}{2^{n/2}\Gamma(n/2)}\, e^{-x/2}$$

which is known as χ^2-distribution χ_n^2 with n degrees of freedom. Its mean and variance can be immediately obtained from the general formulae for the Gamma-distribution:

$$EX_n = n$$
$$\mathrm{Var}(X_n) = 2n.$$

Note that χ_2^2 is the exponential distribution $\exp(0.5)$.

A.4.14   t-distribution

Let $Z \sim N(0,1)$ and $S \sim \chi_n^2$. Assume also that Z and S are independent. Then, the random variable

$$T = \frac{Z}{\sqrt{S/n}} \sim t_n,$$

where t_n is the Student's t-distribution (or simply t-distribution) with n degrees of freedom with the density

$$f_{T_n}(t) = \frac{\Gamma(\frac{n+1}{2})}{\Gamma(\frac{n}{2})\sqrt{\pi n}} \left(1 + \frac{t^2}{n}\right)^{-\frac{n+1}{2}}, \quad t \in \mathbb{R}$$

In particular, $t_1 = Cauchy(0)$. Due to symmetry, $Et_n = 0$ for $n > 1$, while $\mathrm{Var}(t_n) = \frac{n}{n-2}$ for $n > 2$.

The t-distribution naturally appears in statistics for normal samples. For example, let $X_1, ..., X_n$ be i.i.d. $N(\mu, \sigma^2)$. Then, $\bar{X} \sim N(\mu, \sigma^2/n)$ and is independent of $(n-1)s^2 = \sum_{i=1}^n (X_i - \bar{X})^2 \sim \chi_{n-1}^2$ (see Example 1.7). Hence,

$$T = \frac{\sqrt{n}(\bar{X} - \mu)}{s} \sim t_{n-1}$$

A.4.15   F-distribution

Let S_1 and S_2 be two independent χ^2 random variables with n_1 and n_2 degrees of freedom respectively. Then, their normalized ratio

$$F = \frac{S_1/n_1}{S_2/n_2} \sim F_{n_1,n_2},$$

where F_{n_1,n_2} is the F-distribution with n_1 and n_2 degrees of freedom. In particular, $F_{1,n} = t_n^2$. From the definition it is clear that if $F \sim F_{n_1,n_2}$, then $F^{-1} \sim F_{n_2,n_1}$.

It can be shown that

$$\mathrm{E}\,F_{n_1,n_2} = \frac{n_2}{n_2-2}, \quad n_2 > 2$$

$$\mathrm{Var}(F_{n_1,n_2}) = \frac{2n_2^2(n_1+n_2-2)}{n_1(n_2-2)^2)(n_2-4)}, \quad n_2 > 4$$

A.4.16   Multinormal distribution

A.4.16.1   Definition and main properties

Definition A.1 (multinormal distribution) An n-dimensional random vector X has an n-dimensional multinormal distribution $\mathcal{N}_n(\mu, \Sigma)$ if its joint density is given by

$$f_X(x) = \frac{1}{(2\pi)^{n/2}|\Sigma^{1/2}|} e^{-(x-\mu)^\top \Sigma^{-1}(x-\mu)/2}, \tag{A.13}$$

where $\mu \in \mathbb{R}^n$ and $\Sigma \in \mathbb{R}^{n \times n}$ is a self-adjoint positive definite matrix. See Appendix A.3.9 for discussion of random vectors, their moments and linear transformations. In particular, $\mathrm{E}X = \mu$ and $\mathrm{Var}(X) = \Sigma$.

A multivariate normal vector Z has the *standard* multivariate distribution if $\mu = 0$ and $\Sigma = I_n$. In this case

$$f_Z(z) = \prod_{i=1}^p \frac{1}{\sqrt{2\pi}} e^{-z_i^2/2} = \frac{1}{(2\pi)^{n/2}} e^{-\Sigma_{i=1}^p z_i^2/2} = \frac{1}{(2\pi)^{n/2}} e^{-\|z\|^2/2}. \tag{A.14}$$

Proposition A.1 Let Σ be a self-adjoint positive definite matrix. Then, $X \sim \mathcal{N}_n(\mu, \Sigma)$ iff $Z = \Sigma^{-1/2}(X-\mu) \sim \mathcal{N}_n(0, I_n)$.

The proof is an immediate consequence of (A.11).

Random normal vectors have a series of important properties:

Theorem A.1 Let $X \sim \mathcal{N}_n(\mu, \Sigma)$ and split it into two subvectors $X_1 \in \mathbb{R}^{n_1}$ and $X_2 \in \mathbb{R}^{n_2}$, where $n_1 + n_2 = n$. Consider the corresponding splits of the mean vector μ and the variance matrix Σ:

$$X = \begin{pmatrix} X_1 \\ X_2 \end{pmatrix}, \quad \mu = \begin{pmatrix} \mu_1 \\ \mu_2 \end{pmatrix}, \quad \Sigma = \begin{pmatrix} \Sigma_{11} & \Sigma_{12} \\ \Sigma_{21} & \Sigma_{22} \end{pmatrix}.$$

Then,

1. If X_1 and X_2 are uncorrelated ($\Sigma_{12} = 0_{n_1 \times n_2}$), then they are independent.
2. $BX \sim \mathcal{N}(B\mu, B\Sigma B^\top)$ for any $B \in \mathbb{R}^{p \times n}$, $p \le n$, of a full rank.
3. X_1 and X_2 are also normal, where $X_i \sim \mathcal{N}_{n_i}(\mu_i, \Sigma_{ii})$, $i = 1, 2$.

Proof. 1. Let $\Sigma_{12} = 0_{n_1 \times n_2}$. Then,

$$\det(\Sigma) = \det \begin{pmatrix} \Sigma_{11} & 0 \\ 0 & \Sigma_{22} \end{pmatrix} = \det(\Sigma_{11})\det(\Sigma_{22}),$$

$$\Sigma^{-1} = \begin{pmatrix} \Sigma_{11}^{-1} & 0 \\ 0 & \Sigma_{22}^{-1} \end{pmatrix}$$

and

$$(x-\mu)^\top \Sigma^{-1}(x-\mu) = (x_1-\mu_1)^\top \Sigma_{11}^{-1}(x_1-\mu_1) + (x_2-\mu_2)^\top \Sigma_{22}^{-1}(x_2-\mu_2).$$

Hence,

$$f_{\mathbb{X}}(x) = \frac{1}{(2\pi)^{n/2}|\Sigma^{1/2}|} e^{-(x-\mu)^{\top}\Sigma^{-1}(x-\mu)/2}$$

$$= \frac{1}{(2\pi)^{n_1/2}|\Sigma_{11}^{1/2}|} e^{-(x_1-\mu_1)^{\top}\Sigma_{11}^{-1}(x_1-\mu_1)/2} \frac{1}{(2\pi)^{n_2/2}|\Sigma_{22}^{1/2}|} e^{-(x_2-\mu_2)^{\top}\Sigma_{22}^{-1}(x_2-\mu_2)/2}$$

which is the product of the two multivariate marginal normal densities of $X_1 \sim \mathcal{N}(\mu_1, \Sigma_{11})$ and $X_2 \sim \mathcal{N}(\mu_2, \Sigma_{22})$ yielding their independence.

2. If $p = n$, then $BX = B(\mu + \Sigma^{1/2}Z) = B\mu + B\Sigma^{1/2}Z$, where $Z \sim \mathcal{N}_n(0, I_n)$, and the result follows from Proposition A.1.

  If $p < n$, we can always add any $n - p$ orthogonal rows to B that are also orthogonal to its existing p rows, and consider the resulting matrix $\tilde{B} \in \mathbb{R}^{n \times n}$, where $\tilde{B}$ is invertible and $\tilde{B}\tilde{B}^{\top}$ is block diagonal. Part 1 implies then that $BX \sim \mathcal{N}(B\mu, B\Sigma B^{\top})$.

3. Apply Part 2 for $B = (I_{n_1}, 0_{n_1 \times n_2})$.

□

A.4.16.2 *Projections of normal vectors*

In this section we discuss some properties of projections of random normal vectors on subspaces of $\mathbb{R}^n$. Recall that if $\mathcal{S}$ is a subspace of $\mathbb{R}^n$, then the matrix P is an (orthogonal) *projection* onto $\mathcal{S}$ if for every vector $x \in \mathbb{R}^n$, $Px \in \mathcal{S}$ and $(I - P)x \perp \mathcal{S}$. Note that $x = Px + (I - P)x$, and this is the unique representation of x as a sum of two vectors, one in $\mathcal{S}$, and one orthogonal to it.

  Some basic properties of P:

1. If $x \in \mathcal{S}$, then $Px = x$.

2. Hence, for any $x \in \mathbb{R}^n$, $P^2 x = P(Px) = Px$. That is, $P^2 = P$ and P is an idempotent matrix.

3. Since for any $x, y \in \mathbb{R}^n$: $x^{\top}P^{\top}y = (Px)^{\top}y = (Px)^{\top}(Py + (I - P)y) = (Px)^{\top}(Py) = x^{\top}(Py) = x^{\top}Py$, we have $P = P^{\top}$ and P is a self-adjoint matrix.

On the other hand, any self-adjoint idempotent matrix P is a projection matrix, where the corresponding subspace $\mathcal{S}$ is the span of its columns. Note that if P projects on $\mathcal{S}$, then $(I - P)$ projects on its orthogonal complement $\mathcal{S}^c = \{x : x^{\top}y = 0 \text{ for all } y \in \mathcal{S}\}$.

  It is easy to check that for any matrix $A \in \mathbb{R}^{n \times p}$, $p \leq n$ of a full rank (i.e., its columns are linearly independent), the matrix $P = A(A^{\top}A)^{-1}A^{\top}$ is self-adjoint and idempotent, and hence it is a projection matrix.

  There is the following theorem on projections of normal vectors:

Theorem A.4.1 *Let $Z \sim \mathcal{N}_n(0, I_n)$ and $P_1 \in \mathbb{R}^{n_1 \times n}$, $P_2 \in \mathbb{R}^{n_2 \times n}$ are projections on two orthogonal subspaces of $\mathbb{R}^n$ of dimensionalities n_1 and n_2 respectively. Then,*

1. $P_1 Z$ and $P_2 Z$ are independent.

2. $\mathrm{E}\|P_i Z\|^2 = n_i$, $i = 1, 2$.

Proof. Let $B_i \in \mathbb{R}^{n \times n_2}$ be a matrix with orthonormal columns that span the subspace on which P_i projects. Note that $P_i = B_i B_i^{\top}$. To show that $P_1 Z$ is independent of $P_2 Z$ is enough to verify that $B_1^{\top}Z$ is independent of $B_2^{\top}Z$. The latter follows from Part 1 of Theorem A.1 for a matrix $B = \begin{pmatrix} B_1^{\top} \\ B_2^{\top} \\ B_3^{\top} \end{pmatrix}$, where the columns of B_3 are orthogonal to the columns of B_1 and B_2.

  To prove the second claim note that if the vectors $a_1, \ldots, a_k$ form an orthonormal basis for the columns of any projection P on a k-dimensional subspace of $\mathbb{R}^n$, then $P = \sum_{j=1}^{k} a_j a_j^{\top}$. Note that for any vector $x \in \mathbb{R}^n$, $\|Px\|^2 = \|\sum_{j=1}^{k}(a_j^{\top}x)a_j\|^2 = \sum_{j=1}^{k}(a_j^{\top}x)^2$, since $(a_1^{\top}x), \ldots, (a_k^{\top}x)$ are the coefficients of x in the basis $a_1, \ldots, a_k$. By the first claim, $a_1^{\top}Z, \ldots, a_k^{\top}Z$ are independent. Moreover, $a_j^{\top}Z \sim \mathcal{N}(0, 1)$ and hence $\mathrm{E}\|PZ\|^2 = \mathrm{E}\|\sum_{j=1}^{k} a_j a_j^{\top}Z\|^2 = \mathrm{E}\sum_{j=i}^{1} k(a_j^{\top}Z) = k$.

□

Appendix B

Solutions of Selected Exercises

B.1 Chapter 1

Exercise 1.1

1. $T(\mathbf{Y}) = \sum_{i=1}^{n} Y_i \sim Pois(n\lambda)$ and by straightforward calculus similar to that for Bernoulli data in Example 1.5 one has

$$P(\mathbf{Y} = \mathbf{y} | \sum_{i=1}^{n} Y_i = t) = \begin{cases} \frac{P(\mathbf{Y=y})}{P(\sum_{i=1}^{n} Y_i=t)} & \text{if } \sum_{i=1}^{n} y_i = t \\ 0 & \text{if } \sum_{i=1}^{n} y_i \neq t \end{cases} = \begin{cases} \frac{t!}{n^t \prod_{i=1}^{n} y_i!} & \text{if } \sum_{i=1}^{n} y_i = t \\ 0 & \text{if } \sum_{i=1}^{n} y_i \neq t \end{cases}$$

that does not depend on λ. Hence, $T(\mathbf{Y}) = \sum_{i=1}^{n} Y_i$ is a sufficient statistic.

2. Consider the likelihood ratio for any two random samples $\mathbf{Y}_1$ and $\mathbf{Y}_2$ from $Pois(\lambda)$:

$$\frac{L(\lambda; \mathbf{y}_1)}{L(\lambda; \mathbf{y}_2)} = \lambda^{\sum_{i=1}^{n} y_{1i} - \sum_{i=1}^{n} y_{2i}} \frac{\prod_{i=1}^{n} y_{1i}!}{\prod_{i=1}^{n} y_{2i}!}$$

It does not depend on λ iff $\sum_{i=1}^{n} y_{1i} = \sum_{i=1}^{n} y_{2i}$. Thus, according to Theorem 1.2, $T(\mathbf{Y}) = \sum_{i=1}^{n} Y_i$ is a minimal sufficient statistic for λ.

3. (a) Consider an arbitrary function $g(T)$ of $T(\mathbf{Y}) \sim Pois(n\lambda)$. We have

$$Eg(T) = e^{-n\lambda} \sum_{k=0}^{\infty} g(k) \frac{(n\lambda)^k}{k!} = e^{-n\lambda} \sum_{k=0}^{\infty} g_k \frac{\lambda^k}{k!} \ ,$$

where $g_k = g(k)n^k$ and are essentially coefficients of the Taylor expansion of the function $\psi(\lambda) = e^{n\lambda} Eg(T)$ around zero. $Eg(T) = 0$ iff $\psi(\lambda) = 0$, while the latter is zero for all $\lambda > 0$ iff all its Taylor coefficients $g_k = 0$. It obviously implies that $g(k) = 0$, $k = 0, 1, ...$ and, hence, $g(T) \equiv 0$ and $T(\mathbf{Y})$ is complete.

   (b) One can verify (see also Exercise 1.3) that the Poisson distribution $Pois(\lambda)$ belongs to the one-dimensional exponential family and the completeness of $T(\mathbf{Y})$ follows immediately from Theorem 1.5.

Exercise 1.5 Consider any two samples $\mathbf{Y}_1$ and $\mathbf{Y}_2$ of size n from $\mathcal{N}(\mu, a^2\mu^2)$ and the corresponding likelihood ratio

$$\begin{aligned} \frac{L(\mu; \mathbf{y}_1)}{L(\mu; \mathbf{y}_2)} &= \exp\left\{ -\frac{\sum_{i=1}^{n} (y_{1i} - \mu)^2 - \sum_{i=1}^{n} (y_{2i} - \mu)^2}{2a^2\mu^2} \right\} \\ &= \exp\left\{ -\frac{\sum_{i=1}^{n} y_{1i}^2 - \sum_{i=1}^{n} y_{2i}^2 - 2\mu(\sum_{i=1}^{n} y_{1i} - \sum_{i=1}^{n} y_{2i})}{2a^2\mu^2} \right\}. \end{aligned}$$

The likelihood ratio does not depend on μ iff $\sum_{i=1}^{n} y_{1i} = \sum_{i=1}^{n} y_{2i}$ and $\sum_{i=1}^{n} y_{1i}^2 = \sum_{i=1}^{n} y_{2i}^2$ and, therefore, by Theorem 1.2, $(\sum_{i=1}^{n} Y_i, \sum_{i=1}^{n} Y_i^2)$ is a minimal sufficient statistic for μ.

DOI: 10.1201/9781003175407-B

B.2   Chapter 2

Exercise 2.4

1. Let $Y \sim \mathcal{N}(\mu,5^2)$ be an (unobserved) candidate's IQ and X be the number of candidates that successfully pass the test. Evidently, $X \sim \mathcal{B}(100,p)$, where $p = P(Y > 120) = 1 - \Phi\left(\frac{120-\mu}{5}\right)$ or $\mu = 120 - 5z_p$. We already know (see Example 2.1) that the MLE of p is $\hat{p} = \frac{X}{n} = 0.1$. Thus, $\hat{\mu} = 120 - 5z_{\hat{p}} = 113.6$.

2. Let $p_0 = P(Y > 115) = 1 - \Phi(\frac{115-\mu}{5})$ be the chance for the randomly chosen graduate to successfully pass the test with the reduced threshold for the IQ. Then, the MLE $\hat{p}_0 = 1 - \Phi(\frac{115-\hat{\mu}}{5}) = 0.39$.

Exercise 2.5 Note that

$$P(Y = k) = \begin{cases} p + (1-p)e^{-\lambda}, & k = 0 \\ (1-p)e^{-\lambda}\frac{\lambda^k}{k!}, & k > 0 \end{cases} \tag{B.1}$$

1. Let $n_0 = \#\{i : Y_i = 0\}$ be the number of days with zero counts. Then,

$$L(\lambda,p) = \left(p + (1-p)e^{-\lambda}\right)^{n_0} e^{-(n-n_0)}\frac{\lambda^{\Sigma_{i:Y_i>0}Y_i}}{\prod_{i:Y_i>0}Y_i!} = \left(p + (1-p)e^{-\lambda}\right)^{n_0} e^{-(n-n_0)}\frac{\lambda^{\Sigma_{i=1}^n Y_i}}{\prod_{i=1}^n Y_i!} \tag{B.2}$$

and, therefore, the sufficient statistic for (λ,p) is $(\Sigma_{i=1}^n Y_i, n_0)$.

2. From (B.2) it follows that MLE for λ and p do not exist in the closed form.

3. From (B.1) we have $\mu_1 = EY = (1-p)\lambda$ and $\mu_2 = EY^2 = (1-p)(\lambda+\lambda^2) = (1-p)\lambda(\lambda+1)$. Solving two simultaneous equations $(1-p)\lambda = m_1$ and $(1-p)\lambda(\lambda+1) = m_2$ implies

$$\hat{\lambda}_{MME} = \frac{m_2}{m_1} - 1 = \frac{\Sigma_{i=1}^n Y_i^2}{\Sigma_{i=1}^n Y_i} - 1, \quad \hat{p}_{MME} = 1 - \frac{m_1^2}{m_2 - m_1} = 1 - \frac{(\Sigma_{i=1}^n Y_i)^2}{n(\Sigma_{i=1}^n Y_i^2 - \Sigma_{i=1}^n Y_i)}$$

Exercise 2.6 Consider the log-likelihood $l(\theta) = n\ln\theta - \theta\Sigma_{i=1}^n Y_i$ and maximize it w.r.t. θ but subject to the additional constraint $\theta \le \theta_0$. Recall that without the constraint, the MLE $\hat{\theta} = \frac{1}{\bar{Y}}$ (see Example 2.4). Hence, if $\frac{1}{\bar{Y}} \le \theta_0$, it is also the constrained MLE. Otherwise, the solution is evidently achieved at the boundary point $\hat{\theta} = \theta_0$. Combining both cases implies $\hat{\theta}_{MLE} = \min\left(\frac{1}{\bar{Y}}, \theta_0\right)$.

Exercise 2.7 Let p_1 and p_2 be the (unknown) percentages of support for Candidate A in the population among men and women respectively. Thus, $X_1,\ldots,X_n \sim \mathcal{B}(1,p_1)$ and $Y_1,\ldots,Y_m \sim \mathcal{B}(1,p_2)$, where $\Sigma_{i=1}^n X_i = X$, $\Sigma_{j=1}^m Y_j = Y$ and, in addition, it is known that $p_1 \le p_2$. The log-likelihood is

$$l(p_1,p_2) = X\ln p_1 + (n-X)\ln(1-p_1) + Y\ln p_2 + (m-Y)\ln(1-p_2)$$

and we want to maximize it w.r.t. p_1 and p_2 subject to the constraint $p_1 \le p_2$. The unconstrained solutions are evidently $\hat{p}_1 = \frac{X}{n}$ and $\hat{p}_2 = \frac{Y}{m}$. If they satisfy the constraint, i.e. if $\frac{X}{n} \le \frac{Y}{m}$, they are the required MLEs for p_1 and p_2. Otherwise, it is easy to verify that the solution for MLEs is achieved at the boundary point, where $\hat{p}_1 = \hat{p}_2 = \frac{X+Y}{n+m}$ (there is no difference in support for Candidate A between genders). Thus, summarizing, $\hat{p}_1 = \frac{X}{n}$ and $\hat{p}_2 = \frac{Y}{m}$ if $\frac{X}{n} \le \frac{Y}{m}$ and $\hat{p}_1 = \hat{p}_2 = \frac{X+Y}{n+m}$ otherwise.

Exercise 2.8 The first statement can be trivially verified. We have $Var(\alpha T_1 + (1-\alpha)T_2) = \alpha^2 Var(T_1) + (1-\alpha)^2 Var(T_2)$, which is minimized for $\hat{\alpha} = \frac{Var(T_2)}{Var(T_1)+Var(T_2)}$. The resulting estimator $\hat{\alpha}T_1 + (1-\hat{\alpha})T_2$ is the minimal variance unbiased estimator within the family of linear combinations of T_1 and T_2 but there is no guarantee that it is an UMVUE among all possible estimators.

Exercise 2.10 Suppose that T_1 and T_2 are two different UMVUEs of θ, where $ET_1 = ET_2 = \theta$ and $Var(T_1) = Var(T_2) = \sigma^2$, and consider $T = (T_1 + T_2)/2$. Obviously, $ET = \theta$ and T is unbiased. Furthermore,

$$
\begin{aligned}
Var(T) &= \frac{1}{4}Var(T_1 + T_2) = \frac{1}{4}\left(Var(T_1) + Var(T_2) + 2Cov(T_1, T_2)\right) \\
&\leq \frac{1}{4}\left(Var(T_1) + Var(T_2) + 2\sqrt{Var(T_1)Var(T_2)}\right) = \sigma^2 \quad (B.3)
\end{aligned}
$$

On the other hand, T_1 and T_2 are UMVUE so $Var(T) \geq \sigma^2$. The equality in (B.3) holds iff $corr(T_1, T_2) = 1$ and, thus, $T_2 = aT_1 + b$ for some a and b. However, due to unbiasedness, $ET_1 = ET_2 = \theta$ for all θ, which necessarily implies $a = 1$ and $b = 0$ and, therefore, $T_1 = T_2 = T$.

Exercise 2.12

1. By standard probabilistic calculus, the density of $\hat{\theta}_{MLE} = Y_{\max}$ is $f(u) = \frac{n}{\theta}\left(\frac{u}{\theta}\right)^{n-1}$, $0 \leq u \leq \theta$, $E\hat{\theta}_{MLE} = \frac{n}{n+1}\theta$ and $Var(\hat{\theta}_{MLE}) = \frac{n}{(n+1)^2(n+2)}\theta^2$. Thus,

$$
MSE(\hat{\theta}_{MLE}, \theta) = \left(\frac{1}{(n+1)^2} + \frac{n}{(n+1)^2(n+2)}\right)\theta^2 = \frac{2}{(n+1)(n+2)}\theta^2,
$$

   while

$$
MSE(\hat{\theta}_{MME}, \theta) = Var(2\bar{Y}) = \frac{\theta^2}{3n}.
$$

   One can easily check that $MSE(\hat{\theta}_{MLE}, \theta) = MSE(\hat{\theta}_{MME}, \theta)$ for $n = 1, 2$ and $MSE(\hat{\theta}_{MLE}, \theta) < MSE(\hat{\theta}_{MME}, \theta)$ for $n \geq 3$.

2. From the previous paragraph, $E\hat{\theta}_{MLE} = \frac{n}{n+1}\theta \neq \theta$ but then $\hat{\theta}_U = \frac{n+1}{n}\hat{\theta}_{MLE}$ is an unbiased estimator of θ.

3. $\hat{\theta}_{MLE}$ and $\hat{\theta}_U$ are of the form $aY_{\max}$ for $a = 1$ and $a = \frac{n+1}{n}$ respectively. Consider a general estimator of this form: $E(aY_{\max}) = aEY_{\max} = a\frac{n}{n+1}\theta$ and $Var(aY_{\max}) = a^2Var(Y_{\max}) = a^2\frac{n}{(n+1)^2(n+2)}\theta^2$. Thus,

$$
MSE(aY_{\max}, \theta) = \left(\frac{(an - n - 1)^2(n+2) + a^2 n}{(n+1)^2(n+2)}\right)\theta^2
$$

   which is minimized for $\hat{a} = \frac{n+2}{n+1}$.

Exercise 2.17

1. Since X and Y are i.i.d. $\mathcal{N}(0, \sigma^2)$, X^2 and Y^2 are i.i.d. $\sigma^2\chi_1^2$ and, therefore, $R^2 = X^2 + Y^2 \sim \sigma^2\chi_2^2$. The distribution $f_\sigma(r)$ of R can be easily obtained then by standard probabilistic arguments from Section A.3.8. The distribution $f_\sigma(r)$ evidently belongs to the (one-parameter) exponential family with $T(r) = r^2$ and the natural parameter $c(\sigma) = -\frac{1}{2\sigma^2}$.

2. By straightforward calculus $\hat{\sigma}_{MLE}^2 = \frac{\sum_{i=1}^n R_i^2}{2n}$.

3. $ER_i^2 = \sigma^2 E\chi_2^2 = 2\sigma^2$ yielding $E\hat{\sigma}_{MLE}^2 = \sigma^2$. To show that it is UMVUE recall that $f_\sigma(r)$ belongs to the (one-parameter) exponential family with $T(y) = r^2$ and the natural parameter $c(\sigma) = -\frac{1}{2\sigma^2}$. The unbiased estimator $\hat{\sigma}_{MLE}^2$ is a function of the sufficient statistic $T(\mathbf{R}) = \sum_{i=1}^n T(R_i) = \sum_{i=1}^n R_i^2$ and half-heuristic arguments of Section 2.7.4 ($dim(\sigma) = dim(c(\sigma)) = 1$) imply that it is UMVUE. More rigorously, applying Theorem 1.5, one can show completeness of $T(\mathbf{R})$ and then use Lehmann-Scheffé theorem.

Exercise 2.18

1. Let $X_1, \ldots, X_n$ be the number of daily calls made by female clients and $Y_1, \ldots, Y_m$ the number of daily calls made by male clients. Then, X_i's are i.i.d. $Pois(a\lambda)$, Y_j's are i.i.d. $Pois(\lambda)$ and the joint distribution of the data

$$f_\lambda(\mathbf{x}, \mathbf{y}) = e^{-(an+m)\lambda} \lambda^{\sum_{i=1}^{n} x_i + \sum_{j=1}^{m} y_j} \frac{a^{\sum_{i=1}^{n} x_i}}{\prod_{i=1}^{n} x_i! \prod_{j=1}^{m} y_j!}$$

One can easily check that $f_\lambda(\mathbf{x}, \mathbf{y})$ belongs to the one-parameter exponential family of distributions and $\sum_{i=1}^{n} X_i + \sum_{j=1}^{m} Y_j$ is the sufficient statistic for λ.

2. The likelihood $L(\lambda; \mathbf{x}, \mathbf{y}) = f_\lambda(\mathbf{x}, \mathbf{y})$. By straightforward calculus, the MLEs for the average numbers of daily calls are $\hat{\lambda} = \frac{\sum_{i=1}^{n} X_i + \sum_{j=1}^{m} Y_j}{an+m}$ for males and $\widehat{a\lambda} = a\hat{\lambda} = a\frac{\sum_{i=1}^{n} X_i + \sum_{j=1}^{m} Y_j}{an+m}$ for females, respectively.

3. One can immediately verify that the MLEs $\hat{\lambda}$ and $\widehat{a\lambda}$ from the previous paragraph are unbiased estimators for λ and $a\lambda$.

Exercise 2.19

1. By direct calculus, $\hat{\beta}_{MLE} = \frac{\sum_{i=1}^{n} x_i Y_i}{\sum_{i=1}^{n} x_i^2}$. Note that $EY_i = \beta x_i$ and $Var(Y_i) = \sigma^2$. Thus, $E\hat{\beta}_{MLE} = \frac{\sum_{i=1}^{n} x_i EY_i}{\sum_{i=1}^{n} x_i^2} = \beta$ (unbiasedness) and $MSE(\hat{\beta}_{MLE}, \beta) = Var(\hat{\beta}_{MLE}) = \frac{\sum_{i=1}^{n} x_i^2 Var(Y_i)}{(\sum_{i=1}^{n} x_i^2)^2} = \frac{\sigma^2}{\sum_{i=1}^{n} x_i^2}$.

2. $E\tilde{\beta} = \frac{\sum_{i=1}^{n} EY_i}{\sum_{i=1}^{n} x_i} = \beta$ (unbiased estimator) and, therefore, $MSE(\tilde{\beta}, \beta) = Var(\tilde{\beta}) = \frac{\sum_{i=1}^{n} Var(Y_i)}{(\sum_{i=1}^{n} x_i)^2} = \frac{n\sigma^2}{(\sum_{i=1}^{n} x_i)^2} > MSE(\hat{\beta}_{MLE}, \beta)$.

3. $\hat{\sigma}_{MLE}^2 = \frac{\sum_{i=1}^{n} (Y_i - \hat{\beta}_{MLE} x_i)^2}{n}$.

4. $\widehat{\beta/\sigma}_{MLE} = \hat{\beta}_{MLE} / \hat{\sigma}_{MLE}$.

5.

$$I(\beta, \sigma) = \begin{pmatrix} \frac{\sum_{i=1}^{n} x_i^2}{\sigma^2} & 0 \\ 0 & \frac{2n}{\sigma^2} \end{pmatrix}.$$

6. From Theorem 2.4, the Cramer–Rao lower bound for an unbiased estimator of β/σ is

$$\left(\frac{1}{\sigma}, -\frac{\beta}{\sigma^2} \right) I(\beta, \sigma) \begin{pmatrix} \frac{1}{\sigma} \\ -\frac{\beta}{\sigma^2} \end{pmatrix} = \frac{1}{\sum_{i=1}^{n} x_i^2} + \frac{\beta^2}{2n\sigma^2}.$$

Exercise 2.21

1. Evidently, $Y_i \sim \mathcal{B}(k, \xi)$, $i = 1, \ldots, n$, where ξ is the probability of a defective component. The likelihood is

$$L(\xi; \mathbf{y}) = \prod_{i=1}^{n} \binom{k}{y_i} \xi^{y_i} (1 - \xi)^{k - y_i}$$

and $\hat{\xi}_{MLE} = \frac{\sum_{i=1}^{n} Y_i}{nk}$. On the other hand, $p = P(Y = 0) = (1 - \xi)^k$ and, therefore, $\hat{p}_{MLE} = (1 - \hat{\xi}_{MLE})^k = \left(1 - \frac{\sum_{i=1}^{n} Y_i}{nk} \right)^k$.

2. The sufficient statistic $W = \sum_{i=1}^{n} Y_i \sim \mathcal{B}(nk, \xi)$. Since p is the probability of a certain event (a batch contains no defectives), the indicator random variable $T = 1$ if $Y_1 = 0$ and zero otherwise is an unbiased estimator for p. Performing "Rao–Blackwellization," define $T_1 = E(T|W)$. Since

$$T|W \sim \mathscr{B}\left(1, P(Y_1 = 0 | \textstyle\sum_{i=1}^{n} Y_i = W)\right),$$

$$
\begin{aligned}
T_1 &= P(Y_1 = 0 | \sum_{i=1}^{n} Y_i = W) = \frac{P(Y_1 = 0)P(\sum_{i=2}^{n} Y_i = W)}{P(\sum_{i=1}^{n} Y_i = W)} \\
&= \frac{(1-\xi)^k \binom{k(n-1)}{W} \xi^W (1-\xi)^{k(n-1)}}{\binom{nk}{W} \xi^W (1-\xi)^{n-W}} = \frac{\binom{k(n-1)}{W}}{\binom{nk}{W}} = \prod_{j=0}^{k-1} \left(1 - \frac{\sum_{i=1}^{n} Y_i}{n-j}\right).
\end{aligned}
$$

Exercise 2.22 Consider $\psi(\theta; y) = \rho'(\theta; y) = -\alpha \mathbb{I}\{y > \theta\} + (1-\alpha)\mathbb{I}\{y \le \theta\} = -\alpha + \mathbb{I}\{y \le \theta\}$. Solving

$$\sum_{i=1}^{n} \psi(\theta, Y_i) = -n\alpha + \#\{Y_i \le \theta\} = 0$$

yields $\frac{\#\{Y_i \le \hat{\theta}\}}{n} = \alpha$, i.e. $\hat{\theta} = Y_{(\alpha)}$.

B.3   Chapter 3

Exercise 3.2

1. Since Y is a continuous random variable, its cdf $F(\cdot)$ is monotonically increasing and, therefore, there exists a monotonically increasing inverse function $F^{-1}(\cdot)$. Thus, for any $0 \le a \le 1$ we have

$$P(Z < a) = P(F(Y) < a) = P(Y < F^{-1}(a)) = F(F^{-1}(a)) = a$$

and $Z \sim \mathscr{U}(0,1)$.

2. From the results of the previous paragraph, $F_\theta(Y_i) \sim \mathscr{U}(0,1)$ and applying standard probabilistic calculus for the distribution of a function of a random variable, $-\ln F_\theta(Y_i) \sim \exp(1) = \frac{1}{2}\chi_2^2$. Hence, $\Psi(\mathbf{Y}; \theta) = -\sum_{i=1}^{n} \ln F_\theta(Y_i) \sim \frac{1}{2}\chi_{2n}^2$ and is a pivot.

Exercise 3.3

1. We derived the density $f(\cdot)$ of $Y_{\max}$ in Exercise 2.12: $f(u) = \frac{n}{\theta}\left(\frac{u}{\theta}\right)^{n-1}$, $0 \le u \le \theta$. The density $f^*(\cdot)$ of the scaled variable $X = Y_{\max}/\theta$ is, therefore, $f^*(x) = nx^{n-1}$, $0 \le x \le 1$, and X is a pivot.

2. From the results of the previous paragraph, the cdf $F^*(\cdot)$ of X is $F^*(x) = x^n$, $0 \le x \le 1$ and by simple calculus,

$$1 - \alpha = P(\alpha^{\frac{1}{n}} \le X \le 1) = P(Y_{\max} \le \theta \le \alpha^{-\frac{1}{n}} Y_{\max}).$$

Exercise 3.4 The sample mean $\bar{Y} \sim \mathscr{N}\left(\mu, \frac{a\mu^2}{n}\right)$ and, therefore, the standardized variable $Z = \frac{\bar{Y} - \mu}{a\mu/\sqrt{n}} \sim \mathscr{N}(0,1)$ is a pivot. Hence,

$$1 - \alpha = P(-z_{\alpha/2} \le Z \le z_{\alpha/2}) = P\left(\frac{\bar{Y}}{1 + z_{\alpha/2}a/\sqrt{n}} \le \mu \le \frac{\bar{Y}}{1 - z_{\alpha/2}a/\sqrt{n}}\right)$$

if $z_{\alpha/2}a/\sqrt{n} \le 1$ and

$$1 - \alpha = P(-z_{\alpha/2} \le Z \le z_{\alpha/2}) = P\left(\frac{\bar{Y}}{1 - z_{\alpha/2}a/\sqrt{n}} \le \mu \le \frac{\bar{Y}}{1 + z_{\alpha/2}a/\sqrt{n}}\right)$$

otherwise.

Exercise 3.6

1. Let $\mu = EY$ be the average reaction time for cats. The average reaction time for dogs then is $EX = \rho\mu$. Thus, $X_1, ..., X_n \sim \exp(1/(\rho\mu))$ and $Y_1, ..., Y_m \sim \exp(1/\mu)$. In terms of μ the likelihood

$$L(\mu;,\mathbf{y}) = \rho^{-n}\mu^{-(n+m)}e^{-(\rho\sum_{i=1}^n X_i + \sum_{j=1}^m Y_j)/\mu}$$

and by direct calculus, the MLEs $\hat{\mu} = \frac{\rho\sum_{i=1}^n X_i + \sum_{j=1}^m Y_j}{n+m}$ and $\widehat{\rho\mu} = \rho\hat{\mu} = \rho\frac{\rho\sum_{i=1}^n X_i + \sum_{j=1}^m Y_j}{n+m}$.

2. $Y_j \sim \exp(1/\mu)$ and, therefore, $Y_j \sim \mu\exp(1) = \frac{\mu}{2}\chi_2^2$ and $\frac{2}{\mu}\sum_{j=1}^m Y_j \sim \chi_{2n}^2$. Similarly, $\frac{2}{\rho\mu}\sum_{i=1}^n X_i \sim \chi_{2m}^2$. Furthermore, $\frac{2}{\rho\mu}\sum_{i=1}^n X_i$ and $\frac{2}{\mu}\sum_{j=1}^m Y_j$ are independent and, therefore, $\frac{2(\sum_{i=1}^n X_i + \rho\sum_{j=1}^m Y_j)}{\rho\mu} \sim \chi_{2(n+m)}^2$ is a pivot. One then has

$$1 - \alpha = P\left(\chi_{2(n+m),1-\alpha/2}^2 \leq \frac{2(\sum_{i=1}^n X_i + \rho\sum_{j=1}^m Y_j)}{\rho\mu} \leq \chi_{2(n+m),\alpha/2}^2\right)$$

$$= P\left(\frac{2(\sum_{i=1}^n X_i + \rho\sum_{j=1}^m Y_j)}{\rho\chi_{2(n+m),\alpha/2}^2} \leq \mu \leq \frac{2(\sum_{i=1}^n X_i + \rho\sum_{j=1}^m Y_j)}{\rho\chi_{2(n+m),1-\alpha/2}^2}\right)$$

and

$$1 - \alpha = P\left(\frac{2(\sum_{i=1}^n X_i + \rho\sum_{j=1}^m Y_j)}{\chi_{2(n+m),\alpha/2}^2} \leq \rho\mu \leq \frac{2(\sum_{i=1}^n X_i + \rho\sum_{j=1}^m Y_j)}{\chi_{2(n+m),1-\alpha/2}^2}\right).$$

3. Assume now that $\rho = EX/EY$ is unknown. The MLEs for average reaction times for dogs and cats are $\widehat{EX} = \bar{X}$ and $\widehat{EY} = \bar{Y}$ (see Example 2.4), and the MLE for ρ is $\hat{\rho} = \widehat{EX}/\widehat{EY} = \bar{X}/\bar{Y}$. Note that $\frac{2}{EX}\sum_{i=1}^n X_i \sim \chi_{2n}^2$, $\frac{2}{EY}\sum_{j=1}^m Y_j \sim \chi_{2m}^2$ and are independent. Then, $\rho\frac{\bar{Y}}{\bar{X}} = \rho/\hat{\rho} \sim F_{2m,2n}$ (see Appendix A.4.15) is a pivot and

$$1 - \alpha = P\left(F_{2m,2n;1-\alpha/2} \leq \rho/\hat{\rho} \leq F_{2m,2n;\alpha/2}\right) = P\left(\hat{\rho}F_{2m,2n;1-\alpha/2} \leq \rho \leq \hat{\rho}F_{2m,2n;\alpha/2}\right).$$

B.4 Chapter 4

Exercise 4.1

1. Let μ be the average waiting time for food in "FastBurger"'s branches. Mr. Skeptic tests $H_0 : \mu = 50$ (the advertisement is not true) *vs.* $H_1 : \mu = 30$ (the advertisement is true) and rejects the null hypothesis if the waiting time $T < 40$. Then, $\alpha = P_{\mu=50}(T < 40) = 1 - e^{-40/50} = 0.551$, $\beta = P_{\mu=30}(T > 40) = e^{-40/30} = 0.264$.

2. The likelihood ratio $\lambda(t) = \frac{L(30,t)}{L(50,t)} = \frac{50}{30}e^{-(1/30-1/50)t}$ is a decreasing function of t and the test that rejects the null hypothesis for small t is the LRT. According to the Neymann–Pearson lemma, it is the MP test that cannot be improved (in terms of minimizing β) for a given α.

3. Let C be the critical waiting time of Mrs. Skeptic. The corresponding $\alpha_1 = P_{\mu=50}(T < C) = 1 - e^{-C/50}$ and $\beta_1 = P_{\mu=30}(T > C) = e^{-C/30}$ and their sum is minimized when $C = \frac{\ln(50/30)}{1/30 - 1/50} = 38.3$.

4. $\alpha_1 = 1 - e^{-38.3/50} = 0.535 < \alpha$, $\beta_1 = e^{-38.3/30} = 0.279 > \beta$ but $\alpha_1 + \beta_1 = 0.814 < \alpha + \beta = 0.815$.

Exercise 4.4 Let $F_{\theta_0}(\cdot)$ be the cdf of the test statistic $T(\mathbf{Y})$ under $H_0 : \theta = \theta_0$. Then, $F_{\theta_0}(T) \sim \mathcal{U}(0,1)$ (see Exercise 3.2) and $p - \text{value} = 1 - F_{\theta_0}(T) \sim \mathcal{U}(0,1)$.

Exercise 4.5

1. The likelihood ratio is

$$\lambda(\mathbf{y}) = \frac{L(\theta_1; \mathbf{y})}{L(\theta_0; \mathbf{y})} = \begin{cases} \left(\frac{\theta_0}{\theta_1}\right)^n, & Y_{\max} < \theta_0 \\ \infty, & \theta_0 < Y_{\max} < \theta_1 \end{cases}$$

and the MP test (LRT) will reject the null iff $Y_{\max} > C$, where

$$\alpha = P_{\theta_0}(Y_{\max} > C) = 1 - \left(\frac{C}{\theta_0}\right)^n,$$

and, hence, $C = (1 - \alpha)^{1/n} \theta_0$.

2. The power is

$$\pi = P_{\theta_1}\left(Y_{\max} > (1 - \alpha)^{1/n} \theta_0\right) = 1 - \left(\frac{(1 - \alpha)^{1/n} \theta_0}{\theta_1}\right)^n$$

$$= 1 - (1 - \alpha)\left(\frac{\theta_0}{\theta_1}\right)^n.$$

Exercise 4.7

1.

$$\lambda(\mathbf{y}) = \frac{L(\theta_1; \mathbf{y})}{L(\theta_0; \mathbf{y})} = \left(\frac{\theta_0}{\theta_1}\right)^n e^{-\frac{\sum_{i=1}^n y_i^2}{2}\left(\frac{1}{\theta_1} - \frac{1}{\theta_0}\right)}$$

which is an increasing function of $\sum_{i=1}^n y_i^2$. The MP test (LRT) will reject the null hypothesis if $\sum_{i=1}^n y_i^2 \geq C$, where C satisfies $\alpha = P_{\theta_0}(\sum_{i=1}^n Y_i^2 \geq C)$ and, therefore, $C = \theta_0 \chi_{2n,\alpha}^2$.

2. The MP test of the previous paragraph does not depend on a particular $\theta_1 > \theta_0$ and, therefore, is UMP for testing a composite one-sided alternative $H_1 : \theta > \theta_0$. Its power function is

$$\pi(\theta) = P_\theta\left(\sum_{i=1}^n Y_i^2 > \theta_0 \chi_{2n,\alpha}^2\right) = 1 - X_{2n}^2\left(\frac{\theta_0}{\theta}\chi_{2n,\alpha}^2\right),$$

where X_{2n}^2 is the cdf of χ_{2n}^2.

Exercise 4.14

1. The uniform distribution $\mathscr{U}(0,1)$ corresponds to $\theta = 1$.

2. We want to test $H_0 : \theta = 1$ (the uniform distribution $\mathscr{U}(0,1)$) vs. $H_1 : \theta \neq 1$. The likelihood $L(\theta, \mathbf{y}) = \theta^n(\prod_{i=1}^n y_i)^{\theta-1}$ and the MLE $\hat{\theta} = -\frac{n}{\sum_{i=1}^n \ln Y_i}$. The GLR is then

$$\lambda(\mathbf{y}) = \frac{L(\hat{\theta}, \mathbf{y})}{L(1, \mathbf{y})} = \left(-\frac{n}{\sum_{i=1}^n \ln Y_i}\right)^n e^{-\sum_{i=1}^n \ln Y_i \ (n/\sum_{i=1}^n \ln Y_i + 1)}$$

$$= \left(\frac{n}{e}\right)^n \left(-\sum_{i=1}^n \ln Y_i\right)^{-n} e^{-\sum_{i=1}^n \ln Y_i}.$$

The GLRT rejects the null hypothesis iff $\lambda(\mathbf{y}) \geq C$ or, equivalently, in terms of the statistic $T(\mathbf{y}) = -2\sum_{i=1}^n \ln Y_i \overset{H_0}{\sim} \chi_{2n}^2$, iff $T(\mathbf{y}) \leq C_1$ or $T(\mathbf{y}) \geq C_2$, where

$$X_{2n}^2(C_2) - X_{2n}^2(C_1) = 1 - \alpha, \quad C_2 - C_1 = 2n\ln\left(\frac{C_2}{C_1}\right)$$

and X_{2n}^2 is the cdf of the χ_{2n}^2 distribution. There are no closed solutions for C_1 and C_2 but, similar to Example 4.19, they can be approximated by the commonly used quantiles $\chi_{2n;1-\alpha/2}^2$ and $\chi_{2n;\alpha/2}^2$, and the resulting GLRT will not reject the null hypothesis of the uniform distribution iff

$$\chi_{2n;1-\alpha/2}^2 \leq -2\sum_{i=1}^{n} \ln Y_i \leq \chi_{2n;\alpha/2}^2.$$

Exercise 4.16 For a given t the log-likelihood ratio is

$$\ln \lambda_t(\mathbf{Y}) = \ln\left(\frac{p_1}{1-p_1}\frac{1-p_0}{p_0}\right)\sum_{i=1}^{t} Y_i + \ln\frac{1-p_1}{1-p_0}\, t.$$

Thus, the sequential Wald test accepts H_0 if

$$\sum_{i=1}^{t} Y_i < \frac{a}{\ln\left(\frac{p_1}{1-p_1}\frac{1-p_0}{p_0}\right)} + \frac{\ln\frac{1-p_0}{1-p_1}}{\ln\left(\frac{p_1}{1-p_1}\frac{1-p_0}{p_0}\right)}\, t,$$

rejects it if

$$\sum_{i=1}^{t} Y_i > \frac{b}{\ln\left(\frac{p_1}{1-p_1}\frac{1-p_0}{p_0}\right)} + \frac{\ln\frac{1-p_0}{1-p_1}}{\ln\left(\frac{p_1}{1-p_1}\frac{1-p_0}{p_0}\right)}\, t$$

and samples another observation otherwise, where $a = \ln\frac{\beta}{1-\alpha}$ and $b = \ln\frac{1-\beta}{\alpha}$.

Exercise 4.17

1. The p-value for testing the j-th null hypothesis is $p_j = 2(1 - \Phi(|Y_j|/\sigma))$. Following Bonferroni we reject H_{0j} if $p_j \leq \alpha/m$ or, equivalently, if $|Y_j| \geq \sigma z_{\alpha/(2m)}$.

2. Order the p-values $p_{(1)} \leq \ldots \leq p_{(m)}$. The step-down Holm procedure looks for $j^* = \min\{j : p_{(j)} > \frac{\alpha}{m-j+1}\}$ or, equivalently, $j^* = \min\{j : |Y|_{(j)} < \sigma z_{\alpha/(2(m-j+1))}\}$, where $|Y|_{(1)} \geq \ldots \geq |Y|_{(m)}$. It rejects all $j^* - 1$ hypotheses corresponding to the largest gene expression levels (in absolute values) and accept others. Similarly, for the step-up Hochberg procedure $j^* = \max\{j : |Y|_{(j)} > \sigma z_{\alpha/(2(m-j+1))}\}$ and j^* null hypotheses with the largest $|Y|_{(j)}$'s are rejected. Both procedures control FWER at level α.

3. One can easily see that all nulls rejected by Bonferroni will be rejected by Holm, while all nulls rejected by Holm will necessarily be rejected by Hochberg. Hence, Hochberg is the most powerful FWER procedure among those three.

4. Following the Benjamini–Hochberg FDR-controlling procedure, we find $j^* = \max\{j : |Y|_{(j)} \geq \sigma z_{j\alpha/(2m)}\}$ and reject j^* hypotheses with the largest $|Y|_{(j)}$'s. It controls FDR at level α.

5. Bonferroni and Holm do not require the independence assumption and control FWER at level α even if it is violated. Hochberg and Benjamini–Hochberg rely on independence of test statistics and do not guarantee the control of respectively FWER and FDR at the required level α if it does not hold.

B.5   Chapter 5

Exercise 5.1

1. Define $X_i = Y_i^j$. Then, by the WLLN, $m_j = \bar{X} \xrightarrow{p} EX = \mu_j$.

2. $Var(X) = EX^2 - (EX)^2 = EY^{2j} - (EY^j)^2 = \mu_{2j} - \mu_j^2$ and by the CLT, $\sqrt{n}(m_j - \mu_j) = \sqrt{n}(\bar{X} - EX) \xrightarrow{\mathscr{D}} \mathscr{N}(0, \mu_{2j} - \mu_j^2)$.

Exercise 5.4 By the CLT, $\sqrt{n}(\bar{Y}_n - \mu) \xrightarrow{\mathscr{D}} \mathscr{N}(0, \sigma^2)$. Consider $g_1(\mu) = \mu^2$. For $\mu \neq 0$, $g_1'(\mu) = 2\mu \neq 0$ and from Theorem 5.5, $\sqrt{n}(\bar{Y}_n^2 - \mu^2) \xrightarrow{\mathscr{D}} \mathscr{N}(0, 4\mu^2\sigma^2)$. For $\mu = 0$, $\sqrt{n}\bar{Y}_n \xrightarrow{\mathscr{D}} \mathscr{N}(0, \sigma^2)$ and, therefore, $\frac{n}{\sigma^2}\bar{Y}_n^2 \xrightarrow{\mathscr{D}} \chi_1^2$.

Similarly, define $g_2(\mu) = e^{-\mu}$. $g'(\mu) = -e^{-\mu} \neq 0$ for all μ and by Theorem 5.5, $\sqrt{n}(e^{-\bar{Y}_n} - e^{-\mu}) \xrightarrow{\mathscr{D}} \mathscr{N}(0, e^{-2\mu}\sigma^2)$.

Exercise 5.5 We can proceed in (at least) two ways.

1. The asymptotic $100(1-\alpha)\%$ confidence interval for p is $\hat{p}_n \pm z_{\alpha/2}\sqrt{\frac{\hat{p}_n(1-\hat{p}_n)}{n}}$ (see Example 5.12) and since $\rho(p) = \frac{p}{1-p}$ is an increasing function of $p \in (0,1)$, the corresponding asymptotic $100(1-\alpha)\%$ confidence interval for ρ is

$$\left(\frac{\hat{p}_n - z_{\alpha/2}\sqrt{\frac{\hat{p}_n(1-\hat{p}_n)}{n}}}{1 - \hat{p}_n + z_{\alpha/2}\sqrt{\frac{\hat{p}_n(1-\hat{p}_n)}{n}}}, \frac{\hat{p}_n + z_{\alpha/2}\sqrt{\frac{\hat{p}_n(1-\hat{p}_n)}{n}}}{1 - \hat{p}_n - z_{\alpha/2}\sqrt{\frac{\hat{p}_n(1-\hat{p}_n)}{n}}} \right).$$

2. Let $\hat{\rho}_n = \frac{\hat{p}_n}{1-\hat{p}_n}$ be the MLE of ρ. By the CLT, $\sqrt{n}(\hat{p}_n - p) \xrightarrow{\mathscr{D}} \mathscr{N}(0, p(1-p))$ and applying Theorem 5.5, $\sqrt{n}(\hat{\rho}_n - \rho) \xrightarrow{\mathscr{D}} \mathscr{N}(0, \frac{p(1-p)}{(1-p)^4}) = \mathscr{N}(0, \frac{p}{(1-p)^3})$. Substituting the consistent estimator $\hat{p}_n$ in the asymptotic variance instead of the unknown p (see Theorem 5.8) implies the following asymptotic $100(1-\alpha)\%$ confidence interval for ρ:

$$\frac{\hat{p}_n}{1 - \hat{p}_n} \pm z_{\alpha/2}\sqrt{\frac{\hat{p}_n}{(1-\hat{p}_n)^3 n}}.$$

Exercise 5.9

1. $\hat{\tau}_n = Y_{\min}$

2. Show that $\hat{\tau}_n$ has a shifted exponential distribution with the density $f_{n,\tau}(u) = ne^{-n(u-\tau)}$, $u \geq \tau$. Hence, $E\hat{\tau}_n = \frac{1}{n} + \tau$, $Var(\hat{\tau}_n) = \frac{1}{n^2}$ and $MSE(\hat{\tau}_n, \tau) = \frac{2}{n^2} \to 0$ as n increases.

3. It is easy to verify that $\hat{\tau}_n \sim f_{n,\tau}$ yields $n(\hat{\tau}_n - \tau) \sim \exp(1)$.

4. We have $\sqrt{n}(\hat{\tau}_n - \tau) = \frac{1}{\sqrt{n}}n(\hat{\tau}_n - \tau) \sim \frac{1}{\sqrt{n}}\exp(1) \xrightarrow{\mathscr{D}} 0$ as n increases.

5. The asymptotic normality of the MLE $\hat{\tau}_n$ does not hold since one of the regularity conditions of Theorem 5.9 is violated in this case: the support $\text{supp}f_\tau(y) = \{y : y \geq \tau\}$ depends on the unknown parameter τ.

Exercise 5.10

1. In Exercise 3.4 we used the fact that $\bar{Y}_n \sim \mathscr{N}(\mu, \frac{a^2\mu^2}{n})$ to construct an exact $100(1-\alpha)\%$ confidence interval for μ. Now, since $\bar{Y}$ is a consistent estimator of μ, we can substitute it instead of μ in the variance and get an *asymptotic* $100(1-\alpha)\%$ confidence interval for μ:

$$\bar{Y}_n \pm z_{\alpha/2}\frac{a\bar{Y}_n}{\sqrt{n}} = \bar{Y}_n\left(1 \pm z_{\alpha/2}\frac{a}{\sqrt{n}}\right). \tag{B.4}$$

2. A variance stabilizing transformation $g(\mu)$ satisfies $g'(\mu) = \frac{c}{a\mu}$ for a chosen constant c (see Section 5.7). For convenience, take $c = a$ so that $g(\mu) = \ln\mu$. From (5.12), an asymptotic $100(1-\alpha)\%$ confidence interval for $\ln\mu$ is $\ln\bar{Y}_n \pm z_{\alpha/2}\frac{a}{\sqrt{n}}$ and, thus, for μ itself is

$$\bar{Y}_n e^{\pm z_{\alpha/2}\frac{a}{\sqrt{n}}}. \tag{B.5}$$

3. From Exercise 3.4, the exact $100(1-\alpha)\%$ confidence interval for μ is

$$\left(\frac{\bar{Y}_n}{1+z_{\alpha/2}\frac{a}{\sqrt{n}}} \, , \, \frac{\bar{Y}_n}{1-z_{\alpha/2}\frac{a}{\sqrt{n}}} \right) \tag{B.6}$$

(we can assume that $z_{\alpha/2}a/\sqrt{n} \leq 1$ for sufficiently large n).
Using standard asymptotic expansions,

$$\frac{\bar{Y}_n}{1 \mp z_{\alpha/2}\frac{a}{\sqrt{n}}} = \bar{Y}_n \left(1 \pm z_{\alpha/2}\frac{a}{\sqrt{n}} + O(n^{-1}) \right),$$

$$\bar{Y}_n e^{\pm z_{\alpha/2}\frac{a}{\sqrt{n}}} = \bar{Y}_n \left(1 \pm z_{\alpha/2}\frac{a}{\sqrt{n}} + O(n^{-1}) \right)$$

and the three confidence intervals (B.4), (B.5), and (B.6) are asymptotically close.

Exercise 5.11 The log-likelihood for a random Poisson sample is

$$l(\lambda) = -n\lambda + \ln\lambda \sum_{i=1}^{n} Y_i - \sum_{i=1}^{n} \ln Y_i!$$

Hence, the score function $l'(\lambda) = -n + \frac{\sum_{i=1}^{n} Y_i}{\lambda}$ and the Fisher information number $I(\lambda) = -El''(\lambda) = E\left(\frac{\sum_{i=1}^{n} Y_i}{\lambda^2} \right) = \frac{n}{\lambda}$. In addition, the MLE $\hat{\lambda}_n = \bar{Y}_n$.

Thus, the α-level asymptotic Wald test will reject the null hypothesis $H_0 : \lambda = \lambda_0$ if

$$\frac{\sqrt{n}|\bar{Y}_n - \lambda_0|}{\sqrt{\bar{Y}_n}} > z_{\alpha/2}$$

or, equivalently,

$$\frac{n(\bar{Y}_n - \lambda_0)^2}{\bar{Y}_n} > \chi^2_{1;\alpha}$$

The score (Rao) test will reject the null if

$$\frac{|-n + \sum_{i=1}^{n} Y_i/\lambda_0|}{\sqrt{n/\lambda_0}} = \frac{\sqrt{n}|\bar{Y}_n - \lambda_0|}{\sqrt{\lambda_0}} > z_{\alpha/2}$$

Note that under H_0, $\lambda_0 \approx \bar{Y}_n$ and both tests are asymptotically similar.

Finally, the GLRT rejects the null if

$$2\left(l(\bar{Y}_n) - l(\lambda_0) \right) = 2n(\lambda_0 - \bar{Y}_n - \bar{Y}_n \ln(\lambda_0/\bar{Y}_n)) > \chi^2_{1;\alpha}$$

Taylor series expansion yields $\ln\left(\frac{\lambda_0}{\bar{Y}_n} \right) = \ln\left(1 + \frac{\lambda_0 - \bar{Y}_n}{\bar{Y}_n} \right) \approx \frac{\lambda_0 - \bar{Y}_n}{\bar{Y}_n} - \frac{(\lambda_0 - \bar{Y}_n)^2}{2\bar{Y}_n^2}$ and, therefore, the Wilks statistic $2(l(\bar{Y}_n) - l(\lambda_0)) \approx \frac{n(\bar{Y}_n - \lambda_0)^2}{\bar{Y}_n}$ which coincides with the above Wald statistic.

Exercise 5.14 Recall that $\psi(\theta, y) = \rho'(\theta, y) = -\alpha + \mathbb{I}\{y \leq \theta\}$ (see Exercise 2.22). From Theorem 5.13 it follows that under the regularity conditions, $Y_{(\alpha)}$ is a consistent estimator of θ^* satisfying $E_{\theta_0}\psi(\theta^*, Y) = 0$, i.e.

$$E_{\theta_0}\psi(\theta^*, Y) = -\alpha + E_{\theta_0}\mathbb{I}\{Y \leq \theta^*\} = -\alpha + F_{\theta_0}(\theta^*) = 0$$

Hence, $\theta^* = F_{\theta_0}^{-1}(\alpha)$.

Furthermore, note that

$$E_{\theta_0} \psi(\theta^*, Y)^2 = E_{\theta_0} (\alpha - \mathbb{I}\{Y \leq \theta^*\})^2 = \alpha^2 - 2\alpha^2 + \alpha = \alpha(1 - \alpha)$$

and

$$E_{\theta_0} \psi'(\theta^*, Y) = \left(E_{\theta_0} \psi(\theta, Y)\right)'_{\theta=\theta^*} = \left(-\alpha + F_{\theta_0}(\theta)\right)'_{\theta=\theta^*} = f_{\theta_0}(\theta^*)$$

Thus, assuming the required regularity conditions, Theorem 5.14 implies that $\sqrt{n}\left(Y_{(\alpha)} - F_{\theta_0}^{-1}(\alpha)\right)$

$$\xrightarrow{\mathscr{D}} \mathscr{N}\left(0, \frac{\alpha(1-\alpha)}{f_{\theta_0}\left(F_{\theta_0}^{-1}(\alpha)\right)^2}\right).$$

B.6   Chapter 6

Exercise 6.1 Calculate the sequential posterior distribution $\pi^*(\theta|y_1, y_2)$ and compare it with its counterpart $\pi(\theta|y_1, y_2)$ based on the entire data (y_1, y_2). We have

$$\pi(\theta|y_1) = \frac{f(y_1|\theta)\pi(\theta)}{f(y_1)}$$

and

$$\pi^*(\theta|y_1, y_2) = \frac{f(y_2|\theta)\pi(\theta|y_1)}{f(y_2)} = \frac{f(y_2|\theta)f(y_1|\theta)\pi(\theta)}{f(y_2)f(y_1)}. \tag{B.7}$$

Due to the independency of y_1 and y_2, (B.7) implies

$$\pi^*(\theta|y_1, y_2) = \frac{f(y_1, y_2|\theta)}{f(y_1, y_2)} = \pi(\theta|y_1, y_2).$$

Exercise 6.4

1. Let $t_i|\theta \sim \mathscr{U}(0, \theta)$, $i = 1, ..., n$ be the waiting times for a bus ($n = 5$). Then,

$$\pi(\theta|t) \propto L(\theta; t)\pi(\theta) \propto \theta^{-n}\mathbb{I}\{\theta \geq y_{\max}\} \, \theta^{-\alpha-1}\mathbb{I}\{\theta \geq \beta\} = \theta^{-(n+\alpha)-1}\mathbb{I}\{\theta \geq \max(y_{\max}, \beta)\}$$

and, therefore, $\theta|t \sim Pa(n + \alpha, \max(y_{\max}, \beta)) = Pa(12, 14)$ and the Pareto prior is conjugate for uniform data.

2. For the squared error, $\hat{\theta} = E(\theta|t) = \frac{12 \cdot 14}{11} = 15.27$.

3. A 95% HPD for θ is of the form $(14, C)$, where C satisfies $P(14 \leq \theta \leq C|t) = 1 - (14/C)^{12} = 0.95$ implying $C = \frac{14}{0.05^{1/12}} = 17.97$.

4.

$$P(H_0|t) = P(0 \leq \theta \leq 15|t) = P(14 \leq \theta \leq 15|t) = 1 - \left(\frac{14}{15}\right)^{12} = 0.56 > 0.5$$

and H_0 is not rejected.

Exercise 6.5

1.

$$\pi(p|\mathbf{y}) \propto p^n(1-p)^{\sum_{i=1}^n (y_i-1)} p^{\alpha-1}(1-p)^{\beta-1} = p^{n+\alpha-1}(1-p)^{\sum_{i=1}^n y_i-n+\beta-1}.$$

Hence, $p|\mathbf{y} \sim Beta(n + \alpha, \sum_{i=1}^n y_i - n + \beta)$ and

$$\hat{p} = E(p|\mathbf{y}) = \frac{n + \alpha}{\sum_{i=1}^n y_i + \alpha + \beta}.$$

2. By straightforward calculus, $I^*(p) = \frac{1}{p^2(1-p)}$ and Jeffreys prior for p is then $\pi(p) \propto \sqrt{I^*(p)} = \frac{1}{p\sqrt{1-p}}$. The corresponding posterior is

$$\pi(p|\mathbf{y}) \propto p^{n-1}(1-p)^{\sum_{i=1}^n y_i - n - 1/2}$$

which is $Beta(n, \sum_{i=1}^n y_i - n + 1/2)$ and

$$\hat{p} = E(p|\mathbf{y}) = \frac{n}{\sum_{i=1}^n y_i + 0.5}.$$

B.7   Chapter 7

Exercise 7.1

1. By straightforward calculus one can verify that the Bayes estimator that minimizes the posterior expected loss $r(\delta, \mathbf{y}) = E\left(\frac{(\lambda-\delta)^2}{\lambda}\Big|\mathbf{y}\right)$ is $\delta^*(\mathbf{y}) = \left(E(\frac{1}{\lambda}|\mathbf{y})\right)^{-1}$. The posterior distribution $\lambda|\mathbf{y} \sim Gamma(\sum_{i=1}^n y_i + \alpha, n + \beta)$ and, hence,

$$
\begin{aligned}
E\left(\frac{1}{\lambda}\Big|\mathbf{y}\right) &= \int_0^\infty \lambda^{-1} \frac{(n+\beta)^{\sum_{i=1}^n y_i + \alpha}}{\Gamma(\sum_{i=1}^n y_i + \alpha)} \lambda^{\sum_{i=1}^n y_i + \alpha - 1} e^{-(n+\beta)\lambda} d\lambda \\
&= \frac{(n+\beta)^{\sum_{i=1}^n y_i + \alpha}}{\Gamma(\sum_{i=1}^n y_i + \alpha)} \frac{\Gamma(\sum_{i=1}^n y_i + \alpha - 1)}{(n+\beta)^{\sum_{i=1}^n y_i + \alpha - 1}} \\
&= \frac{n+\beta}{\sum_{i=1}^n y_i + \alpha - 1}.
\end{aligned}
$$

Thus, $\hat{\lambda} = \delta^*(\mathbf{y}) = \frac{\sum_{i=1}^n y_i + \alpha - 1}{n + \beta}$. Its risk

$$
\begin{aligned}
R(\lambda, \hat{\lambda}) &= E\left(\frac{(\hat{\lambda} - \lambda)^2}{\lambda}\right) \\
&= \frac{(E\hat{\lambda} - \lambda)^2 + Var(\hat{\lambda})}{\lambda} \\
&= \frac{1}{\lambda}\left(\left(\frac{n\lambda + \alpha - 1}{n + \beta} - \lambda\right)^2 + \frac{n\lambda}{(n+\beta)^2}\right) \\
&= \frac{(\alpha - 1 - \lambda\beta)^2 + n\lambda}{(n+\beta)^2 \lambda}.
\end{aligned}
$$

Since $\hat{\lambda}$ is a unique Bayes estimator and its Bayes risk

$$
\begin{aligned}
\rho(Beta(\alpha,\beta), \hat{\lambda}) &= E_{Beta(\alpha,\beta)}\left(\frac{(\alpha - 1 - \lambda\beta)^2 + n\lambda}{(n+\beta)^2 \lambda}\right) \\
&= \frac{(\alpha - 1)^2 E_{Beta(\alpha,\beta)}(1/\lambda) - 2(\alpha - 1)\beta + \beta^2 E_{Beta(\alpha,\beta)}\lambda + n}{(n+\beta)^2} < \infty
\end{aligned}
$$

for $\alpha > 1$, Theorem 7.2 implies that it is admissible.

2. The Bayes estimator $\hat{p}$ of $p = e^{-\lambda}$ w.r.t. a weighted quadratic loss $L(p,a) = \frac{(p-a)^2}{p^2}$ is $\hat{p} = \frac{E(1/p|\mathbf{y})}{E(1/p^2|\mathbf{y})}$. We have

$$E\left(\frac{1}{p}\Big|\mathbf{y}\right) = \int_0^\infty e^\lambda \frac{(n+\beta)^{\sum_{i=1}^n y_i + \alpha}}{\Gamma(\sum_{i=1}^n y_i + \alpha)} \lambda^{\sum_{i=1}^n y_i + \alpha - 1} e^{-(n+\beta)\lambda} d\lambda$$

$$= \left(\frac{n+\beta}{n+\beta-1}\right)^{\sum_{i=1}^n y_i + \alpha},$$

$$E\left(\frac{1}{p^2}\Big|\mathbf{y}\right) = \int_0^\infty e^{2\lambda} \frac{(n+\beta)^{\sum_{i=1}^n y_i + \alpha}}{\Gamma(\sum_{i=1}^n y_i + \alpha)} \lambda^{\sum_{i=1}^n y_i + \alpha - 1} e^{-(n+\beta)\lambda} d\lambda$$

$$= \left(\frac{n+\beta}{n+\beta-2}\right)^{\sum_{i=1}^n y_i + \alpha}$$

and, therefore,

$$\hat{p} = \left(\frac{n+\beta-2}{n+\beta-1}\right)^{\sum_{i=1}^n y_i + \alpha} = \left(1 - \frac{1}{n+\beta-1}\right)^{\sum_{i=1}^n y_i + \alpha}.$$

3. Both $\hat{\lambda}$ and $\hat{p}$ are consistent estimators since as n increases,

$$\hat{\lambda} = \frac{\bar{Y} + \alpha/n - 1/n}{1 + \beta/n} \xrightarrow{\text{P}} \lambda$$

and

$$\hat{p} = \left(1 - \frac{1}{n+\beta-1}\right)^{n(\bar{Y}+\alpha/n)} \xrightarrow{\text{P}} e^{-\lambda} = p.$$

Exercise 7.4

1. The risk function $R(\hat{\mu}_c, \mu) = (1-c)^2\mu^2 + \sigma^2 c^2$ and, therefore,

$$\sup_{\mu:|\mu|\leq a} R(\hat{\mu}_c, \mu) = \begin{cases} a^2(1-c)^2 + \sigma^2 c^2 & , c \neq 1 \\ \sigma^2 & , c = 1 \end{cases} \tag{B.8}$$

One can easily check that for the case $c \neq 1$, the minimum of (B.8) w.r.t. c is achieved at $\hat{c} = \frac{a^2}{a^2+\sigma^2}$ with the minimal value $\frac{a^2\sigma^2}{a^2+\sigma^2} < \sigma^2$. Hence, $\hat{\mu}_{\hat{c}} = \frac{a^2}{a^2+\sigma^2}Y$ is the minimax linear estimator. Since it is a unique minimax linear estimator, by Theorem 7.1 it is admissible. The minimax risk is $\sup_{\mu:|\mu|\leq a} R(\hat{\mu}_{\hat{c}}, \mu) = \frac{a^2\sigma^2}{a^2+\sigma^2}$.

2. Note that $R(\hat{\mu}_1, \mu) = \sigma^2 > \frac{a^2\sigma^2}{a^2+\sigma^2} = R(\hat{\mu}_{\hat{c}}, \mu)$ for any μ, where $\hat{c} = \frac{a^2}{a^2+\sigma^2}$ from the previous paragraph. Thus, $\hat{\mu}_1 = Y$ is inadmissible (unlike the case with no restrictions on μ—see Example 7.9).

3. Consider the Bayes risk w.r.t. the two-point prior $\pi(a) = \pi(-a) = 0.5$:

$$\rho(\pi, \hat{\mu}_c) = (1-c)^2 E_\pi \mu^2 + \sigma^2 c^2 = (1-c)^2 a^2 + \sigma^2 c^2 \tag{B.9}$$

The corresponding Bayes linear estimator that minimizes (B.9) w.r.t. c is $\hat{\mu}_{\hat{c}} = \frac{a^2}{a^2+\sigma^2}Y$, which is exactly the minimax linear estimator from the previous paragraph and, therefore, the above two-point prior is least favorable.

4. The formula (B.9) for the Bayes risk remains the same for any prior $\pi(\mu)$ with $E_\pi \mu^2 = a^2$ and the extension of the results of the previous paragraph is straightforward.

5. Assume now $\mu \sim \mathcal{U}(-a,a)$. Then, similar to the previous paragraph, $\rho(\pi, \hat{\mu}_c) = (1-c)^2 \frac{a^2}{3} + \sigma^2 c^2$, $\hat{c} = \frac{a^2}{a^2+3\sigma^2}$ and the (unique) Bayes linear estimator w.r.t. the uniform prior is $\hat{\mu}_{\hat{c}} = \frac{a^2}{a^2+3\sigma^2}Y$. The Bayes risk $\rho(\pi, \hat{\mu}_{\hat{c}})$ is evidently finite and, therefore, by Theorem 7.2, $\hat{\mu}_{\hat{c}}$ is admissible.

6. Similar to the unrestricted case (see Example 7.6), for any $c > 1$, $R(\hat{\mu}_c, \mu) > \sigma^2 = R(\hat{\mu}_1, \mu)$, while for any $c < 0$, $R(\hat{\mu}_c, \mu) > \mu^2 = R(\hat{\mu}_0, \mu)$ for any μ that implies their inadmissibility. Inadmissibility of $\hat{\mu}_1 = Y$ has been verified above. The case $c = 0$ corresponds to a trivial zero estimator which has an unimprovable zero risk at $\mu = 0$ and, therefore, is admissible. Consider now $0 < c < 1$. Let $\pi(\mu)$ be any prior for μ on $(-a, a)$ with $m_2 = E\mu^2 < \infty$. The Bayes risk $\rho(\pi, \hat{\mu}_c) = (1 - c)^2 m_2 + \sigma^2 c^2$ is minimized at $\hat{c} = \frac{m_2}{m_2 + \sigma^2}$. In other words, for any $0 < c < 1$, $\hat{\mu}_c$ is a unique Bayes estimator corresponding to the prior $\pi(\mu)$ with $m_2 = \frac{c\sigma^2}{1-c}$ with a finite Bayes risk. Admissibility of $\hat{\mu}_c$ follows then from Theorem 7.2.

Exercise 7.5

1. Let $\delta(\mathbf{y}) = 0$ if H_0 is accepted and $\delta(\mathbf{y}) = 1$ otherwise. The posterior expected loss of a rule δ is then

$$r(\delta_{\pi_0}, \mathbf{y}) = E\left(L(\delta_{\pi_0}, \theta)|\mathbf{y}\right) = \begin{cases} K_1 P(\theta = \theta_1|\mathbf{y}), & \delta_{\pi_0}(\mathbf{y}) = 0 \\ K_0 P(\theta = \theta_0|\mathbf{y}), & \delta_{\pi_0}(\mathbf{y}) = 1 \end{cases}$$

and the resulting Bayes rule

$$\delta_{\pi_0}^*(\mathbf{y}) = \begin{cases} 0, & K_1 P(\theta = \theta_1|\mathbf{y}) < K_0 P(\theta = \theta_0|\mathbf{y}) \\ 1, & \text{otherwise} \end{cases} = \begin{cases} 0, & \frac{f(\mathbf{y}|\theta_1)}{f(\mathbf{y}|\theta_0)} < \frac{K_0}{K_1} \frac{\pi_0}{\pi_1} \\ 1, & \text{otherwise}. \end{cases} \tag{B.10}$$

2. From the frequentist perspective, the above test (B.10) corresponds to the LRT (hence, an MP test) that rejects the null hypothesis if the likelihood ratio $\lambda(\mathbf{y}) = f_{\theta_1}(\mathbf{y})/f_{\theta_0}(\mathbf{y}) \geq C$, where the critical value $C = \frac{K_0}{K_1} \frac{\pi_0}{\pi_1}$.

3. Consider any LRT that rejects the null hypothesis if $\lambda(\mathbf{y}) \geq C$, $C \geq 0$. From the previous paragraph, it is a Bayes test corresponding to $\pi_0 = \frac{CK_1/K_0}{1+CK_1/K_0}$ and, therefore, by Theorem 7.3 for finite $\Theta = \{\theta_0, \theta_1\}$, is admissible.

4. Let $\alpha = P_{\theta_0}(\lambda(\mathbf{y}) \geq C)$ and $\beta = P_{\theta_1}(\lambda(\mathbf{y}) < C)$ be the Type I and Type II error probabilities of the above LRT. Then,

$$R(\theta, \delta^*) = \begin{cases} K_0 \alpha, & \theta = \theta_0 \\ K_1 \beta, & \theta = \theta_1 \end{cases}$$

(see also Example 7.5) that does not depend on θ if C is such that $K_0 \alpha = K_1 \beta$. By Theorem 7.4 it is the minimax test in this case.

5. For the normal data and $\mu_1 > \mu_0$, the LRT can be written in terms of $\bar{Y}$ and rejects $H_0 : \theta = \theta_0$ if $\bar{Y} \geq C^*$ (see Example 4.8). According to the previous paragraph, to get a minimax test, the critical value C^* should be chosen such that

$$25\left(1 - \Phi\left(\frac{C^*}{5/10}\right)\right) = 10\,\Phi\left(\frac{C^* - 2}{5/10}\right)$$

yielding $C^* \approx 1.1$.

Exercise 7.6 The posterior distribution

$$\pi(\tau|y) \propto e^{-(y-\tau)}\mathbb{I}\{\tau \leq y\}e^{-\tau}\mathbb{I}\{\tau \geq 0\} = e^{-y}\mathbb{I}\{0 \leq \tau \leq y\} \propto \mathbb{I}\{0 \leq \tau \leq y\}$$

that corresponds to the uniform distribution $\mathscr{U}(0, y)$. In particular, $\tau|y = 4 \sim \mathscr{U}(0, 4)$ and, hence, $P(A|y = 4) = P(B|y = 4) = P(AB|y = 4) = P(O|y = 4) = 0.25$. The posterior expected loss is then

$$\begin{cases} 0 \cdot 0.25 + 1 \cdot 0.25 + 1 \cdot 0.25 + 3 \cdot 0.25 & = & 1.25, & \delta = AB \\ 1 \cdot 0.25 + 0 \cdot 0.25 + 2 \cdot 0.25 + 3 \cdot 0.25 & = & 1.5, & \delta = A \\ 1 \cdot 0.25 + 2 \cdot 0.25 + 0 \cdot 0.25 + 3 \cdot 0.25 & = & 1.5, & \delta = B \\ 2 \cdot 0.25 + 2 \cdot 0.25 + 2 \cdot 0.25 + 0 \cdot 0.25 & = & 1.5, & \delta = A \end{cases}$$

which is minimized for the decision of blood type AB.

Bibliography

Berger, J. O. (1985). *Statistical Decision Theory and Bayesian Analysis* (2nd ed.). Springer-Verlag, New York.

Bickel, P. J. and Doksum, K. A. (2015). *Mathematical Statistics: Basic Ideas and Selected Topics*, Vol. 1 (2nd ed.). CRC Press, Boca Raton.

Bickel, P. J. and Doksum, K. A. (2016). *Mathematical Statistics: Basic Ideas and Selected Topics*, Vol. 2. CRC Press, Boca Raton.

Box, G. E. P. and Tiao, G. (1973). *Bayesian Inference in Statistical Analysis*. Addison-Wesley, Reading.

Cox, D. R. and Hinkley, D. V. (1974). *Theoretical Statistics*. Chapman and Hall, London.

Draper, N. R. and Smith, H. (1981). *Applied Regression Analysis*. (2nd ed.). John Wiley & Sons, New York.

Gelman, A., Carlin, J. B., Stern, H. S. and Rubin, D. B. (1995). *Bayesian Data Analysis*. Chapman and Hall, London.

Ghosh, J. K., Delampady, M. and Samanta, T. (2006). *An Introduction to Bayesian Analysis. Theory and Methods*. Springer, New York.

Lehmann, E. L. and Romano, J. R. (2005). *Testing Statistical Hypotheses* (3d ed.). Springer, New York.

Siegmund, D. (1985). *Sequential Analysis. Tests and Confidence Intervals*. Springer-Verlag, New York.

Index